Endorsed by the American Welding Society

as satisfying the skills standards set forth by the National Skills Standards Board.

Trainee Guide

Welding

Level 1 - Volume 2

National Center for Construction Education and Research

Wheels of Learning
Standardized Craft Training

Prentice Hall
Upper Saddle River, New Jersey   Columbus, Ohio

Prentice-Hall, Inc.
A Simon & Schuster Company
Upper Saddle River, New Jersey 07458

This information is general in nature and intended for training purposes only. Actual performance of activities described in this manual requires compliance with all applicable operations procedures under the direction of qualified personnel. References in this manual to patented or proprietary devices do not constitute a recommendation for their use.

Printed in the United States of America

10 9 8 7 6 5 4 3 2 1

ISBN: 0-13-982315-8

Prentice-Hall International (UK) Limited, *London*
Prentice-Hall of Australia Pty. Limited, *Sydney*
Prentice-Hall of Canada, Inc., *Toronto*
Prentice-Hall Hispanoamericana, S. A., *Mexico*
Prentice-Hall of India Private Limited, *New Delhi*
Prentice-Hall of Japan, Inc., *Tokyo*
Simon & Schuster Asia Pte. Ltd., *Singapore*
Editora Prentice-Hall do Brasil, Ltda., *Rio de Janeiro*

Preface

This volume is one of many in the *Wheels of Learning* craft training program. This program, covering more than 20 standardized craft areas, including all major construction skills, was developed over a period of years by industry and education specialists. Sixteen of the largest construction and maintenance firms in the U.S. committed financial and human resources to the teams that wrote the curricula and planned the national accredited training process. These materials are industry-proven and consist of competency-based textbooks and instructor guides.

The *Wheels of Learning* was developed by the National Center for Construction Education and Research in response to the training needs of the construction and maintenance industries. The NCCER is a nonprofit educational entity affiliated with the University of Florida and supported by the following industry and craft associations:

Partnering Associations
- American Fire Sprinkler Association
- American Society for Training and Development
- American Vocational Association
- American Welding Society
- Associated Builders and Contractors, Inc.
- Associated General Contractors of America
- Carolinas AGC
- Carolinas Electrical Contractors Association
- Construction Industry Institute
- Merit Contractors Association of Canada
- Metal Building Manufacturers Association
- National Association of Minority Contractors
- National Association of Women in Construction
- National Insulation Association
- National Ready Mixed Concrete Association
- National Utility Contractors Association
- National VoTech Honor Society
- Painting and Decorating Contractors of America
- Portland Cement Association
- Texas Gulf Coast ABC

Some of the features of the *Wheels of Learning* program include:

- A proven record of success over many years of use by industry companies.
- National standardization providing "portability" of learned job skills and educational credits that will be of tremendous value to trainees.
- Recognition: upon successful completion of training with an accredited sponsor, trainees receive an industry-recognized certificate and transcript from NCCER.
- Approval by the U.S. Department of Labor for use in formal apprenticeship programs.
- Well illustrated, up-to-date, and practical information. All standardized manuals are reviewed annually in a continuous improvement process.

Acknowledgments

This manual would not exist were it not for the dedication and unselfish energy of those volunteers who served on the Technical Review Committee. A sincere thanks is extended to:

Michael Goolsby

Ken Kluge

Doyce Redden

John Tate

Ken Teiken

Tom Watson

Morris Weeks

Jack A. Wilson

Contents

The following competencies are required to complete this Volume of AWS Entry Level Welder Certification (Level 1):

Module 09207 - GMAW—PLATE

1. Perform GMAW multi-pass fillet welds on plate in the flat position.
2. Perform GMAW multi-pass fillet welds on plate in the horizontal position.
3. Perform GMAW multi-pass fillet welds on plate in the vertical position.
4. Perform GMAW multi-pass fillet welds on plate in the overhead position.

Module 09209 - FCAW—EQUIPMENT & FILLER METAL

1. Explain Flux Core Arc Welding (FCAW) safety.
2. Identify and explain FCAW equipment.
3. Identify and explain FCAW filler metals.
4. Identify and explain FCAW shielding gasses.
5. Set up FCAW welding equipment.

Module 09210 - FCAW—FILLET AND GROOVE WELDS

1. Perform FCAW multi-pass fillet welds on plate in the 1F (flat) position using flux core carbon steel wire and shielding gas.
2. Perform FCAW multi-pass fillet welds on plate in the 2F (horizontal) position using flux core carbon steel wire and shielding gas.
3. Perform FCAW multi-pass fillet welds on plate in the 3F (vertical) position using flux core carbon steel wire and shielding gas.
4. Perform FCAW multi-pass fillet welds on plate in the 4F (overhead) position using flux core carbon steel wire and shielding gas.
5. Perform FCAW multi-pass groove welds on plate in the 1G (flat) position using flux core carbon steel wire and shielding gas.
6. Perform FCAW multi-pass groove welds on plate in the 2G (horizontal) position using flux core carbon steel wire and shielding gas.
7. Perform FCAW multi-pass groove welds on plate in the 3G (vertical) position using flux core carbon steel wire and shielding gas.
8. Perform FCAW multi-pass groove welds on plate in the 4G (overhead) position using flux core carbon steel wire and shielding gas.

Module 09304 - GTAW—EQUIPMENT & FILLER METALS

1. Explain Gas Tungsten Arc Welding (GTAW) safety.
2. Identify and explain GTAW equipment.
3. Identify and explain GTAW filler metals.
4. Identify and explain GTAW shielding gasses.
5. Set up GTAW welding equipment.

Module 09305 - GTAW—PLATE

1. Pad in all positions with stringer beads using GTAW and carbon steel filler rod.
2. Make multi-pass V-butt open-groove welds on mild steel plate in the 1G (flat) position using GTAW and carbon steel filler rod.
3. Make multi-pass V-butt open-groove welds on mild steel plate in the 2G (horizontal) position using GTAW and carbon steel filler rod.
4. Make multi-pass V-butt open-groove welds on mild steel plate in the 3G (vertical) position using GTAW and carbon steel filler rod.
5. Make multi-pass V-butt open-groove welds on mild steel plate in the 4G (overhead) position using GTAW and carbon steel filler rod.

GMAW-Plate

Module 09207

National
Center for
Construction
Education and
Research

GAS METAL ARC WELDING - (GMAW) - PLATE

Objectives

Upon completion of this module, the trainee will be able to:

1. Pad with GMAW stringer beads using carbon steel wire and carbon dioxide gas.
2. Pad with GMAW weave beads using carbon steel wire and carbon dioxide gas.
3. Perform GMAW multi-pass fillet welds on plate in the 1F (flat) position using carbon steel wire and carbon dioxide gas.
4. Perform GMAW multi-pass fillet welds on plate in the 2F (horizontal) position using carbon steel wire and carbon dioxide gas.
5. Perform GMAW multi-pass fillet welds on plate in the 3F (vertical) position using carbon steel wire and carbon dioxide gas.
6. Perform GMAW multi-pass fillet welds on plate in the 4F (overhead) position using carbon steel wire and carbon dioxide gas.

Prerequisites

Successful completion of the following module(s) is required before beginning study of this module:

- Weld Quality # 09107
- Gas Metal Arc Welding - Equipment and Filler Metals # 09206

Required Student Materials

Each trainee will need:

1. Personal protective equipment
2. Leather welding gloves
3. Welding shield
4. GMAW welding equipment

5. Welding table
6. Carbon steel wire electrode, 0.035 inch diameter
7. Carbon dioxide (CO_2) shielding gas
8. Cutting goggles
9. Chipping hammer
10. Wire brush
11. Needle nose side Cutters
12. Pliers
13. Tape measure
14. Soapstone
15. Scrap steel plate, 1/4 inch to 3/4 inch thick

Each trainee will need access to:

1. Oxyfuel cutting equipment
2. Framing square
3. Grinders

Course Map Information

This course map shows all of the Wheels of Learning modules in the second level of the Welding curricula. The suggested training order begins at the bottom and proceeds up. Skill levels increase as a trainee advances on the course map. The training order may be adjusted by the site Training Program Sponsor.

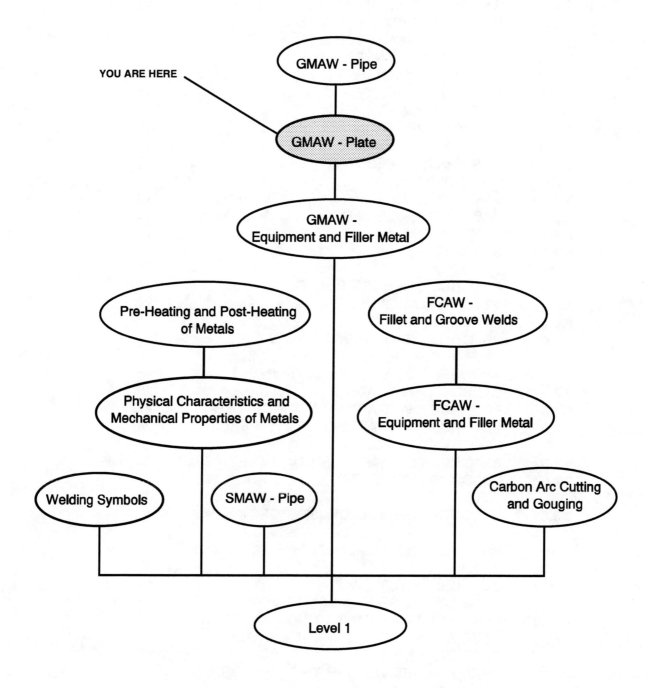

TABLE OF CONTENTS

Trade Terms Introduced In This Module

Arc Transfer. A gas metal arc process where the filler wire (electrode) is fed into contact with the base metal and the arc is formed only momentarily when the shorted electrode melts. This process repeats many times a second. The molten filler metal droplet merges into the molten pool before separation by gravity can take place, making this an excellent process for out-of-position welding.

Buried-Arc Transfer: A GMAW process where the arc length is so short that the wire electrode tip is at or below the plate surface, generally limited to the 1F (flat) position.

cfh: Abbreviation for cubic feet per hour.

DCRP (Direct Current Reverse Polarity): A welding setup using direct current where the electrode holder is attached to the positive cable and the work is attached to the negative cable. Conventional current flows from the electrode to the work; electron flow is from the work to the electrode.

ipm: Abbreviation for inches per minute.

Longitudinal Angle: The angle between the centerline of the welding gun nozzle and a line perpendicular to the axis of the weld. In a pulling angle (trailing), the electrode tip points back toward the weld. In a pushing angle (leading), the electrode tip points ahead of the weld.

Low-Carbon Steel: Also known as mild steel. Steel with a carbon content of 0.05 to 0.30 percent.

Mild Steel: Also know as low-carbon steel. Steel with a carbon content of 0.05 to 0.30.

Out-Of-Position Weld: Out-of-position welds include all welds except the flat position weld. The flat position weld bead is always horizontal and a vertical plane through the root will bisect the weld bead along its entire length.

Penetration: The distance that the weld fusion line extends below the surface of the base metal.

Restart: The point in the weld where one weld bead stops and the continuing bead is started.

Short-Circuiting Transfer: Also called Dip Transfer and Short-Arc. A GMAW process where the electrode wire shorts against the base metal. The welding current repeatedly and rapidly melts the wire establishing the arc which is extinguished when the wire shorts against the base metal.

Spray-Arc Transfer: Also called Spray Transfer, is a GMAW process where the filler metal is melted into fine droplets and carried to the molten pool by the arc.

Stickout: The length of wire electrode that extends past the tip of the gun's contact tube.

Stringer Bead: A weld bead that is made with little or no side-to-side motion of the electrode, usually no more than two to three times the electrode diameter.

Tie-In: Complete fusion of the weld end with the base metal.

Transverse Angle: The angle between the centerline of the welding gun nozzle and the plate surface in a plane normal to the weld axis, or the angle between the centerline of the gun nozzle and a vertical line in a plane normal to the weld axis.

Weave Bead: A weld bead that is made with a wide side-to-side motion of the electrode, generally not more than eight times the electrode diameter.

Weld Coupon: The metal to be welded as a test or practice.

Welding Variables: Items that affect the quality of a weld such as welding voltage, welding current, welding travel rate, welding gun angle and electrode stickout (contact tip-to-work distance).

1.0.0 INTRODUCTION

Gas Metal Arc Welding (GMAW), is an arc welding process that uses continuous solid wire electrode for the filler metal and shielding gas to protect the weld zone. It is a fast and effective method of making high quality welds. Figure 1-1 shows the GMAW process.

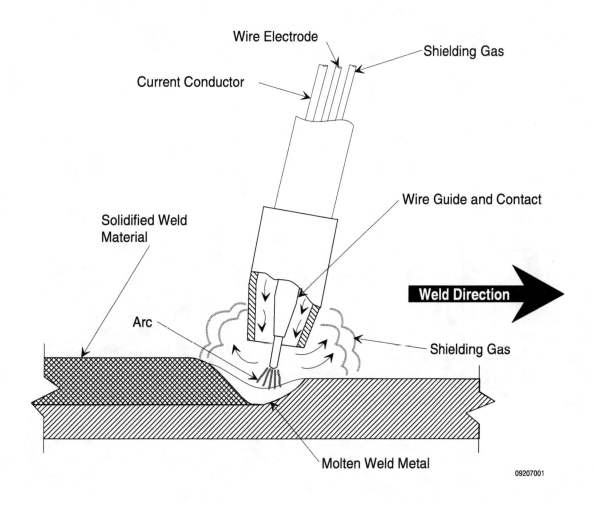

Figure 1-1. GMAW Process.

The GMAW process is commonly used to make fillet welds on mild steel and low-carbon steel plate. This module explains how to set up GMAW equipment and perform fillet welds on steel plate in the 1F, 2F, 3F and 4F welding positions.

Figure 1-2 shows the 1F, 2F, 3F and 4F welding positions.

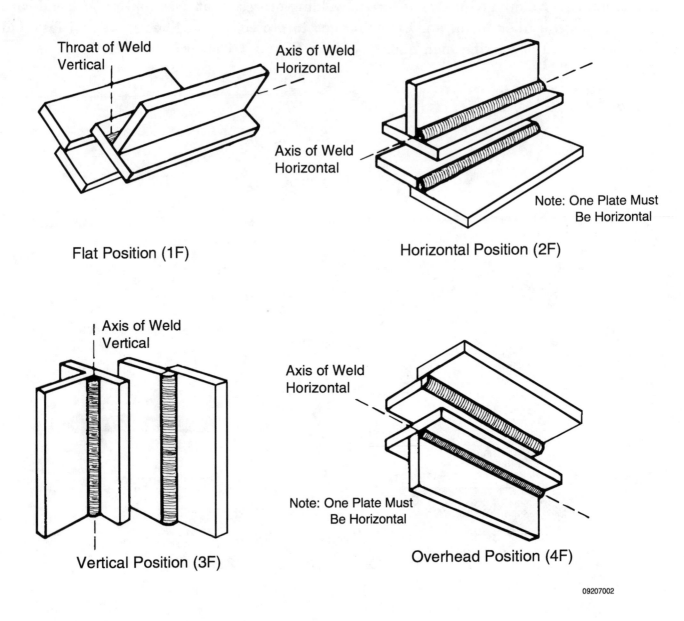

Throat of Weld
Vertical

Axis of Weld
Horizontal

Axis of Weld
Horizontal

Flat Position (1F)

Note: One Plate Must
Be Horizontal

Horizontal Position (2F)

Axis of Weld
Vertical

Axis of Weld
Horizontal

Vertical Position (3F)

Note: One Plate Must
Be Horizontal

Overhead Position (4F)

09207002

Figure 1-2. 1F, 2F, 3F and 4F Welding Positions.

2.0.0   GMAW WELDING EQUIPMENT SET-UP

Before welding can take place the work area has to be made ready, the welding equipment set up and the metal to be welded prepared. The following sections will explain how to prepare the area, set up the equipment and perform GMAW welding of mild and low-carbon steel plate.

2.1.0   PREPARE WELDING AREA

To practice welding, a welding table, bench or stand is needed. The welding surface must be steel and provisions must be made for mounting practice welding coupons out-of-position. A simple mounting for out-of-position coupons is a pipe stand that can be welded vertically to the steel table top. To make a pipe stand, weld a three- or four-foot length of pipe vertically to the table top. Then cut a short section pipe to slide over the vertical pipe. Drill a hole in the slide. Weld a nut over the hole so a bolt can be used to lock the slide in place on the vertical pipe. Weld a piece of pipe or angle horizontally to the slide. The slide can be rotated and adjusted vertically to position the horizontal arm. The weld coupon can be tack-welded to the horizontal arm.

WARNING!   The table must be heavy enough to support the weight of the welding coupon extended on the horizontal arm without falling over. It must also support the coupon during chipping or grinding. Serious injury will result if the table falls onto someone in the area.

Figure 2-1 shows a welding table with out-of-position support.

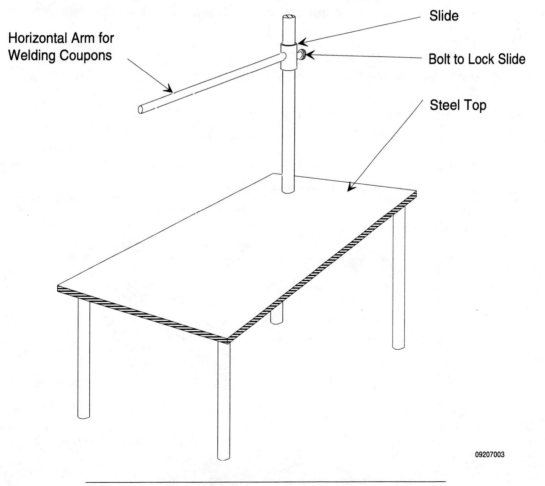

Horizontal Arm for Welding Coupons

Slide

Bolt to Lock Slide

Steel Top

09207003

Figure 2-1. Welding Table With Out-of-position Support

To set up the area for welding follow these steps.

Step 1 Check to be sure the area is properly ventilated. Make use of doors, windows and fans.

Step 2 Check the area for fire hazards. Remove any flammable materials before proceeding.

Step 3 Check the location of the nearest fire extinguisher. Do not proceed unless the extinguisher is charged and you know how to use it.

Step 4 Position a welding table near the welding machine/wire feeder.

Step 5 Set up flash shields around the welding area.

2.2.0 PREPARE WELDING COUPONS

Cut the welding coupons from mild or low-carbon steel 1/4- to 3/4-inch thick. Use a wire brush or grinder to remove any heavy mill scale or corrosion. Prepare welding coupons to practice the welds indicated as follows:

- Running Stringer Beads: The coupons can be any size or shape that can be easily handled.
- Fillet Welds: Cut the metal into four-by-six inch rectangles for the base and three-by-six inch rectangles for the web.

Figure 2-2 shows a mild steel fillet welding coupon.

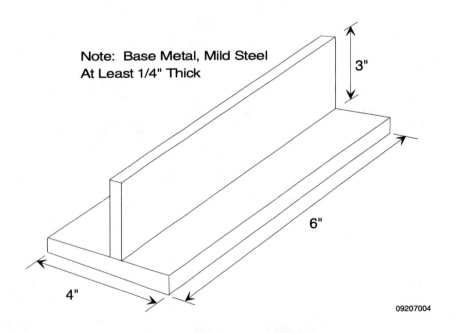

Note: Base Metal, Mild Steel
At Least 1/4" Thick

3"

6"

4"

09207004

Figure 2-2. Mild Steel Fillet Welding Coupon.

Note   Steel for practice welding is expensive and difficult to obtain. Every effort should be made to conserve and not waste the material that is available. Reuse weld coupons until all surfaces have been welded on. Weld on both sides of the joint and then cut the weld coupon apart and reuse the pieces. Use material that can not be cut into weld coupons to practice running beads.

2.3.0   FILLER WIRE

The GMAW process uses continuous wire electrode for filler wire. Filler wire is specified by size (diameter) and metal composition (alloy). Generally, filler wire is selected to be a close chemical match to the base metal. The WPS or job site standards may specify the filler wire to be used. For the welding exercises in this module, 0.035 inch carbon steel wire and carbon dioxide shielding gas will be used.

2.4.0   GMAW WELDING EQUIPMENT

Locate the following GMAW equipment:

- Constant voltage (constant potential) power supply
- Wire feeder
- Gun
- Shielding gas cylinder and flow meter

Figure 2-3 shows a complete GMAW system.

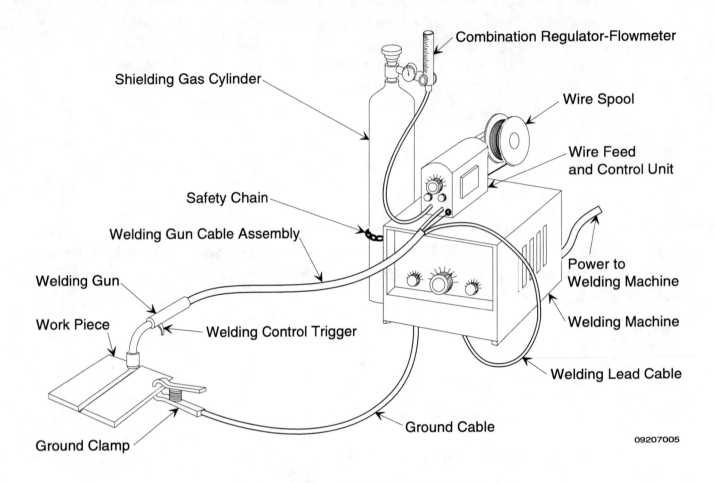

Figure 2-3. Complete GMAW System.

Perform the following steps to set up the GMAW equipment.

Step 1 Verify that the welding machine is a constant voltage type DC power supply.

Step 2 Check to be sure that the welding machine is properly grounded through the primary current receptacle.

Step 3 Verify the location of the primary current disconnect.

Step 4 If the welding machine does not have a wire feeder, locate a compatible wire feeder.

Step 5 Connect the wire feeder power and control cables.

Step 6 Obtain a spool of 0.035 inch mild steel wire.

Step 7 Check the wire feed rollers to be sure they are the correct size for the wire to be run. Change them if they are the wrong size. Follow the manufacturer's instructions for changing feed rollers.

Step 8 Locate a GMAW gun and cable assembly. Check the size of the nozzle contact tip. Change it if it is not correct for the wire size.

Step 9 Connect the gun cable to the wire feeder.

Step 10 Set up the welding machine for reverse polarity (DCRP) welding (positive lead to gun and negative lead to workpiece).

Step 11 Install the wire spool onto the wire feeder. Adjust the spool drag brake and feed the wire through the cable and the gun with the jog control until the proper stick-out is achieved.

Step 12 Locate a cylinder of CO_2 shielding gas and secure it to the welding machine or a nearby structure so that it cannot fall. Install the gas regulator/flowmeter and connect the gas hose to the wire feeder.

Step 13 Fully open the shielding gas cylinder, purge the gun with the purge control on the welding machine or wire feed unit and adjust the gas flowmeter to achieve the specified flow rate.

Step 14 Turn on the welding machine.

Step 15 Set the welding variables as follows:

Wire Size	Wire Speed (ipm)	DCRP Volts	DCRP Amps	Travel Speed (ipm)
0.035 in.	200-300	17-21	130-175	1F & 2F 8-13 3F & 4F 11-16

Note   Shielding Gas: CO_2

Electrical stickout: 1/4-3/8 inch

Use higher end of ranges for 1F

Use lower end of ranges for 3F

Base metal: 1/4-3/4 inch thick mild steel

Step 16 Set the other variables (if applicable).

Refer to the manufacturer's specifications for:

- Slope
- Inductance
- Burn back
- Gas preflow (prepurge)
- Gas postflow (post purge)

3.0.0   GMAW BEAD WIDTH AND PENETRATION

GMAW bead width and penetration depth are affected by many factors. Some factors are equipment related and others are controlled by the welder. These factors include:

- GMAW welding mode (spray-arc or globular transfer)
- Shielding gas type
- Arc voltage
- Electrode stick-out
- Gun travel speed
- Gun position
- Bead type (stringer or weave)

3.1.0   GMAW WELDING MODE

Basic GMAW welding includes three filler metal transfer modes:

- Short-Arc (Short-Circuiting or Dip) Transfer
- Spray-Arc Transfer
- Globular Transfer

3.1.1   Short-Arc Transfer

In short-arc transfer mode, the filler metal is transferred while the wire electrode is shorted to the base metal or weld puddle. As soon as the wire melts, an arc is established. As the wire is again advanced in the gun, the wire shorts and transfers again. Because short-arc transfer mode uses relatively low voltages and welding currents, it produces narrow beads with shallow penetration. It is commonly used for thin materials in all positions and thick materials in vertical and overhead positions.

Figure 3-1 shows the short-arc transfer process.

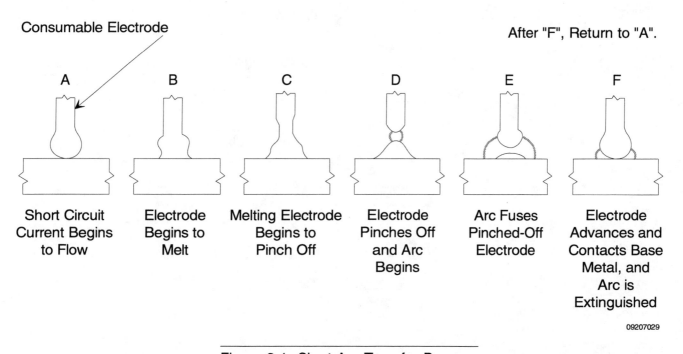

Figure 3-1. Short-Arc Transfer Process.

Circuit inductance is important in short-arc transfer to minimize spattering. Both the amount of weld current and inductance needed for short-arc transfer welding increase with wire size. On welding machines with variable inductance, the inductance is changed with the "slope" or "inductance" control.

3.1.2   Spray-Arc Transfer

In spray-arc transfer mode, the filler metal is transferred as fine droplets propelled by electromagnetic forces. The arc voltages and welding currents are relatively high, producing wider beads and deeper penetrations. It is commonly used for the fast transfer of filler metal on thicker materials in the flat position.

Note   Spray-arc transfer cannot be achieved if the shielding gas contains more than 15 percent carbon dioxide (CO_2).

Figure 3-2 shows the spray-arc transfer process.

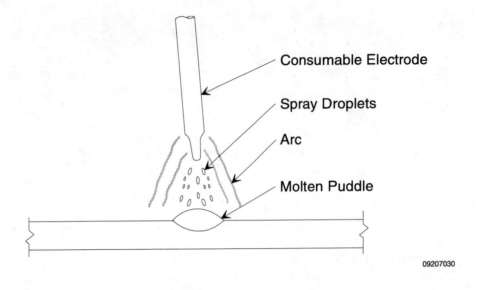

Figure 3-2. The Spray-Arc Transfer Process.

3.1.3   Globular Transfer

In globular transfer, the filler metal is transferred as large molten globs through the arc. The arc voltages and welding currents are higher than for short-arc transfer, but lower than for spray-arc transfer. Weld bead is wider than with short-arc transfer and penetration is shallower than with spray-arc transfer. *Figure 3-3* shows the globular transfer process.

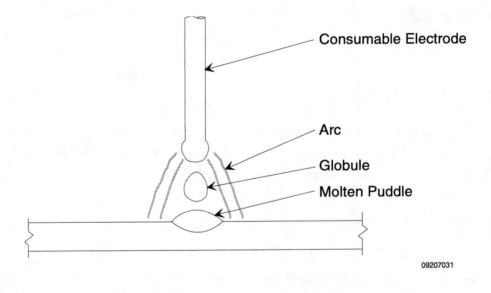

Figure 3-3. The Globular Transfer Process.

3.2.0 SHIELDING GAS TYPE

The choice of shielding gas can have a great effect on arc and weld bead characteristics. Arc stability and temperature change with different types of shielding gases. Also, some shielding gases react with the weld zone to produce alterations in the base metal and filler metal. For more detailed information on shielding gases, refer to Module 09206, *Gas Metal Arc Welding - Equipment and Filler Metals*.

3.3.0 ARC VOLTAGE

A minimum arc voltage is needed to maintain spray-arc transfer. However, penetration is not directly related to voltage. Penetration will increase with voltage for a time, but beyond some optimum voltage, it will actually decrease with further voltage increases.

3.4.0 WELDING AMPERAGE

Weld amperage has a direct effect on weld penetration. Higher weld amperage always produces more heat and deeper penetrations. GMAW power sources do not have an amperage control. The welding power supply is self regulating. It automatically increases or decreases the welding amperage as the wire feed speed is increased or decreased.

3.5.0 GUN TRAVEL SPEED

Gun travel speed affects the weld much as arc voltage does. Deepest penetration is at an optimum travel speed. Any speed faster or slower than the optimum speed results in less penetration.

3.6.0 GUN POSITION

Gun position (longitudinal angle) affects weld penetration greater than either voltage or travel speed. Gun longitudinal angle is the angle the centerline of the gun nozzle or electrode make with a line normal to the axis of the weld.

Longitudinal gun angle can vary anywhere from 25 degrees toward the direction of travel (leading or pushing angles) to 25 degrees opposite to the direction of travel (trailing or pulling angles). Longitudinal angles greater than 25 degrees produce unstable arcs and greater weld spatter. Longitudinal leading (or pushing) angles produce wider beads and shallower penetrations. Longitudinal trailing (or pulling) angles produce narrower beads and deeper penetrations. A 0-degree longitudinal angle (electrode normal to the workpiece) produces a bead whose width and penetration is midway between the pulling and pushing beads. Maximum penetration is achieved with a 25 degree trailing (or pulling) longitudinal angle. A leading (or pushing) longitudinal angle is generally used for thin materials or shallow penetration.

Figure 3-4 shows GMAW longitudinal gun angles.

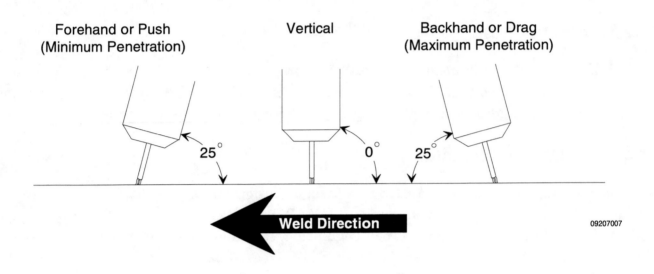

Figure 3-4. GMAW Longitudinal Gun Angles.

3.7.0   WIRE STICK-OUT

Wire stick-out (electrode extension) is the length of the wire that extends beyond the contact tip.

Changing the stickout influences the welding current because it changes the preheating of the wire. As the stickout increases, preheating of the wire increases. Since the welding power source is self-regulating, it does not have to furnish as much welding current to melt the wire so the current output automatically decreases. This results in less penetration and increased deposit rates. Increasing stickout is useful for bridging gaps and compensating for miss-match, but can cause cold-lap (lack of fusion) and a ropy bead appearance. When the stickout is decreased, the power source is forced to increase its current output to burn off the wire. Too little stickout can cause the wire to weld to the contact tip and develop porosity in the weld. Electrical stickout for micro wire (0.030 to 0.045 inch) should be 1/4 to 3/8 inch. The gas nozzle should be adjusted so that it is flush with the contact tip, or is slightly ahead of the contact tip (contact tip slightly recessed).

Figure 3-5 shows electrode stickout and nozzle position.

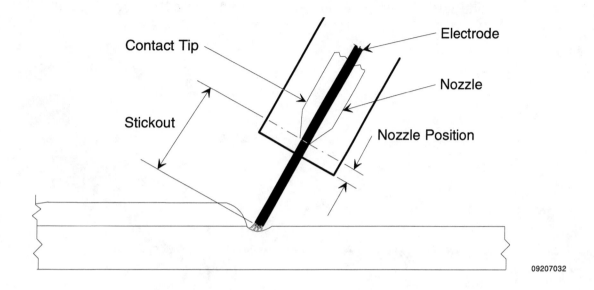

Figure 3-5. Electrode Stickout and Nozzle Position.

3.8.0   GAS NOZZLE CLEANING

With use, weld spatter accumulates on the gas nozzle and contact tip. If it is not occasionally cleaned, the gas nozzle will become blocked, restricting the shielding gas flow and causing porosity in the weld.

Clean the gas nozzle with a reamer or round file, or the tang of a standard file. After cleaning, the nozzle can be sprayed or dipped in a special anti-spatter compound. The anti-spatter helps prevent the spatter from sticking to the nozzle.

CAUTION   Use only anti-spatter material specifically designed for GMAW
             gas nozzles. Other materials may cause porosity in the welds.

3.9.0   BEAD TYPES

There are two basic bead types. They are:

- Stringer beads
- Weave beads

3.9.1 Stringer Beads

Stringer beads are made with little or no side-to-side motion of the gun. *Figure 3-6* shows a stringer bead.

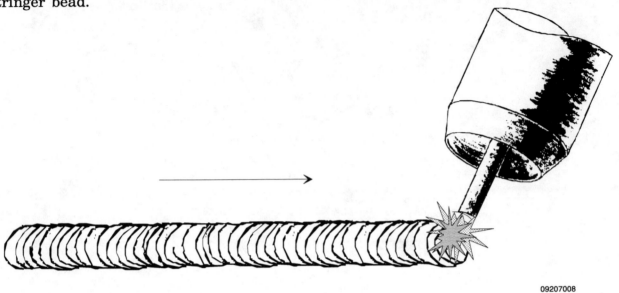

09207008

Figure 3-6. Stringer Bead.

3.9.2 Weave Beads

Weave beads are made with wide side-to-side motions of the electrode. The width of a weave bead is determined by the amount of side-to-side motion. *Figure 3-7* shows a weave bead.

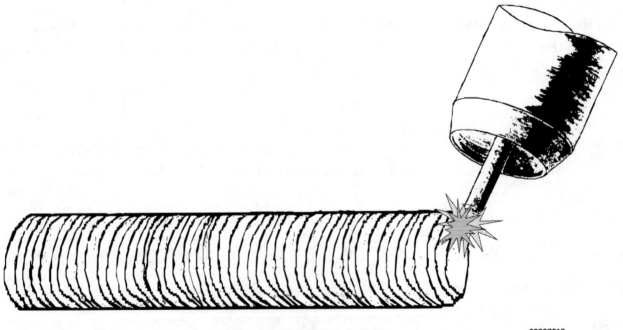

09207010

Figure 3-7. Weave Bead.

When making a weave bead, care must be used at the toes to ensure proper tie-in to the base metal. To ensure proper tie-in at the toes, slow down or pause slightly at the edges. The pause at the edges will also flatten out the weld, giving it the proper profile.

CAUTION The width of weave beads is often specified in the welding code or WPS being used at your site. Do not exceed the widths specified for your site.

4.0.0 PRACTICE STRINGER BEADS

Practice running stringer beads in the flat position. Experiment with different lead angles and stick-outs.

Follow these steps to run stringer beads.

Step 1 Hold the gun at the desired angle with the electrode tip directly over the point where the weld is to begin and pull the gun trigger.

Step 2 Hold the arc in place until the weld puddle begins to form.

Step 3 Slowly advance the arc, while maintaining the gun angle. Stay in the leading edge of the puddle to prevent cold lap.

Step 4 Continue to weld until a bead about 2 to 3 inches long is formed, and then stop by pausing until the crater is filled and then releasing the trigger.

Step 5 Inspect the bead for:

- Straightness of the bead
- Uniform rippled appearance of the bead face
- Smooth flat transition with complete fusion at the toes of weld
- No porosity
- No undercut
- Crater filled
- No cracks

Step 6 Continue practicing stringer beads until you can make acceptable welds every time.

5.0.0 WELD RESTARTS

A restart is the junction where a new weld connects to continue the bead of a previous weld. Restarts are important because an improperly made restart will create a weld defect. A restart must be made so that it blends smoothly with the previous weld and does not stand out. The technique for making a GMAW restart is the same for both stringer and weave beads. Follow these steps to make a restart.

Step 1   Hold the torch at the proper angle while restarting the arc directly over the center of the crater. (The welding codes do not allow arc strikes outside the area to be welded.)

Step 2   Move the electrode tip in a small circular motion over the crater to fill the crater with a molten puddle.

Step 3   As soon as the puddle fills the crater, move to the leading edge of the puddle and continue the stringer or bead pattern being used.

Step 4   Inspect the restart. A properly made restart will blend into the bead, making it hard to detect.

If the restart has undercut, not enough time was spent in the crater to fill it. If undercut is on one side or the other, use more of a side-to-side motion as you move back into the crater. If the restart has a lump, it was overfilled. Too much time was spent in the crater before resuming the forward motion.

Continue to practice restarts until they are correct. Use the same techniques for making restarts whenever performing GMAW.

6.0.0 WELD TERMINATIONS

A weld termination is made at the end of a weld. A termination normally leaves a crater. When making a termination, the welding codes require that the crater must be filled to the full cross-section of the weld. This can be difficult, since most terminations are at the edge of a plate where welding heat tends to build up making filling the crater more difficult.

Follow these steps to make a termination.

Step 1   As you approach the end of the weld, start to bring the gun up to a 0-degree longitudinal angle (gun normal to the bead) and slow forward travel.

Step 2   Stop forward movement about 1/8 inch from the end of the plate and slowly angle the gun to about 10 degrees toward the start of the weld (training (or pulling).

Step 3   Move about 1/8 inch toward the start of the weld and release the trigger when the crater is filled.

CAUTION Do not remove the gun from the weld until the puddle has solidified. The post purge (gas post flow) that continues after the welding stops protects the molten metal. If the gun and shielding gas are removed before the crater has solidified, crater porosity or cracks can occur.

Step 4 Inspect the termination. The crater should be filled to the full cross-section of the weld.

Figure 6-1 shows a terminating a weld.

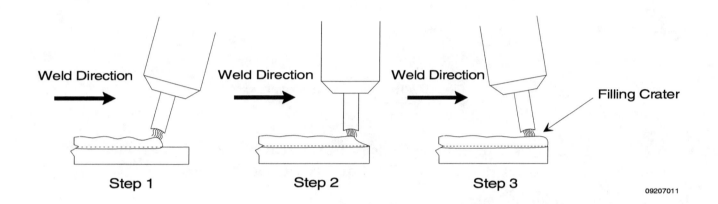

Figure 6-1. Terminating a Weld.

7.0.0   PRACTICE WEAVE BEADS

Practice running weave beads in the flat position. Experiment with different weave motions, lead angles and stick-outs.

Follow these steps to run weave beads.

Step 1 Hold the gun at the desired angle with the electrode tip directly over the point where the weld is to begin and pull the gun trigger.

Step 2 Hold the arc in place until the weld puddle begins to form.

Step 3 Slowly advance the arc in a weaving motion, while maintaining the gun angle.

Figure 7-1 shows the weave motion.

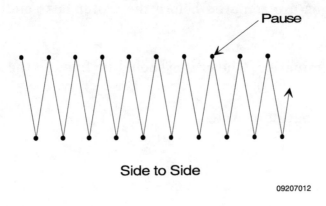

Figure 7-1. The Weave Motion.

Step 4 Continue to weld until a bead about 2 to 3 inches long is formed and then stop by pausing until the crater is filled and releasing the trigger.

Step 5 Inspect the bead for:

- Straightness of the bead
- Uniform rippled appearance of the bead face
- Smooth flat transition with complete fusion at the toes of weld
- No porosity
- No undercut
- Crater filled
- No cracks

Step 6 Continue practicing weave beads until you can make acceptable welds every time.

SELF CHECK REVIEW 1

1. Before welding, how do you check for a fire extinguisher?
2. Why is it important to conserve steel used for practice welding?
3. How is filler wire specified?
4. What type of DC welding machine is best for GMAW?
5. When setting up a GMAW gun, what should be checked for?
6. What is the maximum practical longitudinal gun angle for GMAW?
7. What longitudinal gun angle gives the maximum penetration with GMAW?
8. Which is best for welding sheet metal with GMAW, a leading (pushing) angle or a trailing (pulling) angle?
9. To decrease the bead height and increase the bead width with GMAW, is it better to use a leading (pushing) angle or a trailing (pulling) angle?
10. Why are properly made restarts important?

ANSWERS TO SELF CHECK REVIEW 1

1. Check the location of the nearest fire extinguisher. Do not proceed unless the extinguisher is charged and you know how to use it. (2.1.0)
2. Steel for practice welding is expensive and difficult to obtain. (2.2.0)
3. Filler wire is specified by size (diameter) and metal composition (alloy). (2.3.0)
4. A constant voltage (constant potential) machine is best for GMAW. (2.4.0)
5. The gun contact tip and wire drive rollers should be checked for correct wire size and tension adjustment. (2.4.0)
6. The maximum gun angle is 25 degrees from a normal to the weld bead. (4.6.0)
7. Maximum penetration is achieved with a 25 degree trailing (pulling) angle. (4.6.0)
8. Sheet metal is best welded with a leading (pushing) angle. (4.6.0)
9. Use a leading (pushing)) gun angle to decrease bead height and increase bead width. (4.7.0)
10. An improperly made restart will cause a weld defect. (4.10.0)

8.0.0   OVERLAPPING BEADS

Overlapping beads are made by depositing connective weld beads parallel to one another. The parallel beads overlap forming a flat surface. This is also called padding. Overlapping beads are used to build up a surface and for making multipass welds. Both stringer and weave beads can be overlapped.

Properly overlapped beads when viewed from the end will form a relatively flat surface. *Figure 8-1* shows proper and improper overlapping beads.

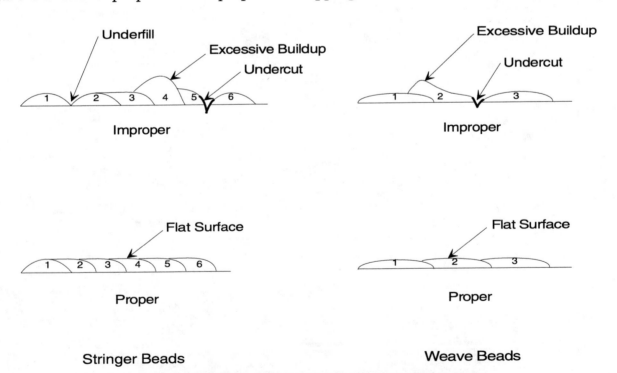

Figure 8-1. Proper and Improper Overlapping Beads.

09207013

WELDING TWO TRAINEE TASK MODULE 09207

8.1.0 PRACTICE OVERLAPPING STRINGER BEADS

Follow these steps to weld overlapping stringer beads using 0.035 inch carbon steel electrode wire and carbon dioxide shielding gas.

Step 1   Mark out a 4 inch square on a piece of steel.

Step 2   Weld a stringer bead along one edge.

Step 3   Position the gun at a transverse angle of 10 to 15 degrees toward the side of the previous bead to get proper tie-in and pull the trigger. *Figure 8-2* shows the transverse gun angle for overlapping beads.

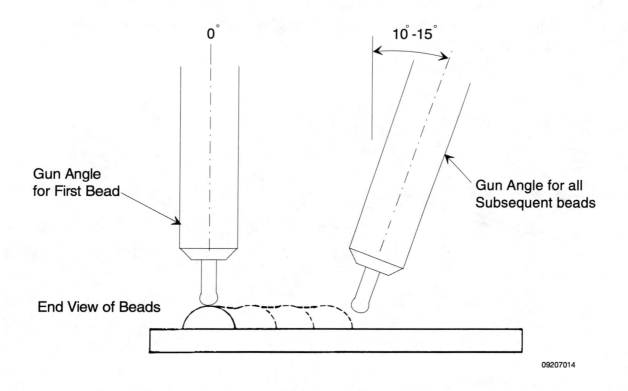

Figure 8-2. Transverse Gun Angle For Overlapping Beads.

Step 5   Continue running overlapping stringer beads until the 4 inch square is covered.

Note   The base metal will get very hot as it is being built up. If necessary, hold it with pliers and cool it with water.

Step 6   Continue building layers of stringer beads, one on top of the other, until the technique is perfected.

8.2.0 PRACTICE OVERLAPPING WEAVE BEADS

Repeat welding overlapping beads using the weaving technique. Remember to angle the electrode toward the previous bead to get good tie-in.

9.0.0 FILLET WELDS

The most common fillet welds are made in lap and tee joints. The weld position is determined by the axis of the weld. The positions for fillet welding are flat or 1F (Fillet), horizontal or 2F, vertical or 3F, and overhead or 4F. Then axis of the weld can rotate plus or minus 15 degrees and the weld can be inclined up to 15 degrees within a position. *Figure 9-1* shows the fillet welding positions.

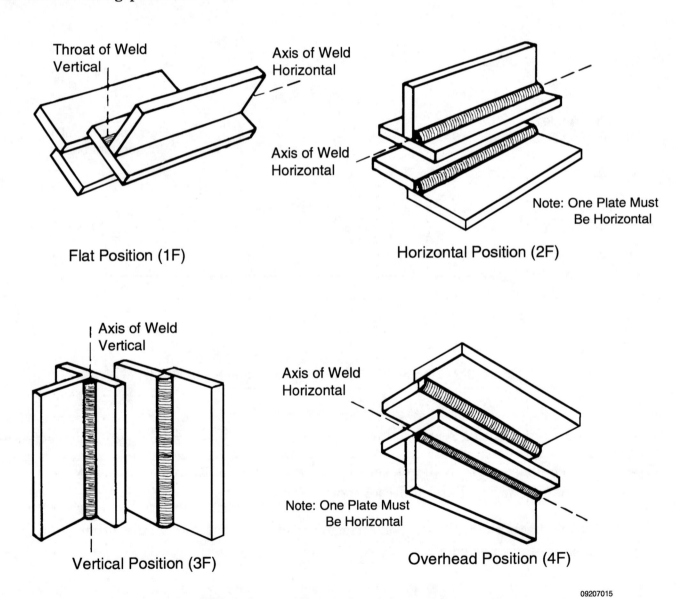

09207015

Figure 9-1. Fillet Welding Positions.

Fillet welds can be concave or convex, depending on the WPS or site quality standards. The welding codes require a fillet weld to have a uniform concave or convex face, although a slightly nonuniform face is acceptable. The convexity of a fillet weld or individual surface bead must not exceed 0.07 times the actual face width or individual surface bead plus 0.06-inch.

A fillet weld is unacceptable and must be repaired if the profile has insufficient throat, excess convexity, excessive undercut, overlap, insufficient leg or inadequate penetration.

Figure 9-2 shows acceptable and unacceptable fillet weld profiles.

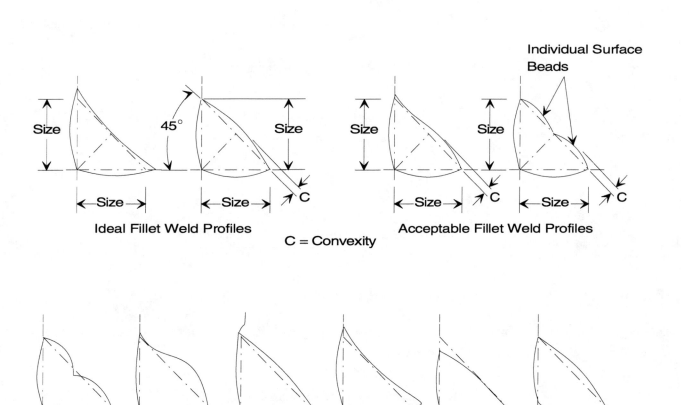

Figure 9-2. Acceptable and Unacceptable Fillet Weld Profiles.

9.1.0 PRACTICE FLAT (1F) POSITION FILLET WELDS

Practice flat fillet welds by welding multiple-pass (six pass) convex fillet welds in a tee joint using 0.035 inch carbon steel electrode wire and carbon dioxide shielding gas. When making flat fillet welds, pay close attention to the gun angle and travel speed. For the first bead, the gun angle is vertical (45 degrees to both plate surfaces). The angle is adjusted for all subsequent beads. The angle is angled 20 degrees so that it points toward the side to be filled. Increase or decrease the travel speed to control the amount of weld metal build-up. *Figure 9-3* shows the gun angles for flat fillet welds.

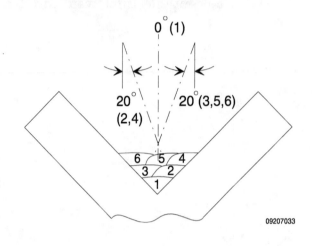

09207033

Figure 9-3. Gun Angles for Flat Fillet Welds.

Follow these steps to make a horizontal fillet weld.

Step 1   Tack two plates together to form a tee joint for the fillet weld coupon. *Figure 9-4* shows the fillet weld coupon.

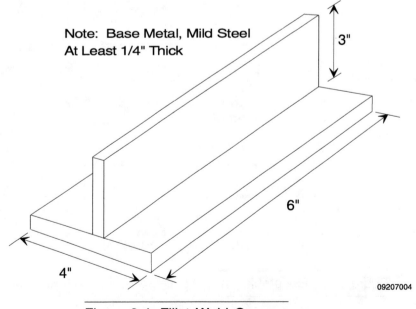

09207004

Figure 9-4. Fillet Weld Coupon.

Step 2 Tack the coupon to the welding table or positioning arm in the 1F (flat) position.

Step 3 Run the first bead along the root of the joint using a transverse gun angle of 45 degrees with a 5- to 10-degree longitudinal gun angle. Use a slight side-to-side oscillation motion.

Step 4 Run the second bead along a toe of the first weld, overlapping about 75 percent of the first bead. Use a transverse gun angle of 70 degrees with a 5- to 10-degree longitudinal drag gun angle and a slight oscillation.

Step 5 Run the third bead along the other toe of the first weld, filling the groove created when the second bead was run. Use a transverse gun angle of 30 degrees with a 5- to 10-degree longitudinal drag gun angle and a slight oscillation.

Figure 9-5 shows weld bead sequences 1, 2 and 3, and the gun angles.

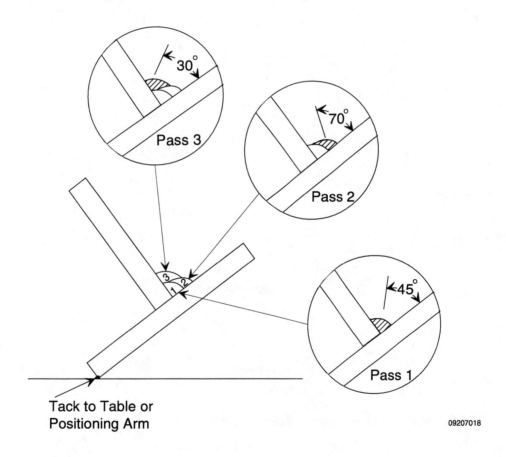

Figure 9-5. Bead Sequences 1, 2 and 3 and Gun Angles.

Step 6 Run the fourth bead along the outside toe of the second weld, overlapping about half the second bead. Use a transverse gun angle of 20 degrees (from the vertical) with a 5- to 10-degree longitudinal drag gun angle and a slight oscillation.

Step 7 Run the fifth bead along the inside toe of the fourth weld, overlapping about half the fourth bead that was run. Use a transverse gun angle of 20 degrees (from the vertical) with a 5- to 10-degree longitudinal drag gun angle and a slight oscillation.

Step 8 Run the sixth bead along the outside toe of the third weld, filling the groove created when the fifth bead was run. Use a transverse gun angle of 20 degrees with a 5- to 10-degree longitudinal gun drag angle and a slight oscillation.

Figure 9-6 shows weld bead sequences 4, 5 and 6 and gun angles.

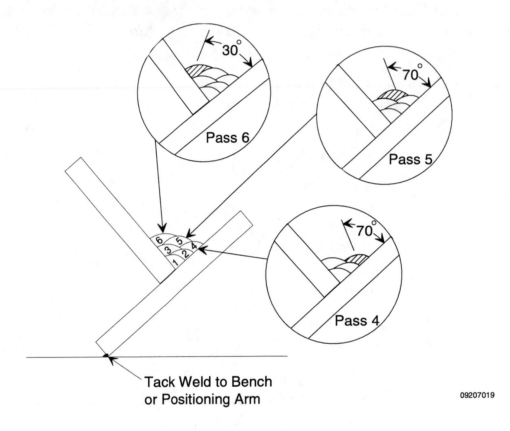

09207019

Figure 9-6. Bead Sequences 4, 5 and 6 and Gun Angles.

Step 9 Inspect the weld. The weld is acceptable if it has:

- Uniform rippled appearance on the bead face
- Craters and restarts filled to the full cross-section of the weld
- Uniform weld size, plus or minus 1/16 inch
- Acceptable weld profile in accordance with AWS D1.1
- Smooth transition with complete fusion at the toes of the weld
- No porosity
- No undercut
- No cracks

9.2.0 PRACTICE HORIZONTAL (2F) POSITION FILLET WELDS

Practice horizontal fillet welds by welding multiple-pass (six pass) convex fillet welds in a tee joint using 0.035 inch carbon steel electrode wire and carbon dioxide shielding gas. When making horizontal fillet welds, pay close attention to the gun angle and travel speed. For the first bead, the gun angle is 45 degrees. The angle is adjusted for all subsequent beads. The angle is raised to deposit beads on the lower side and the angle is decreased to deposit beads on the top edge. Increase or decrease the travel speed to control the amount of weld metal build-up.

Follow these steps to make a horizontal fillet weld.

Step 1   Tack two plates together to form a tee joint for the fillet weld coupon.

Step 2   Clamp the coupon to the welding table or tack it to the positioning arm in the 2F (horizontal) position.

Step 3   Run the first bead along the root of the joint using a transverse gun angle of 45 degrees (from the vertical) with a 5- to 10-degree longitudinal drag angle and a slight oscillation.

Step 4   Run the second bead along the bottom toe of the first weld, overlapping about 75 percent of the first bead. Use a transverse gun angle of 20 degrees (from the vertical) with a 5- to 10-degree longitudinal drag gun angle and a slight oscillation.

Step 5   Run the third bead along the top toe of the first weld, filling the platform created when the second bead was run. Use a transverse gun angle of 70 degrees (from the vertical) with a 5- to 10-degree longitudinal gun angle and a slight oscillation.

Figure 9-7 shows weld bead sequences 1, 2 and 3, and the gun angles.

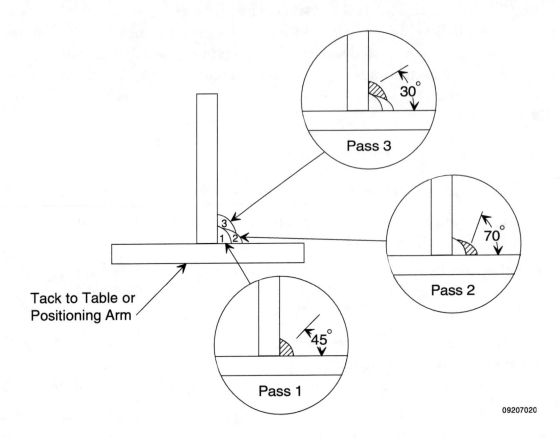

Figure 9-7. Bead Sequences 1, 2 and 3 and Gun Angles.

Step 6 Run the fourth bead along the bottom toe of the second weld, overlapping about half the second bead. Use a transverse gun angle of 20 degrees (from the vertical) with a 5- to 10-degree longitudinal drag gun angle and a slight oscillation.

Step 7 Run the fifth bead along the top toe of the fourth weld, overlapping about half the fourth bead that was run. Use a transverse gun angle of 20 degrees (from the vertical) with a 5- to 10-degree longitudinal drag gun angle and a slight oscillation.

Step 8 Run the sixth bead along the top toe of the third weld, filling the groove created when the fifth bead was run. Use a transverse gun angle of 70 degrees (from the vertical) with a 5- to 10-degree longitudinal drag gun angle and a slight oscillation.

Figure 9-8 shows weld bead sequences 4, 5 and 6 and the gun angles.

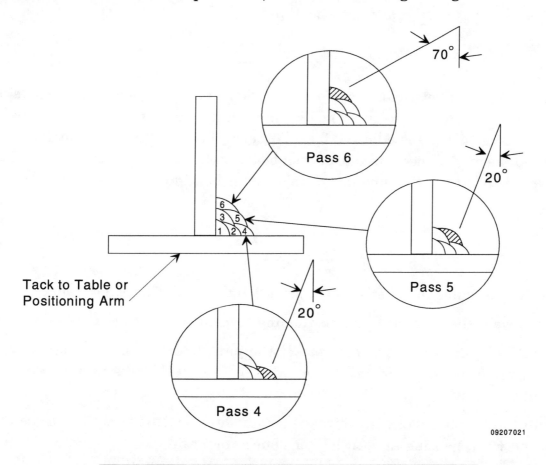

Figure 9-8. Bead Sequences 4, 5 and 6 and Gun Angles.

Step 9   Inspect the weld. The weld is acceptable if it has:

- Uniform rippled appearance on the bead face
- Craters and restarts filled to the full cross-section of the weld
- Uniform weld size, plus or minus 1/16 inch
- Acceptable weld profile in accordance with AWS D1.1
- Smooth transition with complete fusion at the toes of the weld
- No porosity
- No undercut
- No cracks

9.3.0   PRACTICE VERTICAL UP (3F) POSITION FILLET WELDS

Practice vertical up fillet welds by welding multiple-pass convex fillet welds in a tee joint using 0.035 inch carbon steel wire and carbon dioxide shielding gas. When vertical welding, either stringer or weave beads can be used. On the job, the site WPS or Quality Standard will specify which technique to use.

Note   Check with your instructor to see if you should run stringer beads or weave beads or practice both techniques.

When making vertical fillet welds pay close attention to the gun angles and travel speed. For the first bead the transverse gun angle is 45 degrees to either plate (bisects the groove angle). The transverse angle is adjusted for all other welds. It is increased to deposit beads on one side and the angle is decreased to deposit beads on the other side. Increase or decrease the travel speed to control the amount of weld metal build-up.

Follow these steps to make a vertical fillet weld.

Step 1   Tack two plates together to form a tee joint for the fillet weld coupon.

Step 2   Tack-weld the coupon to the positioning arm in the vertical position.

Step 3   Run the first bead along the root of the joint (starting at the bottom) using a transverse gun angle of 45 degrees with a 10- to 15-degree upward angle and a slight oscillation. Pause in the weld puddle to fill the crater.

Step 4   Run the second bead using a weave technique. Use a slow motion across the face of the weld, pausing at each toe for penetration and to fill the crater. Adjust the travel speed across the face of the weld to control the build-up.

Figure 9-9 shows the bead sequence for vertical weave beads.

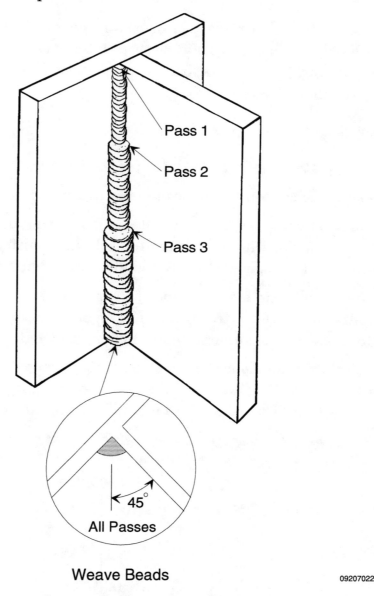

Weave Beads

09207022

Figure 9-9. Bead Sequence For Vertical Weave Beads.

Step 5 Continue to run weld beads as shown in *Figure 9-9*.

Step 6 Inspect the weld. The weld is acceptable if it has:

- Uniform rippled appearance on the bead face
- Craters and restarts filled to the full cross-section of the weld
- Uniform weld size, plus or minus 1/16 inch
- Acceptable weld profile in accordance with AWS D1.1
- Smooth transition with complete fusion at the toes of the weld
- No porosity
- No undercut
- No cracks

9.4.0 PRACTICE OVERHEAD (4F) POSITION FILLET WELDS

Practice overhead fillet welds in a tee joint using stringer beads. Use 0.035 inch carbon steel wire and carbon dioxide shielding gas. Pay close attention to the gun angles and travel speed. The transverse gun angle for the first bead is 45 degrees (from the vertical). The angle is adjusted for all subsequent beads.

Follow these steps to make an overhead fillet weld with stringer beads.

Step 1   Tack two plates together to form a tee joint for the fillet weld coupon.

Step 2   Tack-weld the coupon to the positioning arm in the overhead welding position.

Step 3   Run the first bead along the root of the joint using a transverse gun angle of 0 degrees to the vertical (45 degrees to each plate surface) with a 5- to 10-degree longitudinal drag gun angle and a slight oscillation.

Step 4   Run the second bead along one toe of the first weld, overlapping about 75 percent of the first bead. Use a transverse gun angle of 20 degrees (to the vertical) with a 5- to 10-degree longitudinal gun angle and a slight oscillation.

Step 5   Run the third bead along the other toe of the first weld, filling the groove created when the second bead was run. Use a transverse gun angle of 20 degrees (to the vertical) with a 5- to 10-degree longitudinal gun angle and a slight oscillation.

Figure 9-10 shows overhead bead sequences 1, 2 and 3 and transverse gun angles.

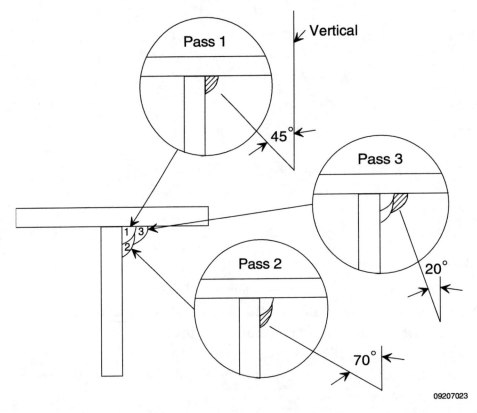

Figure 9-10. Overhead Bead Sequences 1, 2 and 3 and Transverse Gun Angles.

Step 6   Run the fourth bead along the outside toe of the second weld, overlapping about half the second bead. Use a transverse gun angle of 20 degrees with a 5- to 10-degree longitudinal gun angle and a slight oscillation.

Step 7   Run the fifth bead along the inside toe of the fourth weld, overlapping about half the fourth bead that was run. Use a transverse gun angle of 20 degrees with a 5- to 10-degree longitudinal gun angle and a slight oscillation.

Step 8   Run the sixth bead along the outside toe of the third weld, filling the groove created when the fifth bead was run. Use a transverse gun angle of 20 degrees with a 5- to 10-degree longitudinal gun angle and a slight oscillation. *Figure 9-11* shows overhead bead sequences 4, 5 and 6 and transverse gun angles.

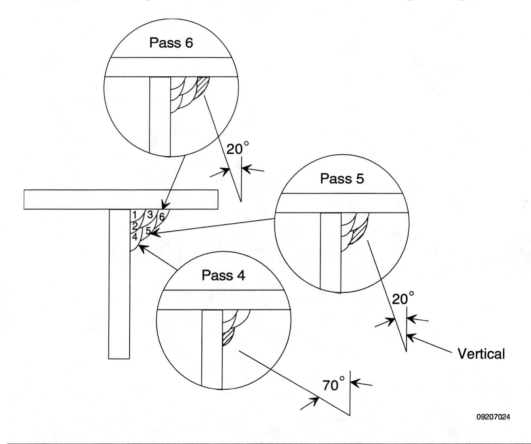

09207024

Figure 9-11. Overhead Bead Sequences 4, 5 and 6 and Transverse Gun Angles.

Step 9   Inspect the weld. The weld is acceptable if it has:

- Uniform rippled appearance on the bead face
- Craters and restarts filled to the full cross-section of the weld
- Uniform weld size, plus or minus 1/16 inch
- Acceptable weld profile in accordance with AWS D1.1
- Smooth transition with complete fusion at the toes of the weld
- No porosity
- No undercut
- No cracks

SELF CHECK REVIEW 2

1. How do you angle the gun to get tie-in when running overlapping stringer beads?
2. What is another name for the horizontal fillet weld position?
3. When making a horizontal fillet weld, what are the gun angles for the first pass?
4. When welding vertically, what determines if you run stringers or weave beads?
5. What are the gun angles for the second pass of an overhead fillet weld?

ANSWERS TO SELF CHECK REVIEW 2

1. Position the gun at a transverse angle of 10 to 15 degrees (from a normal to the plate surface) toward the side of the previous bead. (8.1.0)
2. The horizontal position is also called the 2F position. (9.0.0)
3. The gun is positioned at a 45 degree transverse angle (from the vertical or to each plate) and at a 5- to 10-degree longitudinal angle. (9.2.0)
4. WPS or site quality standards will specify the bead type to use. (9.3.0)
5. Use a 20 degree (from the vertical) transverse angle and a 5- to 10-degree longitudinal drag angle. (9.4.0)

10.0.0 PERFORMANCE QUALIFICATION TASKS

The following tasks are designed to evaluate your ability to run fillet welds with GMAW equipment. Perform each task when you are instructed to do so by your instructor. As you complete each task, show it to your instructor for evaluation. Do not proceed to the next task until directed to do so by your instructor.

10.1.0 MAKE A FILLET WELD IN THE (1F) FLAT POSITION

Using 0.035 inch carbon steel wire, carbon dioxide shielding gas and stringer beads, make a three-pass fillet weld on mild steel plate as shown in *Figure 10-1*.

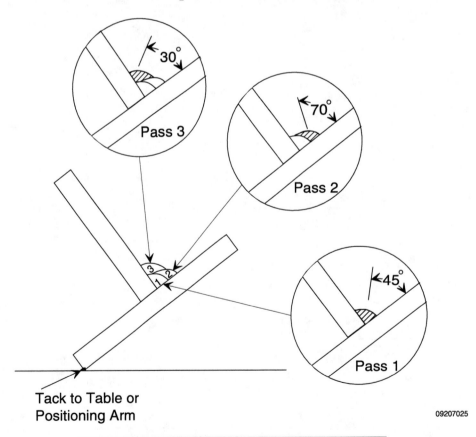

09207025

Figure 10-1. Fillet Weld in the 1F Flat Position.

Criteria For Acceptance:

- Weld beads straight to within 1/8 inch
- Uniform rippled appearance on the bead face
- Craters and restarts filled to the full cross-section of the weld
- Face of the pad flat to within 1/8 inch
- Smooth flat transition with complete fusion at the toes of one bead into the face of the previous bead
- No porosity
- No undercut
- No cracks

10.2.0 MAKE A FILLET WELD IN THE (2F) HORIZONTAL POSITION

Using 0.035 inch carbon steel wire, carbon dioxide shielding gas and stringer beads, make a three-pass fillet weld on mild steel plate as shown in *Figure 10-2*.

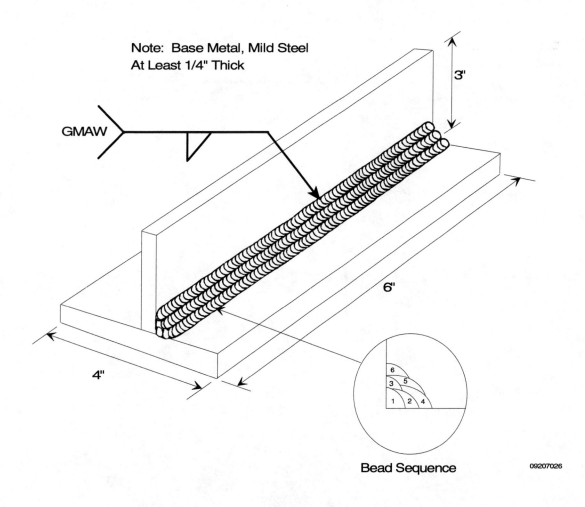

Figure 10-2. Fillet Weld in the 2F Horizontal Position.

Criteria For Acceptance:

- Weld beads straight to within 1/8 inch
- Uniform rippled appearance on the bead face
- Craters and restarts filled to the full cross-section of the weld
- Face of the pad flat to within 1/8 inch
- Smooth flat transition with complete fusion at the toes of one bead into the face of the previous bead
- No porosity
- No undercut
- No cracks

10.3.0   MAKE A FILLET WELD IN THE (3F) VERTICAL POSITION

Using 0.035 inch carbon steel wire, carbon dioxide shielding gas and stringer beads, make a three-pass fillet weld on mild steel plate as shown in *Figure 10-3*.

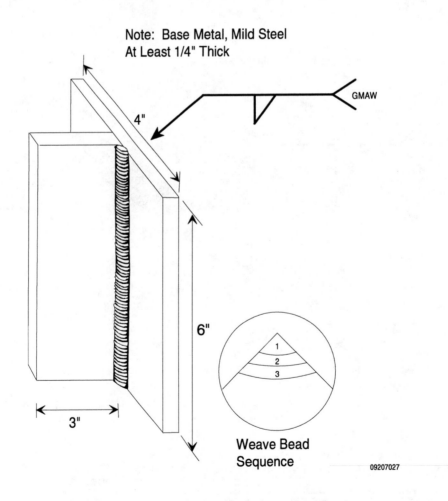

Figure 10-3. Fillet Weld in the 3F Vertical Position.

Criteria For Acceptance:

- Weld beads straight to within 1/8 inch
- Uniform rippled appearance on the bead face
- Craters and restarts filled to the full cross-section of the weld
- Face of the pad flat to within 1/8 inch
- Smooth flat transition with complete fusion at the toes of one bead into the face of the previous bead
- No porosity
- No undercut
- No cracks

10.4.0   MAKE A FILLET WELD IN THE (4F) OVERHEAD POSITION

Using 0.035 inch carbon steel wire, carbon dioxide shielding gas and stringer beads, make a three-pass fillet weld on mild steel plate as shown in *Figure 10-4*.

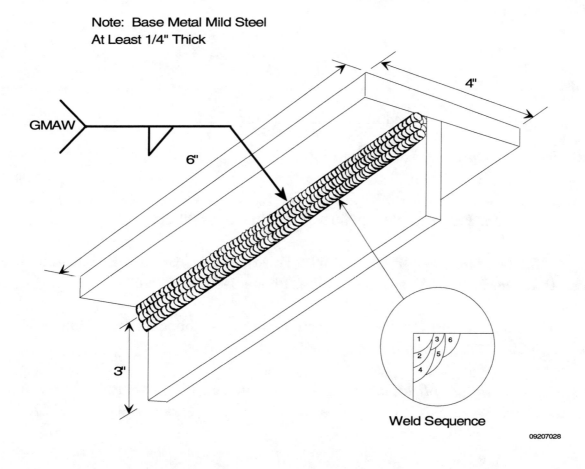

Figure 10-4. Fillet Weld in the 4F Vertical Position.

Criteria For Acceptance:

- Weld beads straight to within 1/8 inch
- Uniform rippled appearance on the bead face
- Craters and restarts filled to the full cross-section of the weld
- Face of the pad flat to within 1/8 inch
- Smooth flat transition with complete fusion at the toes of one bead into the face of the previous bead
- No porosity
- No undercut
- No cracks

SUMMARY

Setting up GMAW equipment, preparing the welding work area, running stringer and weave beads and making acceptable fillet welds in all positions are essential skills a welder must have to perform basic welding jobs or to progress to more difficult welding procedures. Practice these welds until you can consistently produce acceptable welds as defined in the criteria for acceptance.

References

For advanced study of topics covered in this Task Module, the following works are suggested:

OSHA Requirements On Electrical Grounding.

Mig Welding Handbook (P/N 791F18 F-3690-E), L-TEC Welding and Cutting Systems, P.O.Box F-100545, Florence, S.C. 29501-0545, Phone 1-803-669-4411.

Basic Tig & Mig Welding (GTAW & GMAW), (Third Edition) Griffin, Roden and Briggs, Delmar Publishers, Inc., Albany, NY, 1984, Phone 1-800-347-7707.

Welding Skills, Giachino and Weeks, American Technical Publishers Inc., Homewood, IL, 1985, 1-800-323-3471.

Welding Principles and Applications, Jeffus and Johnson, Delmar Publishers, Inc., 2 Computer Drive West, Box 15-015, Albany, N.Y. 12212, Phone 1-800-347-7707.

PERFORMANCE / LABORATORY EXERCISES

1. Practice stringer beads using carbon steel wire and carbon dioxide gas.
2. Practice weave beads using carbon steel wire and carbon dioxide gas.
3. Practice overlapping stringer beads using carbon steel wire and carbon dioxide gas.
4. Practice overlapping weave beads using carbon steel wire and carbon dioxide gas.
5. Practice flat (1F) position fillet welds on steel plate using carbon steel wire and carbon dioxide gas.
6. Practice horizontal (2F) position fillet welds on steel plate using carbon steel wire and carbon dioxide gas.
7. Practice vertical up (3F) position fillet welds on steel plate using carbon steel wire and carbon dioxide gas.
8. Practice overhead (4F) position fillet welds on steel plate using carbon steel wire and carbon dioxide gas.

The NCCER makes every effort to keep these manuals up-to-date and free of technical errors. We appreciate your help in this process. If you have an idea for improving this manual, or if you find an error, a typographical mistake, or an inaccuracy in the *Wheels of Learning*, please write us, using this form or a photocopy. Be sure to include the exact module number, page number, a description of the problem, and the correction, if possible. We'll do our best to correct it in later editions. Thank you for your assistance.

Write: *Wheels of Learning*
National Center for Construction Education and Research
P.O. Box 141104
Gainesville, FL 32614-1104

Fax: 352-334-0932

WHEELS OF LEARNING USER UPDATE

Please let us know if you have found an inaccuracy, error, or other problem in a *Wheels of Learning* manual. Use this form or write us a letter. Please be sure to tell us the exact module name and module number, the page number, and the problem. Thanks for your help.

Craft _____ Module Name _____

Module Number _____ Page Number(s) _____

Description of Problem _____

_____

_____

_____

(Optional) Correction of Problem _____

_____

_____

(Optional) Your Name and Address _____

_____

_____

FCAW-Equipment and Filler Metal

Module 09209

Welder Trainee Task Module 09209

NATIONAL
CENTER FOR
CONSTRUCTION
EDUCATION AND
RESEARCH

FLUX CORE ARC WELDING (FCAW) - EQUIPMENT AND FILLER METALS

OBJECTIVES

Upon completion of this module, the trainee will be able to:

1. Explain Flux Core Arc Welding (FCAW) safety.
2. Identify and explain FCAW equipment.
3. Identify and explain FCAW filler metals.
4. Identify and explain FCAW shielding gasses.
5. Set up FCAW welding equipment.

Prerequisites

Successful completion of the following Task Module(s) is required before beginning study of this Task Module:

- Safety (Common Core)
- Rigging/Material Handling (Common Core)

Required Student Materials

None

Course Map Information

This course map shows all of the *Wheels of Learning* task modules in the second level of the Welding curricula. The suggested training order begins at the bottom and proceeds up. Skill levels increase as a trainee advances on the course map. The training order may be adjusted by the local Training Program Sponsor.

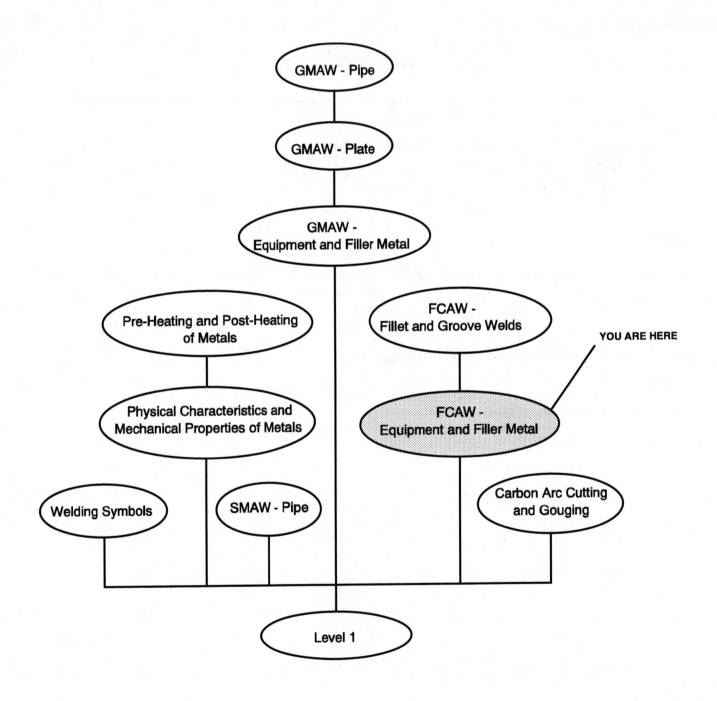

WELDING TWO TRAINEE TASK MODULE 09209

TABLE OF CONTENTS

Trade Terms Introduced in This Module

Alternating Current (AC): Electrical current that reverses its flow at set intervals.

Amperage: A measurement of the rate of flow of electric current.

Arc: Flow of electricity through an air gap which melts the electrode.

Arc Blow: The deflection of an arc from its normal course by a magnetic field.

Arc Flash: Burns to the eyes or skin caused by the arc.

Direct Current (DC): Electrical current that flows in one direction only.

Duty Cycle: The percentage of a ten minute period in which a welding machine can deliver its rated output.

Electrical Stickout: A characteristic of self-shielding electrode guns where the welding current is introduced to the electrode before the end of the contact tip, to preheat the electrode.

Electrode: The filler wire point from which the arc is produced to perform FCAW welding.

Ground: An object that makes an electrical connection with the earth. Also, the welding cable (work lead) attached to the base metal.

Hot Work Permit: Official authorization from a site manager to perform work which may pose a fire hazard.

Out-Of-Position Welding: Any welding position other than the flat position; includes vertical, horizontal, and overhead welding.

Polarity: The direction of flow of electrical current in a direct current welding circuit.

Primary Current: Electrical current received from conventional power lines used to energize welding machines.

Slag: Crusty by-product which forms on the weld bead and is produced by flux or by metallic or non-metallic impurities in the metal.

SMAW: Shielded Metal Arc Welding, an arc welding process using solid wire electrodes with external flux coatings.

Spray-Arc Transfer: GMAW and FCAW welding process where the electrode is held away from the base metal and the filler metal is transferred in the arc as a spray of fine droplets.

Spatter: Metal particles which are expelled into the air during welding.

Stick-Out: The length of electrode that projects from the contact tube, measured from the arc tip to the contact tip.

Voltage: A measurement of electromotive force or pressure that causes current to flow in a circuit.

1.0.0 INTRODUCTION

Flux Core Arc Welding (FCAW) uses an electric arc to melt a flux core (tubular) wire electrode and fuse it with the base metal to form a weld. A wire feeder automatically feeds the flux core wire as it is being consumed in the arc. The equipment used for FCAW is basically the same as that used for Gas Metal Arc Welding (GMAW). The main difference between the two processes is that GMAW uses a solid wire while FCAW uses a tubular flux core wire.

There are two basic FCAW processes, one uses an external shielding gas while the other is self-shielding. Whether or not a shielding gas is used is determined by the type of flux core wire used. When the flux core wire does not require an external shielding gas the process is sometimes called self shielding FCAW or inner shield FCAW, when an external shielding gas is required the process is sometimes called dual shielded FCAW or gas shielded FCAW. Equipment specifically designed for self shielding FCAW does not have provisions for shielding gas (solenoids, pre and post timers) and the guns do not have gas nozzles. Equipment used for gas shielded FCAW welding is the same as that used for GMAW except that the feeder drive assemblies have to be able to handle the core wire which is softer and generally larger in diameter than solid wire. *Figure 1-1* shows the basic FCAW welding process.

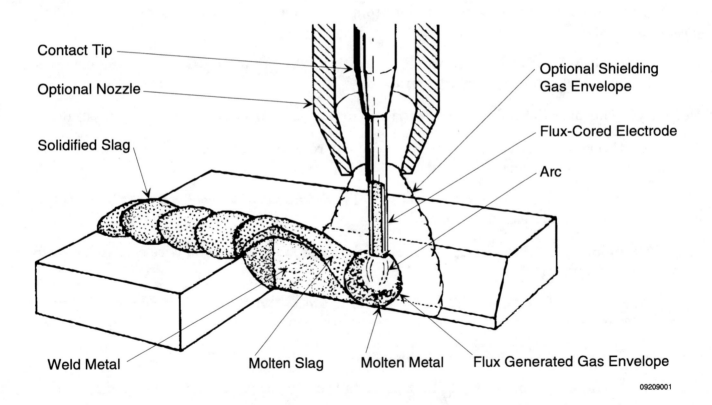

09209001

Figure 1-1. Basic FCAW Process

Flux Core Arc Welding (FCAW) is presently limited to the welding of ferrous metals. These include low- and medium-carbon steels, some low-alloy steels, cast irons and some stainless steels. When critical welds are required, the gas shielded flux core arc welding process is often used because the combination of flux and separate shielding gas produces very high quality welds.

This module provides an overview of the different items of equipment required for gas shielded flux core arc welding. Topics include safety and equipment, power supplies, wire feeders, electrode guns and equipment set-up. Upon completion of this unit, students will be aware of the safety concerns associated with FCAW and will be able to select and set up FCAW equipment safely and efficiently.

Note   All references and procedures covered in this manual will pertain to the gas shielded flux core arc welding process.

2.0.0   FCAW SAFETY

The two major hazards associated with flux core arc welding are heat and arc flash. The intense heat of the process can cause severe burns, and sparks and spatter created during welding are a burn and a fire hazard. The arc gives off ultraviolet and infrared rays which will burn unprotected eyes and skin much as a sunburn does, only much more severely. Severe arc burn to the eyes can cause permanent eye damage. The following sections will explain how FCAW can be safely performed.

2.1.0   PERSONAL PROTECTIVE EQUIPMENT

Because of the potential dangers of FCAW, welders must use personal protective equipment. This equipment includes the proper work clothing, boots, leathers, gloves, safety glasses and special welding helmets. This personal protective equipment is explained in the following sections.

2.1.1   Work Clothing

To avoid burns from sparks and ultraviolet rays, welders must wear the appropriate clothing. Never wear polyester or other synthetic fibers. Sparks or intense heat will melt these materials, causing severe burns. Wool or cotton is more resistant to sparks and therefore should be worn. Dark clothing is also preferred because dark clothes minimize the reflection of arc rays which could be deflected under the welding helmet.

To prevent sparks from lodging in clothing and causing a fire or burns, collars should be kept buttoned and pockets should have flaps that can be buttoned to keep out the sparks. A soft cotton cap worn with the visor reversed will protect the top of your head and keep sparks from going down the back of your collar. Pants should be cuffless and hang straight down the leg. Pant cuffs will catch sparks and catch on fire. Never wear frayed or fuzzy materials. These materials will trap sparks which will catch the clothing on fire. Low-top shoes should never be worn while welding. Sparks will fall into the shoes, causing severe burns. Leather work boots at least eight inches high should be used to protect against sparks and arc flash. *Figure 2-1* shows proper work clothing.

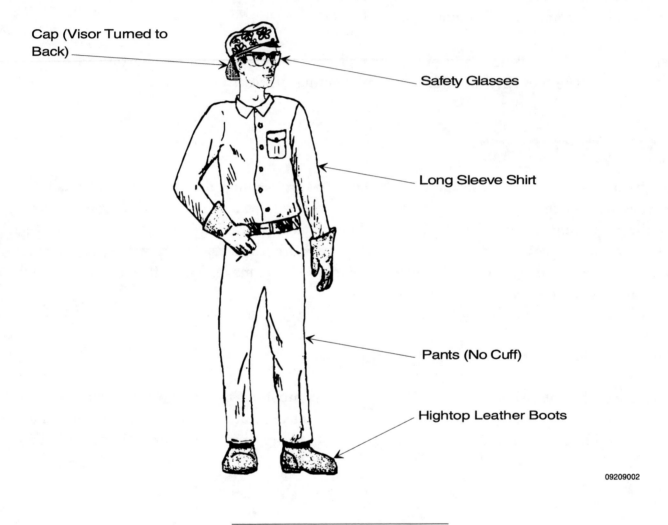

09209002

Figure 2-1. Proper Work Clothing

2.1.2 Leathers

For additional protection, welders often wear leathers over their work clothing. Leather aprons, split leg aprons, sleeves and jackets are flexible enough to allow ease of movement while providing added protection from sparks and heat. Leathers should always be worn when welding out-of-position or when welding in tight quarters.

WELDING TWO TRAINEE TASK MODULE 09209

Figure 2-2 shows the different types of leather protective clothing.

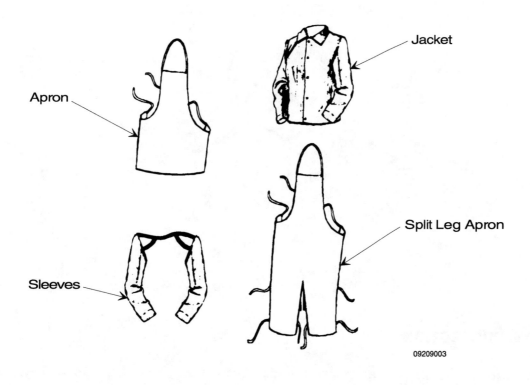

Figure 2-2. Leather Protective Clothing

2.1.3   Welding Gloves

Leather gauntlet-type gloves are designed specifically for welding. They must be worn when performing any type of arc welding to protect against spattering hot metal and the ultraviolet and infrared arc rays. Leather welding gloves are not designed to handle hot metal. Pliers, tongs or some other means should be used to handle hot metal.

CAUTION   Picking up hot metal with leather welding gloves will burn the leather, causing it to shrivel and become hard. Never handle hot metal with leather welding gloves.

Figure 2-3 shows typical leather gauntlet-type welding gloves.

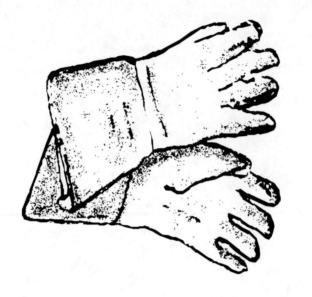

09209004

Figure 2-3. Leather Welding Gloves

2.2.0 EYE PROTECTION

Eyes are delicate and can be easily damaged. The welding arc gives off ultraviolet and infrared rays which will burn unprotected eyes. The burns caused by the arc are called arc flash. Arc flash causes blistering of the outer eye. This blistering feels like sand or grit in the eyes. It can be caused from looking directly at the arc or from receiving reflected glare from the arc.

Another eye hazard associated with flux core arc welding is flying debris. Grinding and surface cleaning operations create many small particles which are propelled in all directions. Slag, the crusty substance which forms on the deposited weld, is removed with a chipping hammer. During chipping, tiny particles of slag fly in all directions. In order to avoid eye injury from flying debris, eye protection must be worn.

2.2.1 Safety Glasses and Goggles

Safety glasses and goggles protect the eyes from flying debris. Safety glasses have impact-resistant lenses. Side shields should be fitted to safety glasses to prevent flying debris from entering from the side. Some safety glasses are equipped with shaded lenses to protect against glare.

WARNING!	Shaded safety glasses will not provide sufficient protection from the arc. Looking directly at the arc with shaded safety glasses will result in severe burns to the eyes and face.

Safety goggles are contoured to fit the wearer's face. Goggles are a very efficient means of protecting the eyes from flying debris. They are often worn over safety glasses to provide extra protection when grinding or performing surface cleaning.

Figure 2-4 shows safety glasses and goggles.

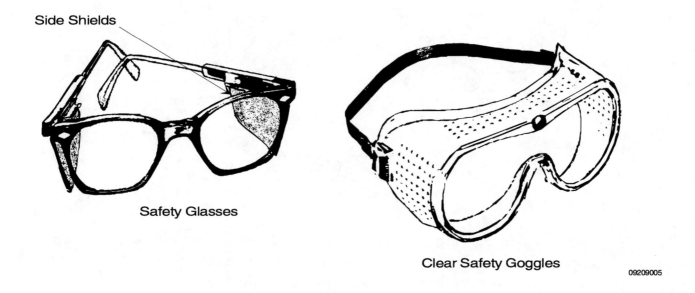

Side Shields

Safety Glasses

Clear Safety Goggles

09209005

Figure 2-4. Safety Glasses and Goggles

2.2.2   Welding Shields

Welding shields (also called welding helmets) provide eye and face protection for welders. Some shields are equipped with handles, but most are designed to be worn on the head. They either connect to helmet-like headgear or attach to a hardhat. Shields designed to be worn on the head have pivot points where they attach to the headgear. They can be raised when not needed. Welding shields are made of dark, non-flammable material. The welder observes the arc through a window that is either 2-1/2 by 4-1/4 inches or 4-1/2 by 5-1/4 inches. The window contains a glass filter plate and a clear glass or plastic safety lens. The safety lens is on the outside to protect the more costly filter plate from damage by spatter and debris. For additional filter plate protection a clear safety lens is sometimes also placed on the inside of the filter plate. On most welding shields the window is fixed in the shield. However some welding shields have a hinge on the 2-1/2 by 4-1/4 inch window. The hinged window containing the filter plate can be raised, leaving a separate clear safety lens. This protects the welder's face from hot slag during weld cleaning.

Figure 2-5 shows typical welding shields.

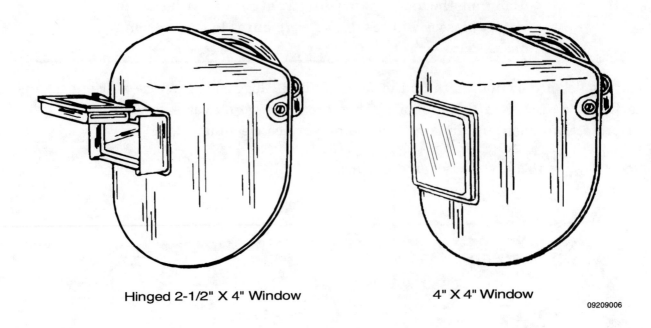

Hinged 2-1/2" X 4" Window   4" X 4" Window

09209006

Figure 2-5. Welding Shields

Filter plates are made in varying shades. The shade required depends on the maximum amount of amperage to be used. The higher the amperage, the more intense the arc and the darker the filter plate must be. Filter plates are graded by numbers. The larger the number, the darker the filter plate. The American National Standards Institute (ANSI) publication Z87.1, *Practices for Occupational and Educational Eye and Face Protection,* provides guidelines for selecting filter plates. The following recommendations are based on these guidelines.

	Welding Current (Amps)	Lowest Shade Number	Comfort Shade Number
Gas Metal Arc and Flux Core Arc Welding	Under 60	7	--
	60-160	10	11
	160-250	10	12
	250-500	10	14

To select the best shade lens, first start with the darkest lens recommended. If it is difficult to see, try a lighter shade lens until there is good visibility. However, do not go below the lowest recommended number.

During normal welding operations the window in the welding shield will become dirty from smoke and weld spatter. To be able to see properly the window will have to be cleaned periodically. The safety lens and filter plate can be easily removed and cleaned in the same manner that safety glasses are cleaned. With use, the outer safety lens will become impregnated with weld spatter. When this occurs, replace it.

2.3.0   EAR PROTECTION

Ear protection is necessary to prevent hearing loss from noise. One source of noise for the welder is pneumatic chipping and scaling hammers. Welders must also protect their ears from flying sparks and weld spatter. This is especially important for out-of-position welding where falling sparks and weld spatter may enter the ear canal, causing painful burns. It is also common for welders who do not protect their ears to suffer from perforated ear drums caused by sparks. Always wear either earmuffs or earplugs for protection. *Figure 2-6* shows earmuffs and ear plugs.

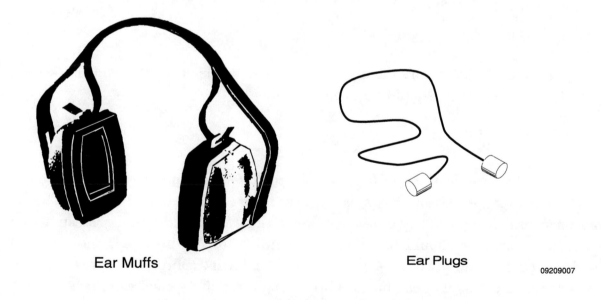

Ear Muffs   Ear Plugs

09209007

Figure 2-6. Earmuffs and Earplugs

2.4.0   VENTILATION

The gases, dust and fumes caused by flux core arc welding can be hazardous if the appropriate safety precautions are not observed. The following section will define these hazards and describe the appropriate safety precautions.

2.4.1 Fume Hazard

Metals heated during FCAW may give off toxic fumes and smoke. These fumes are not considered dangerous as long as there is adequate ventilation. Adequate ventilation can be a problem in tight or cramped working quarters. The following general rules can be used to determine if there is adequate ventilation:

● The welding area should contain at least 10,000 cubic feet of air for each welder
● There should be positive air circulation
● Air circulation should not be blocked by partitions, structural barriers, or equipment

Even when there is adequate ventilation, the welder should try to avoid inhaling welding fumes and smoke. The heated fumes and smoke generally rise straight up from the welding arc. Observe the column of smoke and position yourself to avoid it. A small fan may also be used to divert the smoke, but care must be taken to keep the fan from blowing directly on the arc. The shielding gas must be present at the arc in order to protect the molten metal from the air.

CAUTION   When using FCAW, never point a fan or compressed air stream directly at the arc because it could blow away the protective smoke and shielding gas. Removal of the protective smoke and shielding gas will cause weld defects.

If adequate ventilation cannot be ensured or if the welder cannot avoid the welding fumes and smoke, a respirator should be used.

2.4.2 Respirator

For most FCAW applications, a filter respirator offers adequate protection from dust and fumes. The filter respirator is a mask which covers the nose and mouth. The air intake contains a cloth or paper filter to remove impurities. When additional protection is required, an air-line respirator should be used. It is similar to the filter respirator, except that it provides breathing air through a hose from an external source. The breathing air can be supplied form a cylinder of compressed breathing air, from special compressors that furnish breathing air or from special breathing air filters attached to compressed air lines.

WARNING!   Air line respirators must only be used with pure breathing air sources. Standard compressed air contains oil, which is toxic. Standard compressed air can only be used for breathing if it is cleaned by a special breathing air filter placed in the line just before the respirator.

Figure 2-7 shows respirators.

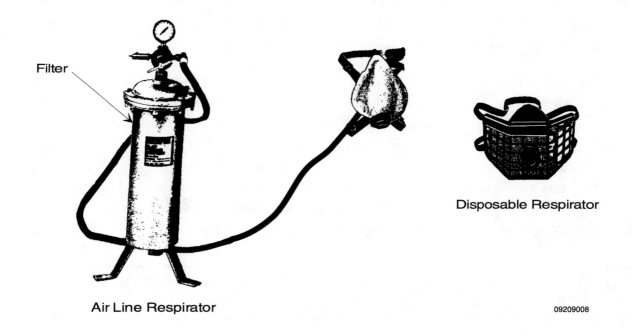

Figure 2-7. Respirators

WARNING! A filter type respirator will not supply oxygen in a low-oxygen environment. If normal oxygen concentration has been reduced because of displacement by shielding gasses, an air-line respirator must be used or suffocation and death could result.

2.4.3 SUFFOCATION

In confined spaces, the shielding gas used with the FCAW process can accumulate and displace the oxygen atmosphere, making suffocation possible. When using the FCAW process in a confined space, the welder must have positive ventilation to remove the shielding gas.

WARNING! When using FCAW in a confined space, positive ventilation must be provided. Failure to provide adequate ventilation could result in suffocation and death.

2.5.0   ELECTRICAL SHOCK HAZARD

Most FCAW welding machines are connected to alternating current (AC) voltages of 208-460 volts. Contact with these voltages can cause extreme shock and possibly death. For this reason never open a welding machine with power on the primary circuit. To prevent electrical shock hazard, the welding machine must always be grounded.

A ground is a conductor that connects an electrical circuit to the earth. Electrical grounding is necessary to prevent the electrical shock which can result from contact with a defective welding machine or other electrical device. The ground provides protection by providing a path from the equipment to ground for stray electrical current created by a short circuit or other defect. The short or other defect will cause the fuse or breaker to open (blow), stopping the flow of current. Without a ground the stray electricity would be present in the frame of the equipment. If someone were to touch the frame the stray current would go to ground through the person, with possible severe injury to the person.

WARNING!   Death by electrocution can occur if equipment is not
                    properly grounded.

To ensure proper electrical grounding, a welding machine which is using single-phase power must have a three-prong outlet plug. Machines which use three-phase power must have a four-prong outlet plug. Welding machines must only be plugged into electrical outlets which have been properly installed by licensed electricians to ensure the ground has been properly connected. *Figure 2-8* shows grounded outlet plugs.

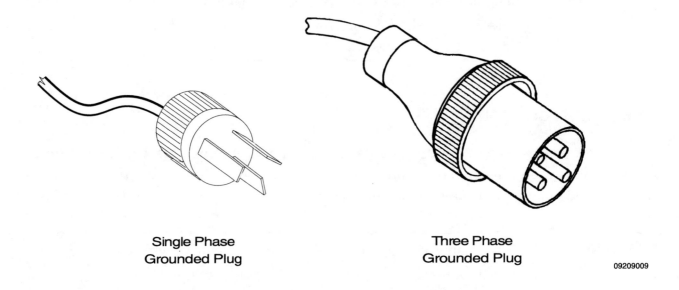

Single Phase
Grounded Plug

Three Phase
Grounded Plug

09209009

Figure 2-8. Grounded Outlet Plugs

2.6.0 AREA SAFETY

Before beginning a welding job, the area must be checked for safety hazards.

Check the area for water. If possible, remove standing water from floors and work surfaces where welding cables or machines are in use in order to avoid injury from slipping or electric shock. When it is necessary to work in a damp environment, make sure that you are well-insulated by wearing rubber boots and gloves. Protect equipment by placing it on pallets above the water and shielding it from overhead leaks. To avoid electric shock, never weld with wet gloves.

Check the area for fire hazards. Welding generates sparks that can fly ten feet or more and can drop several floors. Any flammable materials in the area must be moved or covered. Always have an approved fire extinguisher on hand before starting any welding job.

Other workers in the area must be protected from arc flash. If possible, enclose the welding area with screens to avoid exposing others to harmful rays or flying debris. Inform everyone in the area that you are going to be welding so that they can exercise appropriate caution.

2.7.0 HOT WORK PERMITS AND FIRE WATCH

Most sites require the use of hot work permits and fire watches. Severe penalties are imposed for violation of the hot work permit and fire watch standards.

WARNING!   Never perform any type of heating, cutting or welding until you have obtained a hot work permit and established a fire watch. If you are unsure of the procedure, check with your supervisor. Violation of the hot work permit and fire-watch regulations can result in fires and injury.

A hot work permit is an official authorization from the site manager to perform work which may pose a fire hazard. The permit will include information such as the time, location and type of work being done. The hot work permit system promotes the development of standard fire safety guidelines. Permits also help managers to keep records of who is working where and at what time. This information is essential in the event of an emergency or when personnel need to be evacuated.

Welders should always have a fire watch. During a fire watch, one person (other than the welder) constantly scans the work area for fires. Fire watch personnel should have ready access to fire extinguishers and alarms and know how to use them.

2.8.0   WELDING CONTAINERS

Before welding or cutting containers such as tanks, barrels and other vessels, check to see if they contain combustible and/or hazardous materials or residues of these materials. Such materials include:

- Petroleum products
- Chemicals that give off toxic fumes when heated
- Acids that could produce hydrogen gas as a result of chemical reaction
- Any explosive or flammable residue

To identify the contents of containers check the label and then refer to the MSDS (Material Safety Data Sheet). The MSDS provides information about the chemical or material to help you determine if the material is hazardous. If the label is missing, or if you suspect the container has been used to hold materials other than what is on the label, do not proceed until the material is identified.

WARNING!   Do not heat, cut or weld on a container until its contents have been identified. Hazardous material could cause the container to explode violently.

If a container has held any hazardous materials, it must be cleaned before welding takes place. Clean containers by steam cleaning, flushing with water or washing with detergent until all traces of the material have been removed.

After cleaning the container, fill it with water or an inert gas such as argon or carbon dioxide (CO_2) for additional safety. Air, which contains oxygen, is displaced from inside the container by the water or inert gas. Without oxygen combustion cannot take place. When using water, position the container to minimize the air space. When using an inert gas, provide a vent hole.

Figure 2-9 shows using water in a container to minimize the air space.

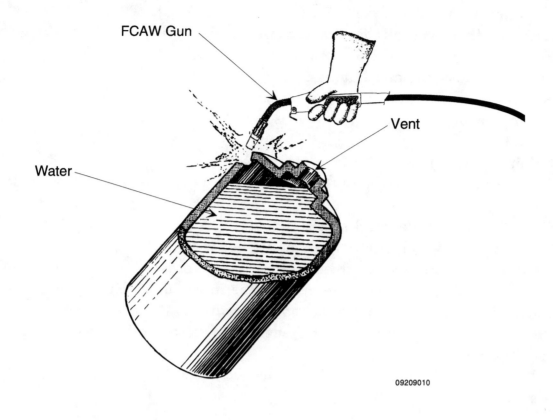

Figure 2-9. Using Water in a Container to Minimize the Air Space

SELF-CHECK REVIEW 1

1. Name the two safety hazards associated with FCAW.
2. Identify what types of fabric should not be worn while welding and explain why.
3. What additional protection is worn over a welders work clothing?
4. What type of eye protection do welders wear?
5. Why is it important to wear ear protection while welding?
6. What should a welder do if it is impossible to provide adequate ventilation in the work area?
7. Why is positive ventilation a concern when performing FCAW in a confined space?
8. Why is the proper electrical grounding of welding equipment necessary?
9. What area checks must be made before welding can take place?
10. How can you protect other workers in the area from hazards while you are welding?
11. What is a hot work permit?
12. After cleaning a container, what else should be done to make it safe for heating, cutting or welding?

ANSWERS TO SELF-CHECK 1

1. Heat and arc flash. (2.0.0)
2. Polyester or other synthetic fabrics because sparks or intense heat will melt these materials, causing severe burns. (2.1.1)
3. Leather aprons, sleeves, and leggings. (2.1.2)
4. Safety glasses, goggles and welding shields. (2.2.1 and 2.2.2)
5. To prevent hearing loss due to excess noise and to keep sparks from entering the ear canal. (2.3.0)
6. Use an air-line respirator. (2.4.2)
7. FCAW shielding gas can displace the oxygen atmosphere and suffocation could result. (2.4.3)
8. To prevent dangerous electrical shock. (2.5.0)
9. Check the area for water, fire hazards and other workers. (2.6.0)
10. Enclose the welding area with screens (if possible) in order to avoid exposing others to harmful rays or flying debris. Inform everyone in the area that you are going to be welding so that they can exercise appropriate caution. (2.6.0)
11. An official authorization from a site manager to perform work which may pose a fire hazard. (2.7.0)
12. It should be filled with water or an inert gas such as argon or carbon dioxide (CO_2) to displace the oxygen. (2.8.0)

3.0.0 CHARACTERISTICS OF WELDING CURRENT

The current produced by a welding machine to perform welding has different characteristics than the current that flows through utility power lines. Welding current has low voltage and high amperage, while the power line current has high voltage and low amperage.

3.1.0 VOLTAGE

Voltage is the measure of the electromotive force or pressure that causes current to flow in a circuit. There are two types of voltage associated with welding current: open circuit voltage and operating voltage. Open circuit voltage is the voltage present when the machine is on but no welding is being done. For FCAW, there are usually ranges of open circuit voltages that can be selected up to about 80 volts. Operating voltage, or arc voltage, is the voltage after the arc is struck. With FCAW, this voltage is generally slightly lower than the open circuit voltage. The arc voltage is typically 2 to 3 volts lower than the open circuit voltage for each 100 amperes of current, but depends on the range selected.

3.2.0 AMPERAGE

Amperage is the electric current flow in a circuit. The unit of measurement for amperage is the Ampere (amp).

In welding, the current flows in a closed loop through two heavy leads (welding cables); the ground lead connects the power supply to the base metal, and the other lead connects the power supply to the consumable wire electrode. During welding, an arc is established between the end of the wire electrode and the work. The arc generates intense heat of 6000 to 10,000°F, melting the base metal and wire electrode and forming the weld. The amount of amperes produced by the welding machine determines the intensity of the arc and the amount of heat available to melt the work and the electrode.

4.0.0   WELDING POWER SUPPLY TYPES

Different types of welding machines (power supplies) are available for FCAW. They include:

- Transformer-Rectifier Welding Machines
- Motor Generator Welding Machines
- Engine-Driven Generator and Alternator Welding Machines

DC welding machines are usually designed to produce either constant current or constant voltage (constant potential) DC welding current. A constant current machine produces a constant current output over a wide voltage range. This is typical of a SMAW welding machine. A constant voltage machine maintains a constant voltage as the output current varies. This is typical of a FCAW or GMAW (MIG) welding machine. The output voltages and output currents of a welding machine can be plotted on a graph to form a curve. These curves show how the output voltage relates to the output current as either changes.

Figure 4-1 shows variable voltage and constant voltage output curves.

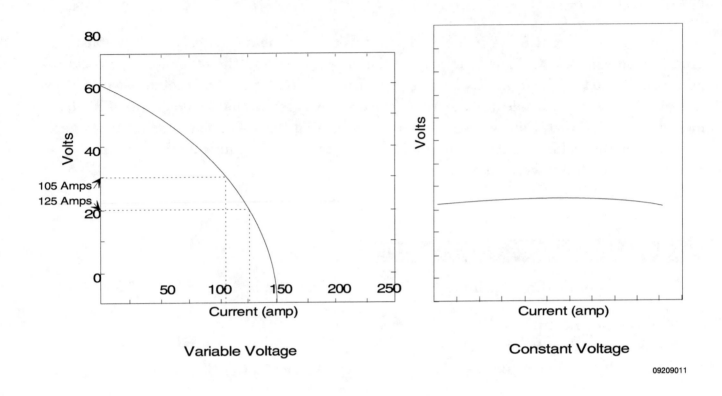

Figure 4-1. Variable Voltage and Constant Voltage Output Curves

4.1.0   TRANSFORMER-RECTIFIER WELDING MACHINES

The transformer-rectifier welding machine uses a transformer to convert the primary current to welding current and a rectifier to change the current from AC to DC. A rectifier is a device for converting AC current to DC current. Transformer-rectifier welding machines can be designed to produce AC and DC welding current or DC current only. Transformer-rectifiers which produce both AC and DC welding current are usually lighter duty than those which produce DC only. Transformer-rectifier welding machines that produce DC welding current only are sometimes called rectifiers. Depending on the size, transformer-rectifier welding machines may require 230-volt single-phase power, 230-volt three-phase power, or 460-volt three-phase power.

WARNING!   Coming into contact with the primary voltage of a welding machine can cause death by electrocution. Ensure that welding machines are properly grounded to prevent injury.

Transformer-rectifiers can be designed to produce variable voltage (constant current) or constant voltage (constant potential) DC welding current. Some will produce either variable or constant voltage by setting a switch. These are referred to as multi-process power sources since they can be used for any welding process including FCAW or GMAW.

Transformer rectifiers used for FCAW have an on/off switch and a voltage control. If the machine has a CV/CC (constant voltage/constant current) switch, it also has an amperage control. When the switch is set to CV for FCAW, the amperage control is disabled. The welding cables (electrode cable and ground cable) are connected to terminals marked ELECTRODE and GROUND or POSITIVE (+) and NEGATIVE (-). They often have selector switches to select DCSP or DCRP. If there is no selector switch, the cables must be manually changed on the machine terminals to select the type of current desired. *Figure 4-2* shows a typical transformer-rectifier welding machine.

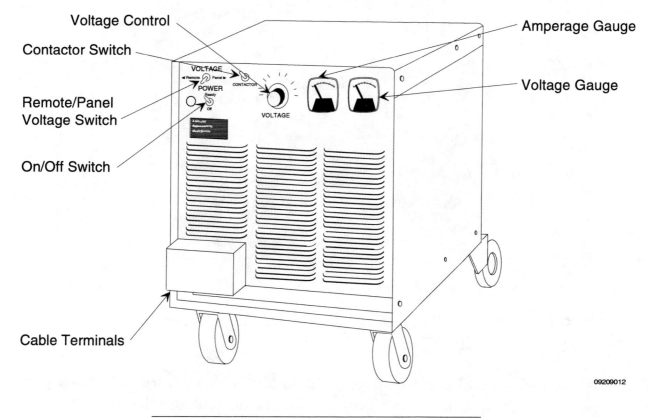

Figure 4-2. Transformer-Rectifier Welding Machine

4.2.0 MOTOR-GENERATOR WELDING MACHINES

Motor-generator welding machines use an electric motor to turn a generator that produces DC welding current. The electric motor requires 440-volt, three-phase primary current. Motor generators can be designed to produce variable voltage, constant voltage or both. When both types of current are available, a selector switch is provided to choose the type of current desired.

Figure 4-3 shows a typical motor-generator welding machine.

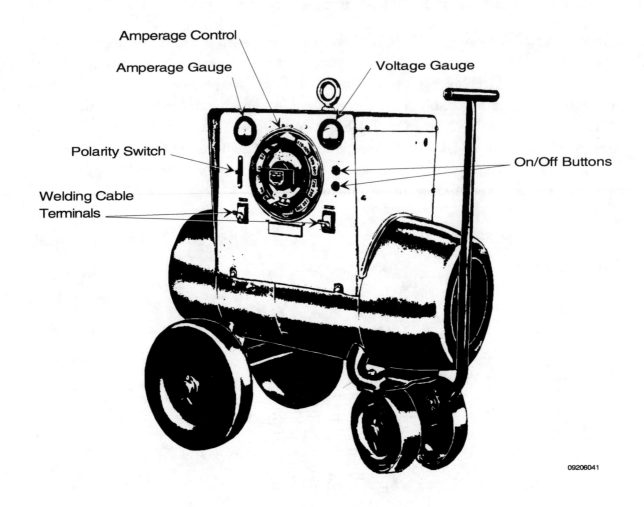

Figure 4-3. Motor-Generator Welding Machine

Motor generators used for FCAW have an on/off switch, voltage control and polarity switch. Some may have voltage and current gauges. They have welding cable terminals marked ELECTRODE and GROUND or POSITIVE (+) and NEGATIVE (-). If the machine has a CV/CC switch, it also has an amperage control. When the switch is set to CV for FCAW, the amperage control is disabled.

Motor generators have fans which pull cooling air through the unit. In dusty conditions, dust builds up inside the unit. This buildup can cause overheating and excessive wear to the armature of the generator. Periodically clean motor generators by blowing them out with compressed air while the unit is running. Blowing out the units while they are running ensures the dust is discharged and not compacted inside the unit.

WARNING!	Motor generators use 460-volt three-phase primary power. Coming into contact with the primary current can cause death by electrocution. When blowing out motor generators, keep the end of the air hose well away from internal parts. Never remove guards.

4.3.0   ENGINE-DRIVEN GENERATOR AND ALTERNATOR WELDING MACHINES

Welding machines can also be powered by gasoline or diesel engines. The engine can be connected to a generator or to an alternator. Engine-driven generators produce DC welding current. Engine-driven alternators produce AC current which is fed through a rectifier to produce DC welding current.

The size and type of engine used depends on the size of the welding machine. Single-cylinder engines are used to power small rectifier alternators while six-cylinder engines are used to power larger generators.

To produce welding current, the generator or alternator must turn at a required rpm (revolutions per minute). The engines powering alternators and generators have governors to control the engine speed. Most governors will have a welding speed switch. The switch can be set to idle the engine when no welding is taking place. When the electrode is touched to the base metal, the governor will automatically increase the speed of the engine to the required rpm for welding. After about 15 seconds of no welding the engine will automatically return to an idle. The switch can also be set for the engine to run continuously at the welding speed. *Figure 4-4* shows an example of an engine-driven generator welding machine.

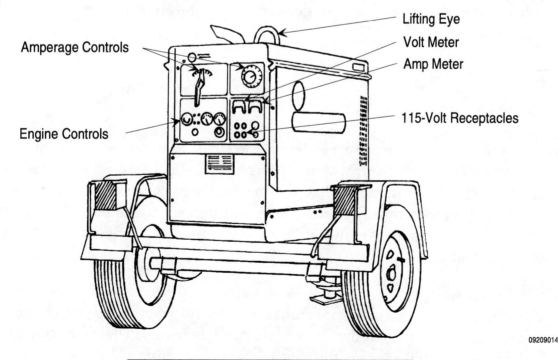

09209014

Figure 4-4. Engine Driven Welding Machine

Engine-driven generators and alternators often have an auxiliary power unit to produce 115 VAC for lighting, power tools, and other electrical equipment. When 115 VAC is required, the engine-driven generator or alternator must run continuously at the welding speed.

When used for FCAW, engine driven generators must be set to run continuously. The wire feeder requires 115 VAC to operate and welding current is required as soon as the wire makes contact with the work. If welding current is not available when the wire touches the work, the wire feeder may jamb.

Engine-driven generators and alternators have engine controls and welding current controls. The engine controls vary with the type and size of engine, but normally include:

- Starter
- Voltage Gauge
- Temperature Gauge
- Fuel Gauge
- Hour Meter

Engine driven generators and alternators used for FCAW have a voltage control. There may be a polarity switch or you may have to manually change the welding cables at the welding current terminals to change welding polarity. If the machine has a CV/CC switch, it also has an amperage control. When the switch is set to CV for FCAW, the amperage control is disabled.

The advantage of engine-driven generators and alternators is that they are portable and can be used in the field where electricity is not available for other types of welding machines. The disadvantage is that engine-driven generators and alternators are costly to purchase, operate and maintain.

4.4.0   POWER SUPPLY RATINGS

The rating (size) of a welding machine is determined by the amperage output of the machine at a given duty cycle. The duty cycle of a welding machine is based on a ten-minute period. It is the percentage of ten minutes that the machine can continuously produce its rated amperage without overheating. For example, a machine with a rated output of 300 amps at 60 percent duty cycle can deliver 300 amps of welding current for six minutes out of every ten without overheating.

The duty cycle of a welding machine will be 10 percent, 20 percent, 30 percent, 40 percent, 60 percent, or 100 percent. A welding machine having a duty cycle of 10 percent to 40 percent is considered a light- to medium-duty machine. Most industrial, heavy-duty machines for manual welding will be 60 percent duty cycle. Machines designed for automatic welding operations are 100 percent duty cycle.

With the exception of 100 percent duty cycle machines, the maximum amperage that a welding machine will produce is always higher than its rated capacity. A welding machine rated 300 amps at 60 percent duty cycle will generally put out a maximum of 375 to 400 amps. But, since the duty cycle is a function of its rated capacity, the duty cycle will decrease as the amperage is raised over 300 amps. Welding at 375 amps with a welding machine rated 300 amps at 60 percent duty cycle will lower the duty cycle to about 30 percent. If welding continues for more than three out of ten minutes, the machine will overheat.

Note　　　Most welding machines have a heat-activated circuit breaker which will shut off the machine automatically when it overheats. The machine cannot be turned back on until it has cooled.

If the amperage is set below the rated amperage, the duty cycle increases. Setting the amperage at 200 amps for a welding machine rated 300 amps at 60 percent duty cycle will increase the duty cycle to 100 percent. *Figure 4-5* shows the relationship between amperage and duty cycle.

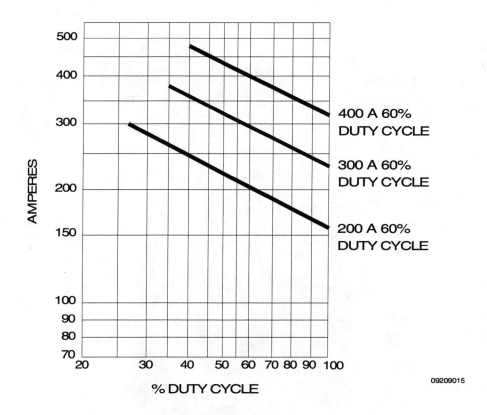

09209015

Figure 4-5. Amperage and Duty Cycle

5.0.0 FCAW EQUIPMENT

The typical FCAW system is composed of several interconnected pieces of equipment. They include:

- DC constant voltage welding machine (power supply)
- Wire feeder
- Welding gun (also called welding torch)
- Shielding gas supply

Heavy duty water cooled guns may also be equipped with a closed loop water cooling system.

Most FCAW systems have a separate wire feeder which pushes the wire from the feeder through the gun cable and gun (push system). Other systems may incorporate the wire feeder into the welding gun (pull system), or do both (push-pull system). There are some systems that have synchronized wire feeders located at intervals along very long feeder conduits. *Figure 5-1* shows several FCAW wire feeder systems.

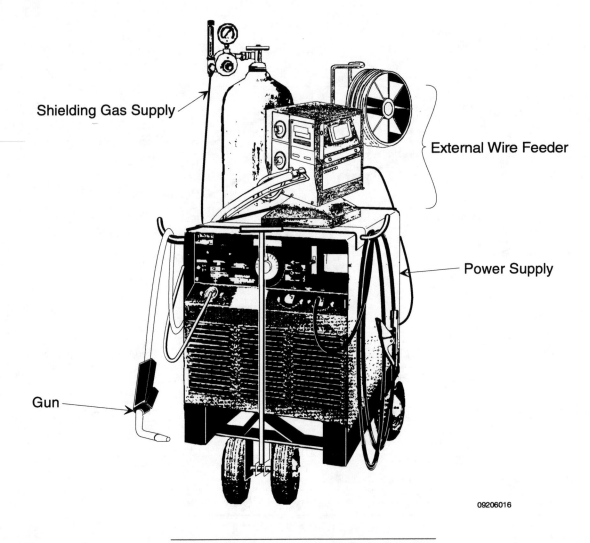

09206016

Figure 5-1. FCAW Wire Feeder Systems

5.1.0   FCAW METAL TRANSFER PROCESS

The FCAW process uses an electric arc to melt the filler and base metals and transfer fine droplets of filler metal to the weld puddle. The electro-magnetic force that propels the metal spray is strong, but the process produces very little spatter. The flux within the wire's core contains ingredients to perform several functions:

- Ionizers to stabilize the arc
- Deoxidizers to purge the weld of gasses and slag
- Additional metals and elements to enhance the quality of the weld metal
- Ingredients to generate shielding gas
- Ingredients to form a protective slag over the weld bead

The gas shield generated by flux in the wire core, as well as the external gas shield, protects the arc and molten weld metal. The slag cover produced by the flux helps to support the weld metal permitting the process to be used for making high quality welds in all positions. With larger diameter flux core wires, deposition rates of 25 pounds per hour can be achieved. *Figure 5-2* shows the FCAW process.

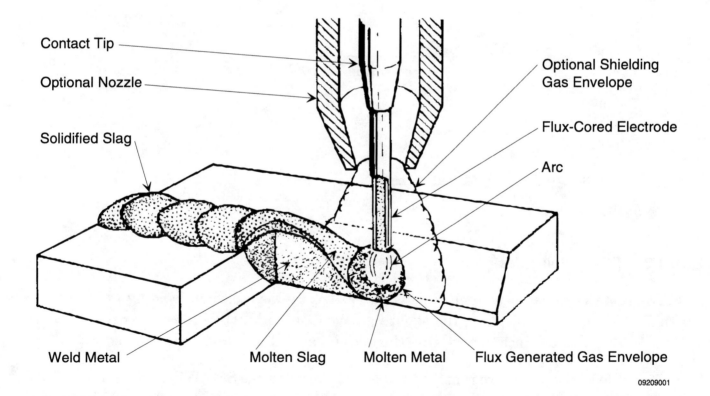

Contact Tip

Optional Nozzle

Solidified Slag

Optional Shielding Gas Envelope

Flux-Cored Electrode

Arc

Weld Metal   Molten Slag   Molten Metal   Flux Generated Gas Envelope

09209001

Figure 5-2. FCAW Process.

5.1.1 FCAW Weld Penetration

FCAW welds made with carbon dioxide shielding have much deeper penetration than either SMAW (shielded metal arc welding) or FCAW without shielding gas. For this reason, edge preparation is often not needed for double-welded butt joints on plate up to 5/8 inch thick. When making FCAW fillet welds using carbon dioxide shielding, a smaller size fillet weld will have as much strength as a larger fillet made with SMAW because of the deep penetration achieved with FCAW and the carbon dioxide shielding.

FCAW fillet welds made with only self-shielding wire are not as strong as the carbon dioxide shielded FCAW welds because they don't penetrate as deeply. *Figure 5-3* shows SMAW and FCAW-CO$_2$ shielded fillet welds.

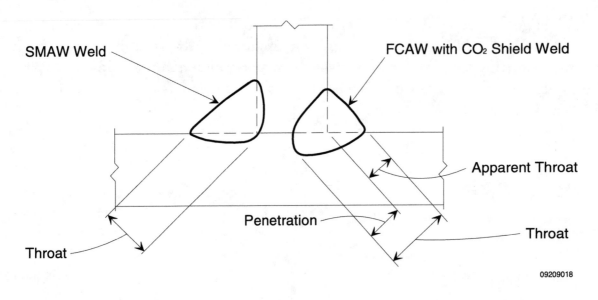

Figure 5-3. SMAW and FCAW-CO$_2$ Shielded Fillet Welds.

5.1.2 FCAW Joint Design

FCAW uses the same basic joint designs as SMAW. However, when welding groove welds with FCAW and CO$_2$ shielding, the square groove (square butt) joint can be used for plate thicknesses up to 5/8 inch. Above this thickness, beveled joints should be used.

Because FCAW electrode wire is much thinner than comparable SMAW electrodes and FCAW penetration with carbon dioxide is superior to SMAW, the included angles of V-joints can be reduced to one-half the angle required for SMAW. For example, a (SMAW) included angle of 60 degrees can be reduced to 30 degrees when using FCAW with CO$_2$ shielding. This can save approximately 50 percent on filler metal and considerable welding time when compared with using SMAW.

CAUTION   Always refer to your site WPS or Quality Standard for specific information on joint requirements. Information in this manual is provided as a general guideline only.

Figure 5-4 shows joint designs for FCAW with CO_2 shielding.

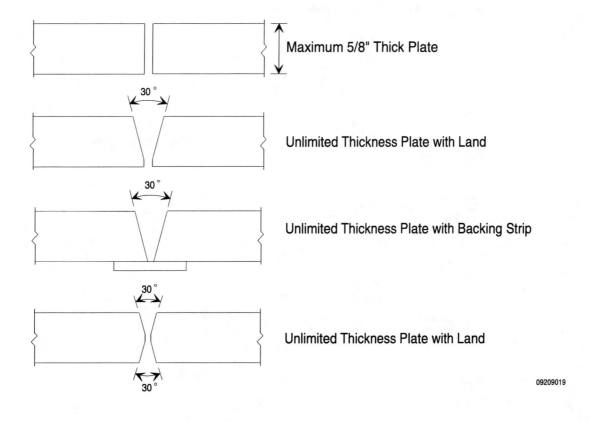

Figure 5-4. Joint Designs for FCAW with CO_2 Shielding.

5.2.0   FCAW POWER SUPPLIES

FCAW is performed with a constant voltage (constant potential) DC power supply (welding machine). In constant voltage welding machines, the open circuit voltage (no arc struck) and the welding voltage are nearly the same.

Power supplies designed for FCAW often contain controls to adjust arc voltage, slope, and inductance and have gauges to monitor weld current and voltage.

5.2.1 Slope

The constant voltage power supplies used for FCAW are not truly constant in their voltage output because the voltage always drops some as the current (amperage) increases. This voltage to amperage relationship forms a slight curve (or sloping line) when plotted as a graph. The general slope angle of the volt-ampere curve is known as the "slope." This slope is adjustable on some power supplies. A flatter (more horizontal) slope is best for FCAW. *Figure 5-5* shows flat slope and steep slope curves.

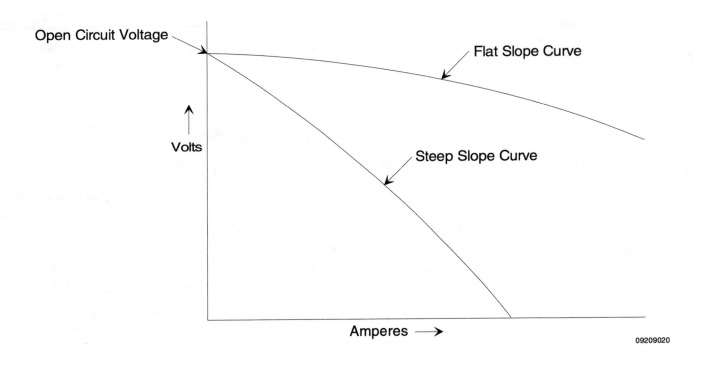

Figure 5-5. Flat Slope and Steep Slope Volt-Ampere Curves

5.2.2 Inductance

Constant voltage power sources are self-regulating in that they automatically produce current as it is required. Inductance controls the rate at which the current output rises. With low circuit inductance, the current rise is very fast and can cause the wire to explode or spatter. When the circuit inductance is high, the current rises more slowly. Some inductance is desirable in FCAW because it prevents explosive arc starts by slowing down the current rise rate. Many FCAW power supplies have a control to adjust the inductance.

5.2.3 Arc Blow

When performing FCAW, arc blow may sometimes be a problem. Arc blow is the deflection of the arc from its normal course because of the attraction or repulsion of the arc's magnetic field with the weld current's magnetic field in the base metal. The amount and direction of arc deflection will depend upon the relative position, direction and density of the base metal's magnetic field. Arc blow can result in excess spatter and weld defects. In FCAW it can be minimized by:

● Relocating the ground clamp
● Changing the weld angle of the gun

5.3.0 WELDING CABLE

Cables used to carry welding current are designed for maximum strength and flexibility. The conductors inside the cable are made of fine strands of copper wire. The copper strands are covered with layers of rubber reinforced with nylon or dacron cord. *Figure 5-6* shows a cutaway section of welding cable.

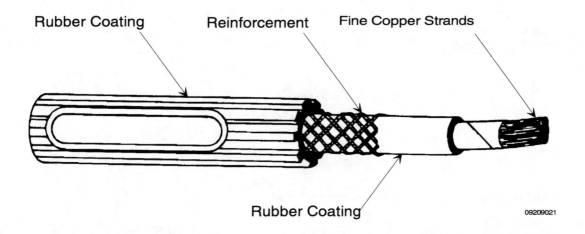

Rubber Coating   Reinforcement   Fine Copper Strands

Rubber Coating   09209021

Figure 5-6. Welding Cable

The size of a welding cable is based on the number of copper strands it contains. Large diameter cable has more copper strands and can carry more welding current. Typically the smallest cable size is number 8 and the largest is number 3/0 (three-O).

When selecting welding cable size, the amperage load as well as the distance the current will travel must be considered. The longer the distance the current has to travel, the larger the cable must be to reduce voltage drop and heating caused by the electrical resistance in the welding cable. When selecting welding cable, use the rated capacity of the welding machine for the cable amperage requirement. For the distance, measure both the electrode and ground leads and add the two lengths together. To identify the size welding cable required, refer to a "Recommended Welding Cable Size Table" furnished by most welding cable manufacturers. *Figure 5-7* shows a typical welding cable size table.

RECOMMENDED CABLE SIZES FOR MANUAL WELDING

Machine Size In Amperes	Duty Cycle (%)	Copper Cable Sizes for Combined Lengths of Electrodes Plus Ground Cable				
		Up to 50 Feet	50 - 100 Feet	150 - 150 Feet	150 - 200 Feet	200 - 250 Feet
100	20	#8	#4	#3	#2	#1
180	20	#5	#4	#3	#2	#1
180	30	#4	#4	#3	#2	#1
200	50	#3	#3	#2	#1	#1/0
200	60	#2	#2	#2	#1	#1/0
225	20	#4	#3	#2	#1	#1/0
250	30	#3	#3	#2	#1	#1/0
300	60	#1/0	#1/0	#1/0	#2/0	#3/0
400	60	#2/0	#2/0	#2/0	#3/0	#4/0
500	60	#2/0	#2/0	#3/0	#3/0	#4/0
600	60	#3/0	#3/0	#3/0	#4/0	* * *
650	60	#3/0	#3/0	#4/0	* *	* * *

* * Use Double Strand of #2/0
* * * Use Double Strand of #3/0

09209022

Figure 5-7. Welding Cable Size Table

5.3.1 Welding Cable End Connections

Welding cables must be equipped with the proper end connections or terminals to be used efficiently.

CAUTION If the end connection is not tightly secured to the cable, the connection will overheat and oxidize. An overheated connection will cause variations in the welding current and permanent damage to the connector and/or cable. Check connections for tightness and repair loose or overheated connections.

Lugs are used at the end of the welding cable to connect the cable to the welding machine current terminals. The lugs come in various sizes to match the welding cable size and are mechanically crimped onto the welding cable.

Quick disconnects are also mechanically connected to the cable ends. They are insulated and serve as cable extensions for splicing two lengths of cable together. Quick disconnects are connected or disconnected with a half twist. When using quick disconnects, care must be taken to ensure that they are tightly connected to prevent overheating or arcing in the connector.

Figure 5-8 shows typical lugs and quick disconnects used as welding cable connectors.

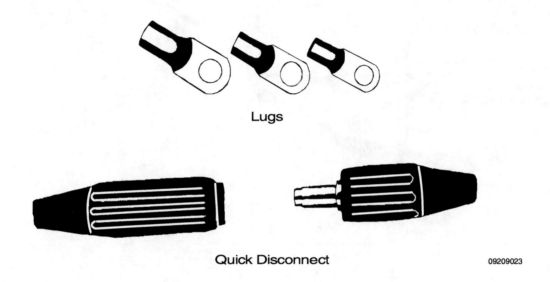

Lugs

Quick Disconnect

09209023

Figure 5-8. Lugs and Quick Disconnects

The ground clamp provides the connection between the end of the ground cable and the workpiece. Ground clamps are mechanically connected to the welding cable and come in a variety of shapes and sizes. The size of a ground clamp is the rated amperage that it can carry without overheating. When selecting a ground clamp be sure it is rated at least the same as the rated capacity of the power source on which it will be used.

Figure 5-9 shows examples of various styles of ground clamps.

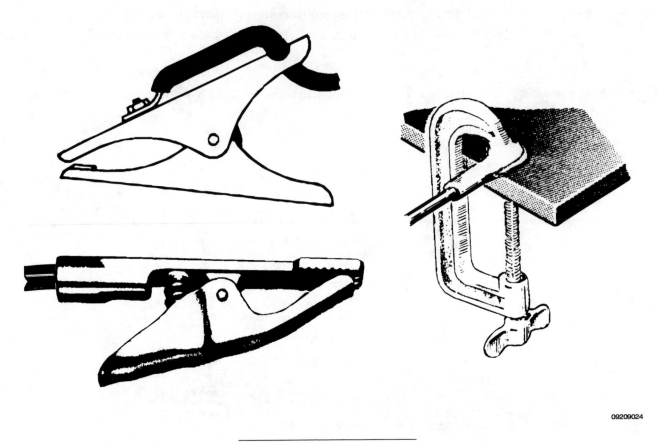

Figure 5-9. Ground Clamps

5.4.0   FCAW WIRE FEEDERS

FCAW wire feeders are identical to GMAW wire feeders, except they are typically designed to handle the larger diameter flux core wires. The wire feeder pulls the electrode wire from a spool and pushes it through the wire conduit (gun cable) and the gun. It consists of a wire spool holder with a drag brake, an electric motor that drives either one or two sets of opposing slotted rollers and various controls. The wire feeder also contains the shielding gas connections and control solenoid, and if the torch is water cooled, optional cooling water flow control solenoid. Controls that are typically located on a wire feeder include:

- **Wire spool drag brake adjustment** A manual device used to adjust the drag on the wire spool to prevent the uncontrolled unwinding of the wire.
- **Wire speed** A variable control used to set the wire feed speed. Since GMAW power sources are self regulating (produce the required current on demand), increasing or decreasing the wire feed speed will cause the welding machine to automatically increase or decrease its current output.

WELDING TWO TRAINEE TASK MODULE 09209

- **Wire jog** A manual switch used to start and stop the wire feeder without energizing the welding current.
- **Wire burnback** An adjustable time delay which causes the welding current to continue after the trigger is released. This consumes the length of wire that continues to feed after the trigger was released so that the proper stickout is maintained to start the next weld.
- **Gas purge** A manual switch used to operate the gas solenoid to purge the shielding gas system prior to actual welding. Also used to adjust the flowmeter.
- **Gas Preflow** An adjustable time delay that prevents the welding current solenoid from closing until the shielding gas is flowing.
- **Gas Postflow** An adjustable time delay used to keep shielding gas flowing for a short period after the gun trigger is released.

Figure 5-10 shows a FCAW wire feeder unit.

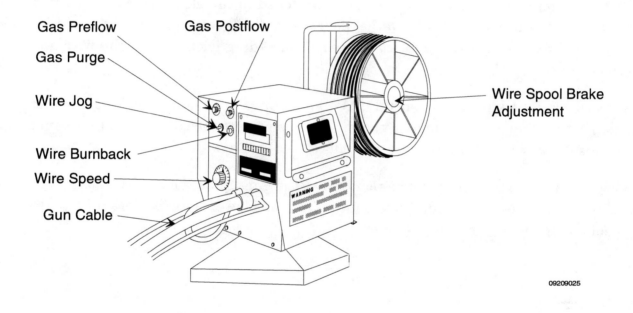

Figure 5-10 FCAW Wire Feeder Unit.

The wire feeder grips the wire in the grooves of opposing grooved rollers. There are usually two sets of opposed rollers. Each roller contains a groove to grip one side of the wire without crushing it.

The rollers are interchangeable to match different size wires, and must be changed when ever the wire size exceeds the range specified for the rollers. Also, rollers are often specifically designed for either solid wire which is hard or core wire which is soft. Always check the manufactures specifications for the wire feeder rollers to be sure the rollers match the wire size being run and that they are designed for core wire. The roller specifications are generally stamped on the side of the rollers.

The tension of the rollers against the wire is also adjustable generally with a spring loaded thumb screw directly above the top roller. It should be set high enough not to slip but not so high that it scores or distorts the wire.

5.5.0 FCAW GUNS

FCAW guns are similar to GMAW guns but are often designed to handle larger diameter wire. The FCAW gun supports and guides the wire electrode. It also contains the remote switch to start and stop the welding current and the shielding gas flow when used. A gas nozzle on the end of the gun directs the shielding gas around the arc. The electrode wire extends through the contact tip located at the end of the gun. The contact tip provides the electrical connection from the welding machine electrode lead to the electrode wire. The contact tip is replaceable and must always be matched to the electrode wire size. If it is too small, the wire will jam. If it is too large, poor contact will cause damaging arcing between the tip and the wire. It is also subject to wear from sliding friction with the wire and must be replaced periodically.

FCAW guns designed for self shielded FCAW do not have provisions to deliver shielding gas. This makes the guns nozzles smaller in diameter. To protect the contact tip, these guns have a fiber insulator which threads onto the end of the gun nozzle.

FCAW generates smoke which can be removed at the gun tip with a special smoke extraction nozzle. The nozzle is connected to a vacuum pump or blower by a flexible tube to draw off the smoke and thereby improve visibility.

Figure 5-11 shows a FCAW guns with and without smoke extractor.

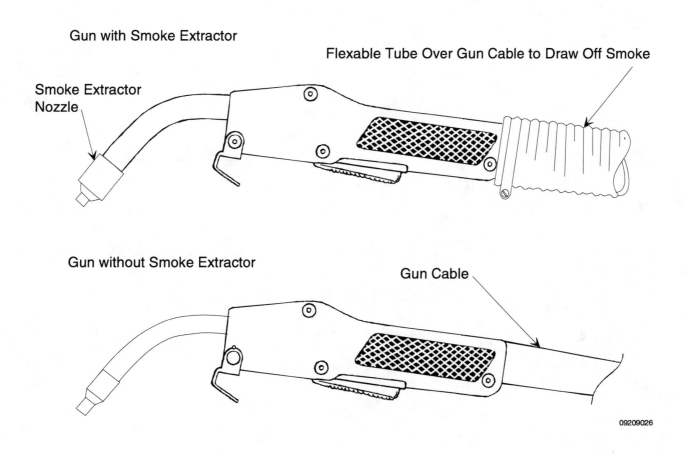

Gun with Smoke Extractor

Flexable Tube Over Gun Cable to Draw Off Smoke

Smoke Extractor Nozzle

Gun without Smoke Extractor

Gun Cable

09209026

Figure 5-11. FCAW Guns With and Without Smoke Extractor.

Heavy duty guns are water cooled. The cooling water flow may be controlled by a solenoid in the wire feeder, or by a separate dedicated cooling system. Cooling water is piped to and from the gun through flexible tubing integrated into the gun cable. Some cooling systems use domestic water to cool the gun. Others use a closed loop system which recirculates demineralized water or special cooling fluid that will not corrode internal gun surfaces or plug the cooling passages with mineral scale.

Figure 5-12 shows a closed loop cooling system.

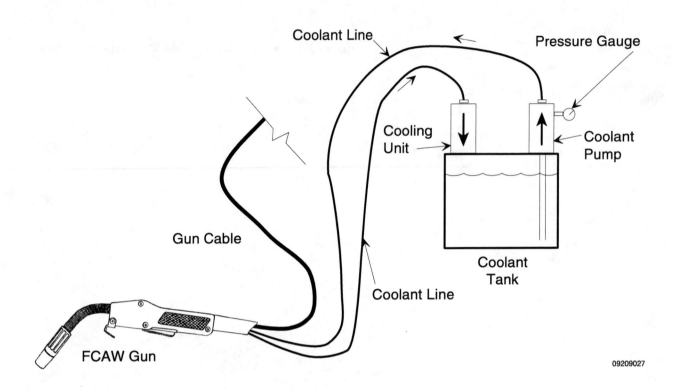

Figure 5-12. Closed Loop Cooling System.

5.6.0 SHIELDING GAS SUPPLY

Some FCAW core wires require the use of a shielding gas. The shielding gas is directed around the electrode end and the weld zone by the nozzle of the gun. The shielding gas used depends on the filler wire and base metal being welded. The most common shielding gas for FCAW carbon and low alloy steels is carbon dioxide (CO_2). For FCAW high alloy or stainless steels an argon carbon dioxide mixture is generally used. Both gasses are available in pressurized cylinders of various sizes and the carbon dioxide is available in bulk liquid. Check your site WPS or Quality Standards for the type of shielding gas to use. If these are not available, refer to the filler metal manufactures specifications.

5.6.1 Cylinder Safety

Cylinders must always be handled with great care because they contain high pressure gas. When a valve is broken off of a high pressure cylinder, the gas escapes with explosive force and can severely injure personnel with the blast and blown debris. The cylinder itself can become a powerful uncontrolled missile. When transporting and handling cylinders, always abide by the following rules:

- The safety cap should always be installed over the valve except when the cylinder is secured and connected for use.
- When in use, the cylinder must always be secured against falling. It should be chained or secured to the welding machine or to a post, beam or pipe.
- Always use a cylinder cart to transport a cylinder.
- Cylinders should never be hoisted by slings. They can slip out of the sling. Always use a hoisting basket or similar device.
- Always crack the cylinder valve to clean out the nozzle before attaching the flow meter.
- Always open the cylinder valve slowly. Once open, the valve should be completely opened to prevent leakage from around the valve stem.

WARNING!   Do not remove the protective cap unless the cylinder is secured. The cylinder could fall over and the valve could break off. The cylinder would then shoot like a rocket and could cause severe injury or death to anyone in its path.

A regulator-flow meter is required to meter the shielding gas to the gun at the proper flow rate. A typical regulator-flow meter consists of a preset pressure regulator with cylinder valve spud, that is joined to a flow metering (needle) valve and a flow rate gauge. The pressure regulator is usually equipped with a tank pressure gauge for monitoring the cylinder pressure. The metering valve is used to adjust the gas flow rate to the gun nozzle. The flow gauge shows the gas flow rate in cubic feet per minute or liters per second.

Figure 5-13 shows a shielding gas regulator-flow meter.

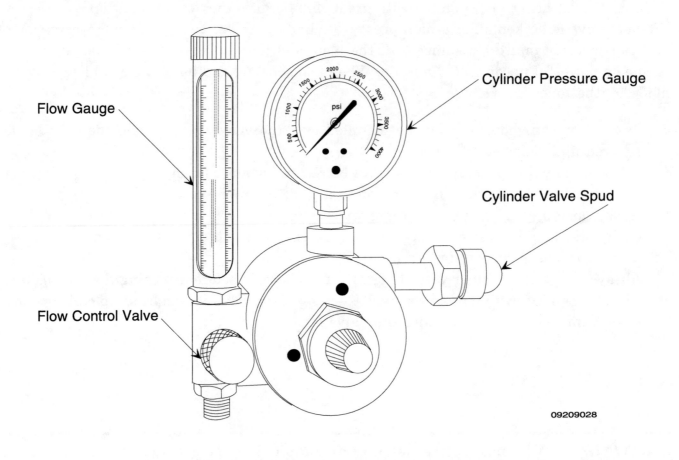

Figure 5-13. Shielding Gas Regulator-Flow Meter.

Gas flow to the gun is started and stopped by an electric solenoid valve that is usually located inside the control unit or the wire feeder unit. The solenoid is controlled by the operator with the trigger switch on the gun. The same trigger switch simultaneously starts and stops the welding current and wire feed.

5.6.3   Shielding Gas Flow Rate

Gas flow rate from the gun tip is important because it affects the quality and the cost of a weld. Too low a flow rate will not shield the weld zone adequately and will result in a poor quality weld. Excessive gas flow wastes expensive gasses and can cause turbulence at the weld. The turbulence can pull atmosphere into the weld zone causing porosity in the weld. Welding specifications specify shielding gas flow rates.

Typically, the flow rate for carbon dioxide used with FCAW is 35 to 45 cfm.

Other factors that may affect the shielding gas flow rate include:

Drafts: The flow rate must be increased in a drafty location to maintain the gas shield around the weld zone.

Welding current: High welding currents require higher flow rates.

Nozzle size: Larger nozzles require higher flow rates.

Joint type: Welds on flat surfaces require higher flow rates than welds in grooves or fillets.

Welding speed: Fast advance speeds require higher flow rates than slower advance speeds.

Weld position: Vertical and horizontal position welds require higher flow rates than flat or overhead welds.

6.0.0   FCAW FILLER METALS

Flux Core Arc Welding (FCAW) is presently limited to the welding of ferrous metals. These include low- and medium-carbon steels, some low-alloy steels, cast irons and some stainless steels.

The FCAW electrode is a continuous flux core wire electrode for filler material. The wire is made by forming a thin filler metal strip into a U-shape, filling the U with the flux material, squeezing the strip shut around the flux, and then drawing the tube through dies to size the wire and compact the flux. During manufacture, the wire is carefully inspected for defects. Flux core wires are manufactured in a range of sizes from 0.045 inch to 5/32 inch. The finished flux-core wire is coiled on all the standard spool sizes and forms.

Note   All flux-core wires are considered to be low hydrogen. For this reason, proper electrode control procedures must be followed to keep the wire dry.

By varying the power settings on the welding machine, smaller size electrode wire (0.045 and 1/16") can be used both for thin metals and for multipass welds on metals of unlimited thickness in all positions.

Several industry organizations and the U.S. Government publish specification standards for flux-core welding wire. The most common is the American Welding Society (AWS). The purpose of the AWS specification is to set standards that all manufacturers must follow when manufacturing welding consumables. This ensures consistency for the user regardless of who manufactured the product. The three AWS specifications presently available for flux core electrodes are:

- AWS A5.20 for carbon steel electrodes
- AWS A5.22 for corrosion-resistant chromium and chromium-nickel electrodes
- AWS A5.29 for low-alloy steel electrodes

Each specification defines standards for the:

- Classification, identification and marking of the electrode wire
- Chemical composition of the deposited weld metal
- Chemical composition of the flux filling
- Mechanical properties of the deposited weld metal

The classification number format is different for each AWS classification.

Figure 6-1 shows examples of the AWS specification formats for flux core electrodes.

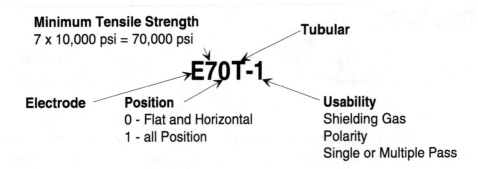

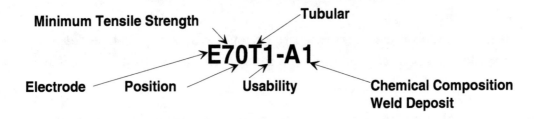

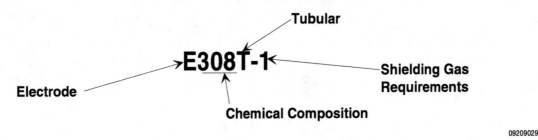

09209029

Figure 6-1. AWS Specification Formats for Flux Core Electrodes.

Table 6-2 shows the carbon steel electrode welding characteristics as indicated by specification number suffix.

Table 6-2

CARBON STEEL ELECTRODE WELDING CHARACTERISTICS.

AWS Classification	Welding Current	Shielding	Single or Multiple Pass
EXXT-1	DCEP/DCRP	CO_2	Multiple Pass
EXXT-2	DCEP/DCRP	CO_2	Single Pass
EXXT-3	DCEP/DCRP	None	Single Pass
EXXT-4	DCEP/DCRP	None	Multiple Pass
EXXT-5	DCEP/DCRP	CO_2	Multiple Pass
EXXT-6	DCEP/DCRP	None	Multiple Pass
EXXT-7	DCEN/DCSP	None	Multiple Pass
EXXT-8	DCEN/DCSP	None	Multiple Pass
EXXT-10	DCEN/DCSP	None	Single Pass
EXXT-11	DCEN/DCSP	None	Multiple Pass
EXXT-G	*	*	Multiple Pass
EXXT-GS	*	*	Single Pass

* Note: This characteristic is as agreed upon by the supplier and the purchaser.

09209035

Table 6-3 lists carbon steel electrode classes and uses.

Table 6-3

CARBON STEEL ELECTRODE CLASSES AND USES

Electrode Class	Shielding Gas Required	Single or Multiple Pass	Weld Positions
E60T-7	None	Either	Flat & Horizontal
E60T-8	None	Either	Flat & Horizontal
E70T-1	CO_2	Either	Flat & Horizontal Fillets
E70T-2	CO_2	Single Pass	Flat & Horizontal Fillets
E70T-3	None	Single Pass	Flat & Horizontal Thin Gauge
E70T-4	None	Multiple	Flat & Horizontal
E70T-5	None/CO_2	Multiple	All (Globular Transfer)
E70T-6	None	Multiple	All (Globular Transfer)

E70T-G: May be used with or without gas shielding and is limited to multipl-pass work. Required to meet tension tests, but not chemical, radiographic, bend or impact tests.

E70T-GS: May be used with or without gas shielding and is limited to multipl-pass work. Required to meet tension tests, but not chemical, radiographic, bend or impact tests.

09209036

SELF-CHECK REVIEW 2

1. How does the power line current and voltage supplied to a welding machine differ from the current and voltage supplied to the welding cables?
2. What is arc voltage?
3. What determines the intensity of an arc and the amount of heat available for welding?
4. Is a constant current or a constant voltage (potential) welding machine typically used for FCAW?
5. What functions does the flux in the wire core perform?
6. What allows FCAW to be used successfully in all positions?
7. Why is a fillet weld made with FCAW and carbon dioxide shielding gas stronger than an equal sized fillet weld made by SMAW?
8. Would more or less inductance prevent explosive arc starts?
9. What two factors affect the selection of a welding cable size?
10. What grips the electrode wire in a wire feeder?
11. What part of the gun conducts welding current to the FCAW wire electrode?
12. Is a shielding gas system always required with FCAW equipment?
13. What controls the rate of shielding gas flow?
14. What types of self-shielding flux-core electrodes are considered to be low-hydrogen electrodes?
15. Why must carbon dioxide shielding gas be dry (welding grade)?

ANSWERS TO SELF-CHECK 2

1. Power line voltage is much higher and the current is much lower than welding voltage and current. (3.0.0)
2. Arc voltage is the welding voltage after the arc is established. (3.1.0)
3. The welding amperes produced by the welding machine. (3.2.0)
4. A constant voltage machine is used for FCAW. (4.0.0)
5. The flux stabilizes the arc, purges gases and slag from the weld, improves the weld metal quality, generates shielding gas and forms slag over the weld bead. (5.1.0)
6. The slag cover. (5.1.0)
7. Because the FCAW with carbon dioxide shielding has much greater penetration. (5.1.1)
8. More inductance slows current build-up and prevents explosive arc starts. (5.2.2)
9. The maximum welding current and the cable length. (5.3.0)
10. Sets of opposing grooved rollers. (5.4.0)
11. The gun contact tip. (5.5.0)
12. The shielding gas system is not required with self-shielding flux-core electrodes. (5.5.0)
13. The gas regulator-flowmeter. (5.6.2)
14. All FCAW electrode wire is considered to be low-hydrogen electrode. (6.0.0)
15. Moisture in the shielding gas causes porosity and cracks in the weld. (7.0.0)

8.0.0 WELDING EQUIPMENT SET-UP

In order to weld safely and efficiently the welding equipment must be properly set up. The following sections will explain the steps for setting up welding equipment. *Figure 8-1* shows a simplified diagram of a FCAW system.

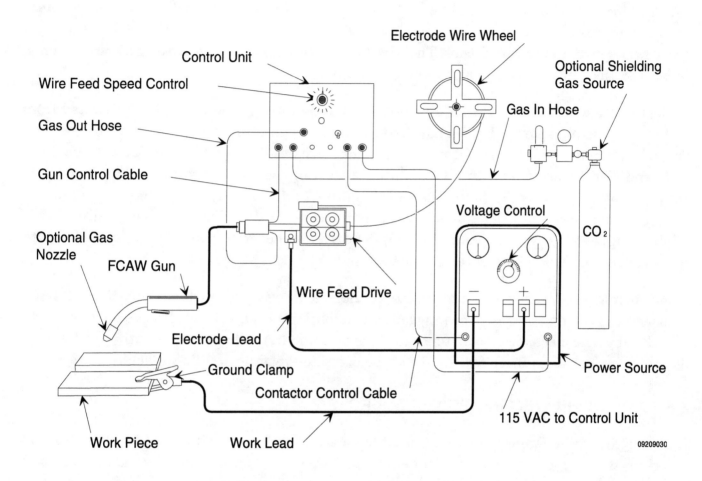

Figure 8-1. Simplified Diagram of a FCAW System.

8.1.0 SELECTING A FCAW POWER SUPPLY

To select a welding machine the following factors must be considered.

- FCAW requires a constant voltage (constant potential) power source.
- The type welding current required (AC or DC). FCAW almost always operates with DC and electrode positive (or reverse polarity DCRP), with the exception of a few special applications which use AC.
- Required operating amperages range from 150 amps with a 0.045 inch electrode to 650 amps or more with a 1/8 inch electrode. Larger electrodes would require even higher amperages.

- The primary power requirements. Are there AC electrical receptacles to plug a welding machine into or do engine driven generators/alternators need to be used?

8.2.0 LOCATING THE FCAW EQUIPMENT

Because of the short length of most FCAW gun cables, the wire feeder and power supply must be located near the work to be performed. The power source must also be located near the wire feeder because of the limited length of the contactor/control cable which runs from the power source to the wire feeder. The wire feeder is often located on top of the power supply, or on the floor or a bench near the power supply.

Select a site where the FCAW equipment will not be in the way but will be protected from welding or cutting sparks. There should be good air circulation to keep the machine cool and the environment should be free from explosive or corrosive fumes and as free as possible from dust and dirt. Welding machines have internal cooling fans which will pull these materials into the welding machine if they are present. Also dust and dirt can collect on the wire causing weld contamination and clogging of the liner in the gun cable. The site should also be free of standing water or water leaks. If an engine-driven generator or alternator is used, locate it so it can be easily refueled and serviced.

There should also be easy access to the site so the machine can be started, stopped or adjusted as needed. If the machine is to be plugged into an outlet, be sure the outlet has been properly installed by a licensed electrician to ensure it is grounded. Also be sure to identify the location of the electrical disconnect for the outlet before plugging the welding machine into it.

8.3.0 MOVING WELDING POWER SOURCES

Large engine-driven generators are mounted on a trailer frame and can be easily moved by a pickup truck or tractor using a trailer hitch. Other types of welding machines will have a skid base or be mounted on steel or rubber wheels. When moving welding machines that are mounted on wheels by hand, use care. Some machines will be top-heavy and may fall over in a tight turn or if the floor or ground is uneven or soft.

CAUTION   Secure or remove the wire feeder and gas cylinder before attempting to move the power supply.

WARNING!   If a welding machine starts to fall over do not attempt to hold it. Welding machines are very heavy and severe crushing injuries can occur if a welding machine falls on someone.

WELDING TWO TRAINEE TASK MODULE 09209

Most welding machines have a lifting eye. The lifting eye is used to move machines mounted on skids or to lift any machine. Before lifting a welding machine check the equipment specifications for the weight. Be sure the lifting device and tackle will handle the weight of the machine. When lifting a welding machine always use a shackle. Never attempt to lift a machine by placing the lifting hook in the machine eye. Also, before lifting or moving a welding machine be sure the welding cables are secure. *Figure 8-2* shows lifting a welding machine.

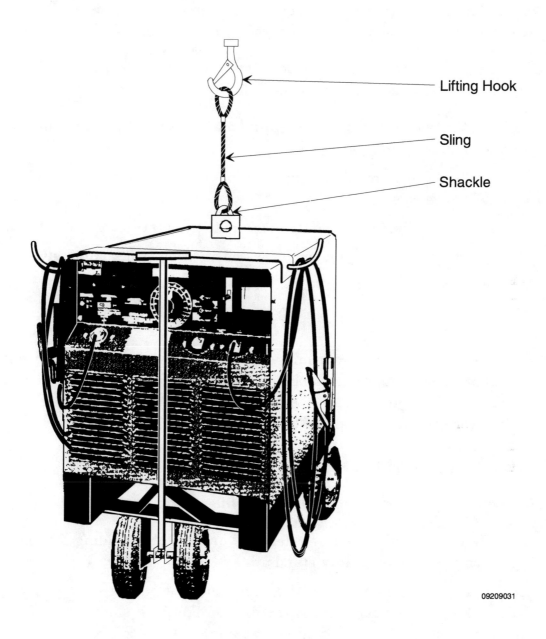

Lifting Hook

Sling

Shackle

09209031

Figure 8-2. Lifting A Welding Machine

8.4.0 CONNECTING THE SHIELDING GAS

The shielding gas generally connects to the wire feeder unit. When shielding gas is required, prepare the shielding gas system by performing the following steps:

Step 1 Identify the shielding gas required by referring to the WPS (Welding Procedure Specification) or site quality standard.

Step 2 Locate a cylinder of the correct gas or mixture and securely chain or tie it to the welding machine or a solid structural member to prevent it from falling over.

CAUTION Do not remove the cylinder's protective valve cap until the cylinder is secured and cannot fall over.

Step 3 Remove the cylinder's protective cap and momentarily crack open the cylinder valve to blow out any dirt.

Step 4 Install the pressure regulator and flow meter assembly to the cylinder. Leave the cylinder valve closed.

Step 5 Connect the gas hose to the flow meter and to the gas connection on the wire feeder.

Step 6 Slowly crack open the cylinder valve and then open it completely.

Step 7 Adjust the gas flow with the flow meter valve to the specified flow rate.

Note: Some flow meters are equipped with several scales of different calibrations around the same sight tube for monitoring the flows of different types (densities) of gases. Be sure to rotate the scales or read the correct side for the gas type in use.

8.5.0 SELECT AND INSTALL FILLER WIRE

Use the following steps to select and install the filler wire:

Step 1 Identify the filler required by referring to the WPS (Welding Procedure Specification) or site quality standard.

Step 2 Locate a spool of the required wire and mount it on the wire feeder spool holder. Lock the holder and set the drag brake for a slight drag. The brake control is usually located on the end of the spool axle.

Step 3 Check (and change if necessary) the wire feeder drive wheels and gun contact tip to make sure they are the correct size for the wire being loaded. Adjust the wheel tension if necessary. For detailed instructions on changing or adjusting the wire feeder drive wheels or changing the gun contact tip, refer to the specific manufacturer's instructions.

Step 4   Feed the wire into the feeder wheels and then use the "jog" control to feed it through the gun cable and the gun contact tip.

8.6.0   ATTACH THE GROUND CLAMP

The ground clamp must be properly located on the work piece to prevent damage by welding current to any associated or surrounding equipment. If the electrical welding current travels through a bearing, seal, valve, contacting surface or delicate electronic or electrical device, it could cause severe damage from heat and arcing, This would require that these items be replaced. Carefully check the area to be welded and position the ground clamp so the welding current will not pass through any contacting surfaces or delicate or sensitive equipment. If in doubt, ask your supervisor for assistance before proceeding.

CAUTION   Welding current can severely damage bearings, seals, valves or contacting surfaces. Position the ground clamp to prevent welding current from passing through them.

CAUTION   Welding current passing through electrical or electronic equipment will cause severe damage. Before welding on any type of mobile equipment, the ground lead at the battery must be disconnected to protect the electrical system.

WARNING!   If welding near a battery, the battery must be removed. Batteries produce hydrogen gas which is extremely explosive. A welding spark could cause the battery to explode showering the area with battery acid. Never weld near a battery.

CAUTION   The slightest spark of welding current can destroy electronic or electrical equipment. Have an electrician check the equipment and if necessary isolate the system before welding.

Ground clamps must never be connected to pipes carrying flammable or corrosive materials. The welding current could cause overheating or sparks, resulting in an explosion or fire.

The ground clamp must make a good electrical contact when it is connected. Dirt and paint will inhibit the connection and cause arcing resulting in overheating of the ground clamp and variations in the welding current which can cause defects in the weld. Clean the surface before connecting the ground clamp. If the ground clamp is damaged and does not close securely onto the surface, replace it.

8.7.0 ENERGIZE THE POWER SUPPLY

Electrically powered welding machines are energized from AC utility power circuits. They may be plugged into an electrical outlet or hard wired into a connection or switch box by an electrician. The electrical requirements, (primary current) will be on the equipment specification tag displayed prominently on the machine. Most machines will require single-phase 230-volt current or 460-volt three-phase current. Machines requiring single-phase 230-volt power will have a three-prong plug. Machines requiring three-phase 460-volt power will have a four-prong plug. *Figure 8-3* shows the welding machine electrical plugs.

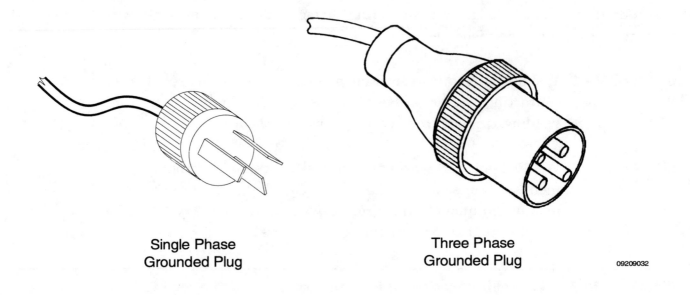

Single Phase
Grounded Plug

Three Phase
Grounded Plug

09209032

Figure 8-3. Welding Machine Electrical Plugs.

If a welding machine does not have a power plug, an electrician must connect it. The electrician will add a plug or hard-wire the machine directly into an electrical box.

WARNING!   Never use a welding machine until you identify the electrical disconnect. In the event of an emergency, you must be able to quickly turn off the power to the welder at the disconnect.

8.8.0 STARTING ENGINE-DRIVEN GENERATORS OR ALTERNATORS

Before welding can take place with an engine-driven generator or alternator the engine must be checked and then started. As with a car engine, the engine powering the generator or alternator must also requires routine maintenance.

8.8.1 Pre-Start Checks

Many sites will have pre-start check lists which must be completed and signed prior to starting or operating an engine-driven generator or alternator. Check with your supervisor. If your site has such a check list, complete and sign it. If your site does not have a pre-start check list, perform the following checks before starting the engine.

- Check the oil using the engine oil dipstick. If the oil is low, add the appropriate grade oil for the time of year.
- Check the coolant level in the radiator if the engine is liquid-cooled. If the coolant level is low, add coolant.

CAUTION   Do not add plain water to radiators that contain antifreeze. Antifreeze not only protects radiators from freezing in cold weather. It also has rust inhibitors and additives to aid cooling. If the antifreeze is diluted, it will not function properly. If the weather turns cold the system may freeze, causing damage to the radiator, engine block and water pump.

- Check the fuel. The unit may have a fuel gauge or a dipstick. If the fuel is low, add the correct fuel (diesel or gasoline) to the fuel tank. The type of fuel required should be marked on the fuel tank. If it is not marked, contact your supervisor to verify the fuel required and have the tank marked.

CAUTION   Adding gasoline to a diesel engine or diesel to a gasoline engine will cause severe engine problems. It can also cause a fire hazard. Always be sure to add the correct fuel to the fuel tank.

- Check the battery water level unless the battery is sealed. Add de-mineralized water if the battery water level is low.
- Open the fuel shut-off valve if the equipment has one. If there is a fuel shut-off valve it will be located in the fuel line between the fuel tank and the carburetor.
- Record the hours from the hour meter if the equipment has one. An hour meter records the total number of hours the engine runs. This information is used to determine when the engine needs to be serviced. The hours will be displayed on a gauge similar to an odometer.
- Clean the unit. Use a compressed air hose to blow off the engine and generator or alternator. Use a rag to remove heavier deposits that cannot be removed with the compressed air.

| *WARNING!* | Always wear eye protection when using compressed air to blow dirt and debris from surfaces. NEVER point the nozzle at yourself or anyone else. |

Note   Cleaning may not be required on a daily basis. Clean the unit as required.

8.8.2   Starting the Engine

Most engines will have an on/off ignition switch and a starter. They may be combined into a key switch similar to the ignition on a car. To start the engine, turn on the ignition switch and press the starter. Release the starter when the engine starts. The engine speed will be controlled by the governor. If the governor switch is set for idle the engine will slow to an idle after a few seconds. If the governor is set to welding speed the engine will continue to run at the welding speed. For FCAW, set the governor switch to welding speed. The welding current as well as 115 volt AC auxiliary current must be available continuously.

Small engine-driven alternators may have an on/off switch and a pull cord. These are started by turning on the ignition switch and pulling the cord similar to starting a lawn mower.

Engine-driven generators and alternators should be started about five to ten minutes before they are needed for welding. This will allow the engine to warm up before a welding load is placed on it.

8.8.3   Stopping the Engine

If no welding is required for thirty or more minutes, stop the engine by turning off the ignition switch. If you are finished with the welding machine for the day, also close the fuel valve if there is one.

8.8.4   Preventive Maintenance

Engine-driven generators and alternators require regular preventive maintenance to keep the equipment operating properly. Most sites will have a preventive maintenance schedule based on the hours that the engine operates. In severe conditions such as very dusty or cold weather, maintenance may have to be performed more frequently.

CAUTION   To prevent equipment damage, perform preventive maintenance as recommended by the site procedures or manufacturer's maintenance schedule in the equipment manual.

The responsibility for performing preventive maintenance will vary by site. Check with your supervisor to determine who is responsible for performing preventive maintenance.

When performing preventive maintenance follow the manufacturer's guidelines in the equipment manual. Typical tasks to be performed as a part of preventive maintenance include:

- Changing the oil
- Changing the gas filter
- Changing the air filter
- Checking/changing the antifreeze
- Greasing the undercarriage
- Re-packing the wheel bearings

9.0.0   TOOLS FOR WELD CLEANING

The tools used for weld cleaning include hand-held tools such as chipping hammers, wire brushes and pliers, and power tools such as pneumatic weld flux chippers and needle scalers.

9.1.0   HAND TOOLS

Wire brushes are used to clean welds and to remove paint or surface corrosion. Wire brushing will remove light to medium corrosion, but will not remove tight corrosion. Tight corrosion must be removed by filing, grinding, or sand blasting.

Chipping hammers are used to remove cutting and welding slag. The head of a chipping hammer has a point at one end and a chisel at the other. When chipping hammers become dull they can be sharpened on a grinder. When sharpening chipping hammers use care not to overheat the head. The head is hardened by tempering. Overheating will remove the temper, causing the head to become soft. If the temper is removed from a chipping hammer head it will mushroom and wear out much faster than a tempered head. Prevent overheating by plunging the chipping hammer's head into a pail of water every few seconds while grinding.

Welders should also have pliers to handle hot metal. Hot metal should not be handled with leather welding gloves. The heat will cause the leather to shrivel and become hard.

Figure 9-1 shows wire brushes, chipping hammers and pliers.

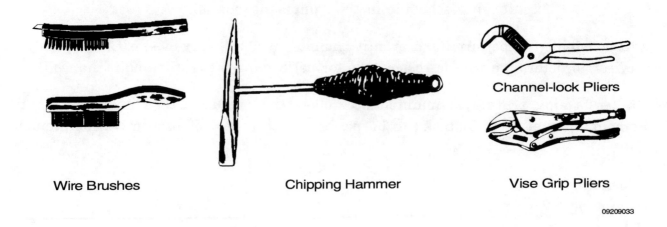

Figure 9-1. Wire Brushes, Chipping Hammer and Pliers

9.2.0 WELD FLUX CHIPPERS AND NEEDLE SCALERS

Weld flux chippers and needle scalers are pneumatically powered. They are used by welders to clean surfaces and to remove slag from cuts and welds. Weld flux chippers and needle scalers are also excellent for removing paint or hardened dirt but are not very effective for removing surface corrosion. Weld flux chippers have a single chisel, and needle scalers have about 18 to 20 blunt steel needles approximately ten inches long. Most weld flux chippers can be converted to needle scalers with a needle scaler attachment. *Figure 9-2* shows a weld flux chipper and a needle scaler.

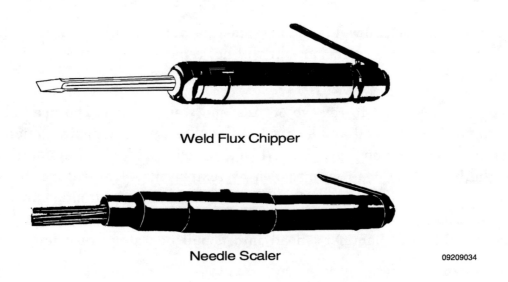

Figure 9-2. Weld Flux Chipper and Needle Scaler

SELF-CHECK REVIEW 3

1. What four factors must be considered when selecting a FCAW power supply?
2. What must be done before attempting to move a FCAW welding machine?
3. What should be checked to make certain the correct electrode wire type and shielding gas (if required) are selected?
4. When placing a ground clamp, how can you avoid damaging bearings, seals and other heat-damageable surfaces?
5. What should you locate before using an electrically powered welding machine?
6. What should you always check before starting an engine-driven welding machine?
7. Why should you never fuel a diesel engine with gasoline or a gasoline engine with diesel fuel?

ANSWERS TO SELF-CHECK 3

1. DC welding current, constant voltage (constant potential) output, maximum welding current required, and availability of electric service to power the welding machine. (8.1.0)
2. The shielding gas cylinder and wire feeder must be removed or safely secured before attempting to move the machine. (8.3.0)
3. The Welding Procedure Specification or site quality standard. (9.4.0 and 9.5.0))
4. Place the clamp so that no bearings, seals or other easily damaged items are in the path of the welding current (between the weld zone and the ground clamp). Batteries should be disconnected or removed and electronics should be isolated. (8.6.0)
5. The electrical disconnect for the electric service supplying power to the welding machine. (8.7.0)
6. The engine lube oil level and coolant levels should always be checked before starting the engine. Operating an engine with insufficient oil or coolant can destroy an engine. (8.8.1)
7. Operating a diesel engine on gasoline or a gasoline engine on diesel fuel will cause severe engine problems. (8.8.1)

SUMMARY

Flux core arc welding can be dangerous if the proper safety precautions are not followed. Before proceeding be sure you understand and can follow the safety precautions presented in this module. Before FCAW can take place the appropriate equipment must be selected and set up. By following the recommendations in this module, you will be able to select and safely set up the appropriate welding equipment.

REFERENCES

For advanced study of topics covered in this Task Module, the following works are suggested:

Basic Tig & Mig Welding (GTAW & GMAW), Griffin, Roden and Briggs, Delmar Publishers, Inc., 2 Computer Drive West, Box 15-015, Albany, N.Y. 12212, Phone 1-800-347-7707.

Modern Welding Technology (1979), Howard B. Cary, Prentice-Hall, Inc., Englewood Cliffs, N.J. 07632. Phone 201-767-5937

OSHA Requirements On Electrical Grounding.

Welding Principles and Applications, Jeffus and Johnson, Delmar Publishers, Inc., 2 Computer Drive West, Box 15-015, Albany, N.Y. 12212, Phone 1-800-347-7707.

Welding Skills, Giachino and Weeks, American Technical Publishers Inc, Homewood, IL, 1985, 1-800-323-3471.

PERFORMANCE / LABORATORY EXERCISES

None

The NCCER makes every effort to keep these manuals up-to-date and free of technical errors. We appreciate your help in this process. If you have an idea for improving this manual, or if you find an error, a typographical mistake, or an inaccuracy in the *Wheels of Learning*, please write us, using this form or a photocopy. Be sure to include the exact module number, page number, a description of the problem, and the correction, if possible. We'll do our best to correct it in later editions. Thank you for your assistance.

Write: *Wheels of Learning*
National Center for Construction Education and Research
P.O. Box 141104
Gainesville, FL 32614-1104

Fax: 352-334-0932

WHEELS OF LEARNING USER UPDATE

Please let us know if you have found an inaccuracy, error, or other problem in a *Wheels of Learning* manual. Use this form or write us a letter. Please be sure to tell us the exact module name and module number, the page number, and the problem. Thanks for your help.

Craft _____   Module Name _____

Module Number _____   Page Number(s) _____

Description of Problem _____

_____

_____

(Optional) Correction of Problem _____

_____

_____

(Optional) Your Name and Address _____

_____

FCAW-Fillet and Groove Welds

Module 09210

NATIONAL
CENTER FOR
CONSTRUCTION
EDUCATION AND
RESEARCH

FLUX CORE ARC WELDING (FCAW) - FILLET AND GROOVE WELDS

Objectives

Upon completion of this module, the trainee will be able to:

1. Perform FCAW multi-pass fillet welds on plate in the 1F (flat) position using flux core carbon steel wire and shielding gas.
2. Perform FCAW multi-pass fillet welds on plate in the 2F (horizontal) position using flux core carbon steel wire and shielding gas.
3. Perform FCAW multi-pass fillet welds on plate in the 3F (vertical) position using flux core carbon steel wire and shielding gas.
4. Perform FCAW multi-pass fillet welds on plate in the 4F (overhead) position using flux core carbon steel wire and shielding gas.
5. Perform FCAW multi-pass groove welds on plate in the 1G (flat) position using flux core carbon steel wire and shielding gas.
6. Perform FCAW multi-pass groove welds on plate in the 2G (horizontal) position using flux core carbon steel wire and shielding gas.
7. Perform FCAW multi-pass groove welds on plate in the 3G (vertical) position using flux core carbon steel wire and shielding gas.
8. Perform FCAW multi-pass groove welds on plate in the 4G (overhead) position using flux core carbon steel wire and shielding gas.

Prerequisites

Successful completion of the following module(s) is required before beginning study of this module:

● *Flux Core Arc Welding - Equipment and Filler Metals*, Module 09209

Required Student Materials

Each trainee will need:

1. Personal protective equipment
2. Leather welding gloves
3. Leather jacket or sleeves

4. Welding shield
5. FCAW welding power supply with wire feeder, shielding gas equipment, and electrode gun adapted to 0.045 inch diameter wire
6. Welding table with out-of-position arm
7. Electrode wire, 0.045 inch diameter dual shield flux core carbon steel, (Class E71T-1) or as specified by the instructor
8. Shielding gas, 75/25 argon-carbon dioxide mixture, or as specified by the instructor
9. Cutting goggles
10. Chipping hammer
11. Wire brush
12. Needle nose side cutters
13. Pliers
14. Tape measure
15. Soapstone
16. Scrap steel plate, 3/8 inch to 2 inches thick

Each trainee will need access to:

1. Oxyfuel cutting equipment
2. Framing square
3. Grinders
4. Chippers
5. Wire brushes

Course Map Information

This course map shows all of the *Wheels of Learning* task modules in the second level of the Welding curricula. The suggested training order begins at the bottom and proceeds up. Skill levels increase as a trainee advances on the course map. The training order may be adjusted by the local Training Program Sponsor.

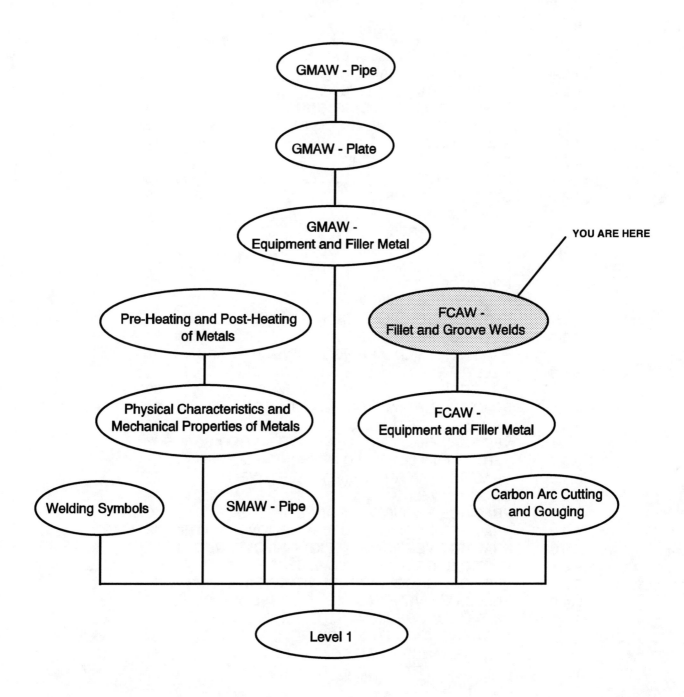

3

TABLE OF CONTENTS

Trade Terms Introduced In This Module

DCRP (Direct Current Reverse Polarity): The electrical connection of welding cables of a direct current welding machine where the machine's positive cable is connected to the electrode holder and the negative cable is connected to the workpiece. In this configuration, the conventional current flows from the electrode to the workpiece, while the actual electron flow is from the workpiece to the electrode.

Globular Transfer: A GMAW/FCAW process where the filler metal is melted into large drops which are sequentially carried to the molten pool by the arc.

Low-Carbon Steel: Also known as mild steel. Steel with a carbon content of 0.05 to 0.30 percent.

Mild Steel: Also know as low-carbon steel. Steel with a carbon content of 0.05 to 0.30.

Out-Of-Position Weld: Out-of-position welds include all welds except the flat position weld. The flat position weld bead is always horizontal and a vertical plane through the root will bisect the weld bead along its entire length.

Penetration: The distance that the weld fusion line extends below the surface of the base metal.

Restart: The point in the weld where one weld bead stops and the continuing bead is started.

Spray-Arc Transfer: Also called Spray Transfer. A GMAW/FCAW welding process where the filler metal is melted into fine droplets and carried to the molten pool by the welding arc.

Stickout: Also called Electrical Extension. In FCAW and GMAW, the length of wire electrode that extends beyond the contact tube of the welding gun.

Stringer Bead: A weld bead that is made with little or no side-to-side motion of the electrode, usually no more than two to three times the electrode diameter.

Tie-In: Complete fusion of the weld end with the base metal.

Weave Bead: A weld bead that is made with a wide side-to-side motion of the electrode, generally not more than eight times the electrode diameter.

Weld Coupon: The metal to be welded as a test or practice.

1.0.0 INTRODUCTION

The Flux Core Arc Welding (FCAW) process is a semiautomatic process that uses an electric arc to melt and fuse a continuous flux-core wire electrode with the base metal to form a weld. The wire electrode is consumed in the process. Shielding gas and smoke from the flux core wire electrode protects the molten weld metal from contamination by the atmosphere. A slag formed by flux components stabilizes and protects the weld bead. The shielding gas can be generated entirely from the flux in the wire core (self-shielding), or by both the flux in the wire core and an external shielding gas (dual shield). The external shielding gas usually promotes much deeper weld penetration. In all types of flux-core wires, the flux contains agents for degassing the weld, stabilizing the arc and forming a protective slag over the weld bead. In addition, other elements are often added to the flux core to alloy with the base metal and improve its chemical and/or mechanical characteristics.

Figure 1-1 shows the FCAW dual-shield process.

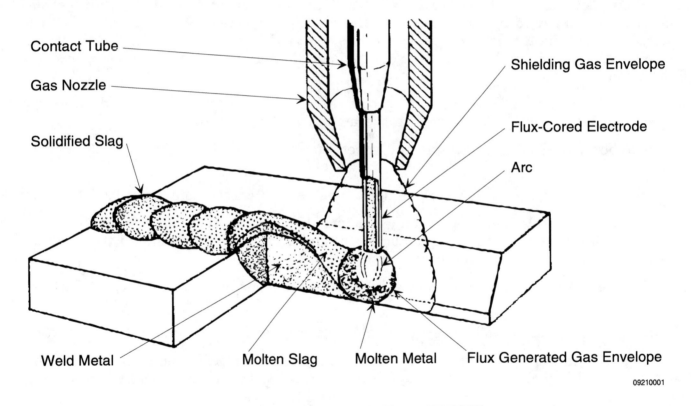

09210001

Figure 1-1. FCAW Dual-Shield Process.

The equipment used for FCAW dual-shield and FCAW self-shielding is basically the same except for the components needed for the gas used by the dual-shield process. This additional equipment includes a gun equipped with a gas nozzle, gas hoses, gas control solenoid valve, gas regulator/flowmeter and a shielding gas cylinder (gas source). A special gun is typically used with the FCAW self-shielding process since an external shielding gas is not needed. The gun nozzle is smaller in diameter and has a solid fiber insulation around the contact tip instead of the gas nozzle found on a dual-shield gun.

Figure 1-2 is a simple diagram of the FCAW dual-shield equipment.

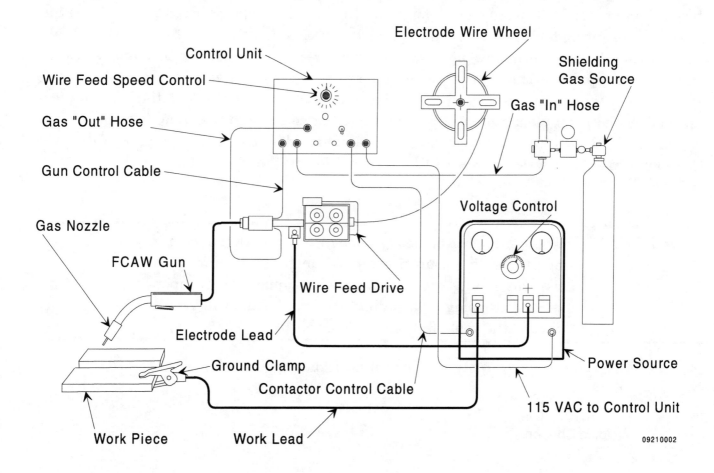

Figure 1-2. FCAW Dual-Shield Equipment.

The FCAW process is basically limited to welding low-carbon steels, some low-alloy steels and some stainless steels. FCAW is an excellent process for making fillet and groove welds on mild steel and low-carbon steel plate. This module explains how to set up FCAW equipment and perform fillet welds on steel plate in the 1F, 2F, 3F and 4F weld positions and groove welds on steel plate in the 1G, 2G, 3G and 4G positions, using dual-shield wire (flux-core wire and an external shielding gas).

2.0.0   FCAW WELDING EQUIPMENT SET-UP

Before welding can take place the work area has to be made ready, the welding equipment set up and the metal to be welded prepared. The following sections will explain how to prepare the area, set up the equipment and make FCAW fillet and groove welds on low-carbon steel plate.

2.1.0 PREPARE WELDING AREA

To practice welding, a welding table, bench or stand is needed. The welding surface must be steel and provisions must be made for mounting practice welding coupons out-of-position. A simple mounting for out-of-position coupons is a pipe stand that can be welded vertically to the steel table top. To make a pipe stand, weld a three- to four-foot length of pipe vertically to the table top. Then cut a short section pipe to slide over the vertical pipe. Drill a hole in this slide. Weld a nut over the hole so that a bolt can be used to lock the slide in place on the vertical pipe. Weld a piece of pipe or angle horizontally to the slide, to form the arm which will hold the coupons. The slide can be rotated and adjusted vertically to position the horizontal arm. The weld coupon will be tack-welded to the horizontal arm.

WARNING!   The table must be heavy enough to support the weight of the welding coupon extended on the horizontal arm without falling over. It must also support the coupon during chipping or grinding. Serious injury will result should the table fall onto someone.

Figure 2-1 shows a welding table with out-of-position support.

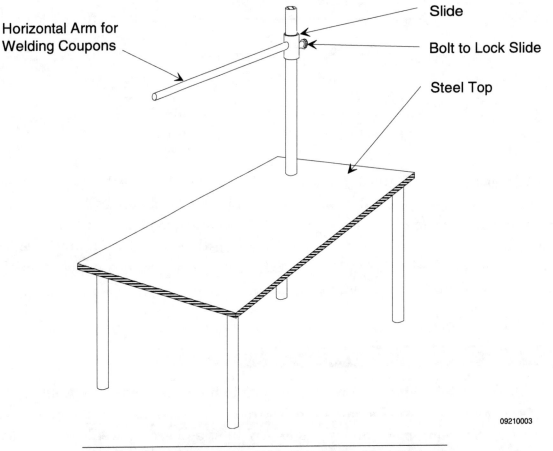

Horizontal Arm for Welding Coupons

Slide

Bolt to Lock Slide

Steel Top

09210003

Figure 2-1. Welding Table With Out-Of-Position Support

To set up the area for welding follow these steps.

Step 1   Check to be sure the area is properly ventilated. Make use of doors, windows and fans.

Step 2   Check the area for fire hazards. Remove any flammable materials before proceeding.

Step 3   Check the location of the nearest fire extinguisher. Do not proceed unless the extinguisher is charged and you know how to use it.

Step 4   Position a welding table near the welding machine.

Step 5   Set up flash shields around the welding area.

2.2.0   PREPARE WELDING COUPONS

The welding coupons should be mild or low-carbon steel 1/4- to 3/4-inch thick. Use a wire brush or grinder to remove any heavy mill scale or corrosion. Prepare welding coupons to practice the welds indicated as follows:

- Running Stringer Beads: The coupons can be any size or shape that can be easily handled.
- Fillet Welds: Cut the metal into four-by-six inch rectangles for the base and three-by-six inch rectangles for the web.
- Butt Joints: Cut a pair of three-by-seven inch rectangles for each butt joint. Each of the pieces should have a 30-degree bevel along one of the long sides. Each bevel should have a 1/16 inch land ground on it. (Most codes allow a land size of 1/16 to 1/8 inch. The land size can be adjusted as needed within these limits. Always refer to your site WPS or quality standards for specific joint configurations. Information in this manual is provided as a general guideline only.

Figure 2-2 shows fillet and open V-butt weld coupon.

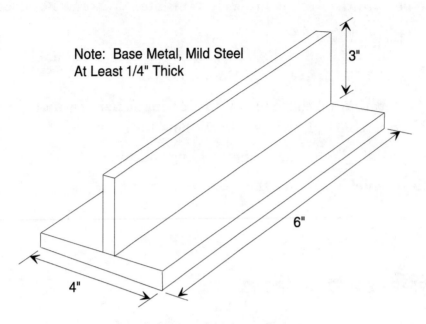

Note: Base Metal, Mild Steel
At Least 1/4" Thick

3"

6"

4"

Fillet Weld Coupons

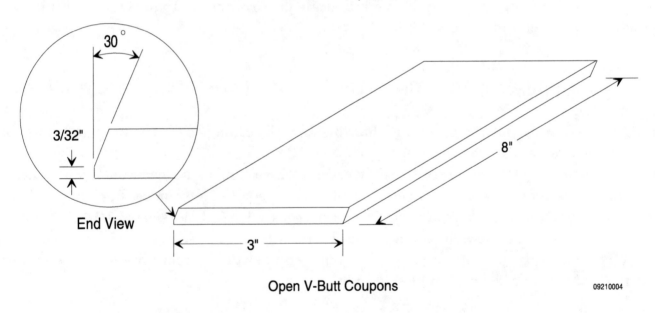

30°

3/32"

End View

8"

3"

Open V-Butt Coupons

09210004

Figure 2-2. Fillet and Open V-Butt Weld Coupon.

Note   Steel for practice welding is expensive and difficult to obtain. Every effort should be made to conserve and not waste the material that is available. Reuse weld coupons until all surfaces have been welded on. Weld on both sides of the joint and then cut the weld coupon apart and reuse the pieces. Use material that cannot be cut into weld coupons to practice running beads.

2.3.0   FILLER WIRE

The FCAW process uses continuous flux core wire electrode for filler wire. The wire can be self-shielding or dual shield (used with an external shielding gas). Filler wire is specified by size (diameter) and AWS classification number. Generally, filler wire is selected to be a close chemical match to the base metal. The WPS or job site standards may specify the filler wire to be used.

For the welding exercises in this module, 0.045 inch flux core dual shield carbon steel wire (such as E70T-1) will be used for all weld types and weld positions.

Note   Depending on local conditions, another wire type and size may be substituted. Check with your instructor for the wire type and size to use.

2.4.0   FCAW WELDING EQUIPMENT

Locate the following FCAW welding equipment.

- CV power source
- FCAW wire feeder
- FCAW Gun
- Gas cylinder and regulator/flow meter

Figure 2-3 is a simplified diagram of a complete FCAW dual-shield system.

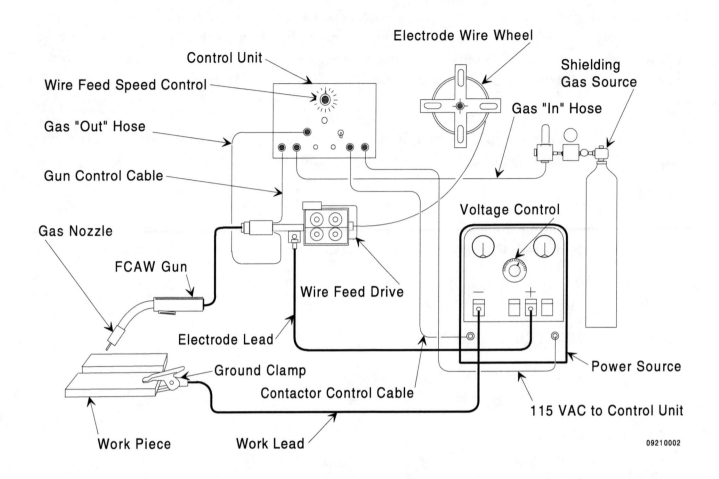

Figure 2-3. Complete FCAW Dual-Shield System.

Perform the following steps to set up the FCAW equipment.

Step 1 Verify that the welding machine is a constant voltage type DC power supply.

Step 2 Check to be sure that the welding machine is properly grounded through the primary current receptacle.

Step 3 Verify the location of the primary current disconnect.

Step 4 If the welding machine does not have a wire feeder, locate a compatible wire feeder.

Step 5 Connect the wire feeder power and control cables.

Step 6 Obtain a spool of 0.045 inch flux core dual shield mild steel wire.

Step 7 Check the wire feed rollers to be sure they are the correct size for the wire to be run. Change them if they are the wrong size. Follow the manufacturer's instructions for changing feed rollers.

Step 8 Locate a FCAW gun and cable assembly. Check the size of the nozzle contact tip. Change it if it is not correct for the size wire to be used.

Step 9 Connect the gun cable to the wire feeder.

Step 10 Set up the welding machine for reverse polarity (DCRP) welding (positive lead to gun and negative lead to workpiece).

Step 11 Turn on the welding machine.

Step 12 Install the wire spool onto the wire feeder. Adjust the spool drag brake and feed the wire through the cable and the gun until the proper stick-out is achieved.

Step 13 Locate a cylinder of 75/25 argon CO_2 (75 percent argon / 25 percent CO_2) shielding gas mixture and secure it to the welding machine or a nearby structure so that it cannot fall. Install the gas regulator-flowmeter and connect the gas hose to the wire feeder.

Note Alternative shielding gas may be used depending on local site conditions and available wire requirements.

Step 14 Fully open the shielding gas cylinder and then purge the gun with the purge control on the welding machine or wire feed unit, and adjust the gas flowmeter to achieve the specified flow rate.

Step 15 Set the welding variables as follows:

Wire	DCRP	
Size	Volts	Amps
0.045 in.	20-25	125-225

Note Shielding Gas Mixture: 75% Argon, 25% CO_2 at 35 cfh
Electrical stickout: 3/8 to 3/4 inch
Use higher end of ranges for flat welds
Use lower end of ranges for vertical welds
Base metal: 1/4-3/4 inch thick mild steel

Step 16 Set the other variables (if applicable).

Refer to the manufacturer's specifications for:

- Slope
- Inductance
- Burn back
- Gas preflow (prepurge)
- Gas postflow (post purge)

3.0.0 ARC BLOW

When current flows, magnetic fields are created. The magnetic fields tend to concentrate in corners, deep grooves or the ends of the base metal. When the arc approaches these concentrated magnetic fields it is deflected from the intended path. Arc blow can cause defects such as excessive weld spatter and porosity. If arc blow occurs, try one or more of these methods to control it.

- Change the position of the ground clamp. This will change the flow of welding current, affecting the way the magnetic fields are created.
- Change the angle of the gun. Reducing or even reversing the angle of the gun can compensate for the arc blow.

4.0.0 FCAW BEAD AND PENETRATION FACTORS

FCAW bead size, shape and penetration are affected by a number of factors. Some factors are equipment related and others are welder controlled. The factors include:

- Welding voltage
- Welding amperage
- Welding travel rate
- Welding gun angle
- Contact tip-to-work distance (stickout)
- Type of bead (stringer or weave)

4.1.0 WELDING VOLTAGE

Arc length is determined by voltage, which is set at the power supply. Arc length is the distance from the wire electrode tip to the base metal or molten pool at the base metal. If voltage is set too high, the arc will be too long and could weld the end of the contact tip.

Too high a voltage will also cause porosity and excessive spatter. Voltage must be increased or decreased as wire feed speed (welding amperage) is increased or decreased. Voltage must be set in relation to wire feed speed (welding amperage). Set the voltage so that the arc length is just above the surface of the base metal. *Figure 4-1* shows the arc length.

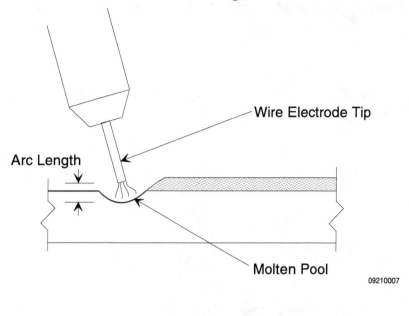

09210007

Figure 4-1. Arc Length.

4.2.0   WELDING AMPERAGE

With a constant potential power source, the electrode feed speed controls the welding amperage. The welding power supply provides the amperes necessary to melt the wire electrode, while maintaining the selected welding voltage. When the wire electrode feed speed is increased, the welding amperes and deposition rate also increase. This results in higher welding heat, deeper penetration and larger beads.

When the wire electrode feed speed is decreased, the welding amperage automatically decreases. With lower welding amperage and less heat, the deposition rate drops and the weld beads are smaller with less penetration.

4.3.0   WELDING TRAVEL RATE

Weld travel rate is the relative speed of the electrode tip to the base metal, in the direction of the weld. Travel speed has a great effect on penetration and bead size. Slower travel speeds build bigger beads with deeper penetration. Faster travel speeds build narrower beads with less penetration.

4.4.0   LONGITUDINAL WELDING GUN ANGLE

The longitudinal welding gun angle is the angle that the center line of the electrode makes with a line normal to the weld axis. Gun angles can be pushing (leading), neutral (vertical) or pulling (lagging, dragging or trailing). *Figure 4-2* shows longitudinal welding gun angles.

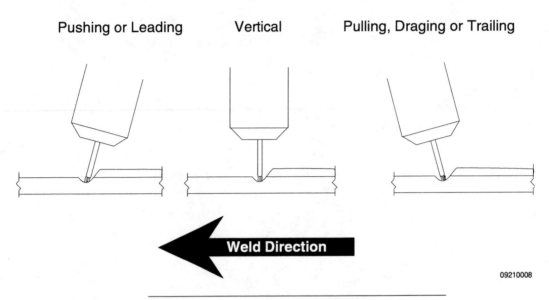

Pushing or Leading   Vertical   Pulling, Draging or Trailing

Weld Direction

09210008

Figure 4-2. Longitudinal Welding Gun Angles.

In a pushing (leading) gun angle, the electrode tip proceeds the gun in the direction of the weld. The electrode tip and gas shield are directed ahead of the weld bead resulting in a narrower, higher bead with less penetration. The maximum practical push or drag angle is about 55 degrees from a normal to the weld bead.

In the neutral or vertical gun angle, the electrode is normal to the axis of the weld bead. Penetration is moderate.

In a pulling (lagging) gun angle, the electrode tip trails the gun in the direction of the weld. The electrode tip and gas shield are directed back into the weld bead resulting in a wider, flatter bead with more penetration. Maximum penetration is achieved with a gun angle of approximately 25 degrees from a normal position to the weld bead.

4.5.0   CONTACT TIP-TO-WORK DISTANCE (STICKOUT)

The contact tip-to-work distance is the distance from the contact tip to the bottom of the crater, which includes the arc length. The electrode stickout is the length of electrode wire that extends from the contact tip.

When the stickout is increased, the constant potential power supply continues to supply the current necessary to maintain the preset voltage. However, because more voltage is lost in the electrode length, arc voltage is lower and irregular arc action and spatter may occur when the stickout is too long. Penetration is decreased as stickout is increased.

With shorter stickout and less voltage drop in the wire, higher arc voltage is available. If the stickout is too short, spatter will increase and rapidly build up in the nozzle and on contact tip.

Note   The stickout for dual-shielding FCAW with 0.045-inch wire is 3/8 to 3/4 inch, depending on the welding position.

Figure 4-3 shows stickout range for 0.045 inch carbon steel dual-shield wire.

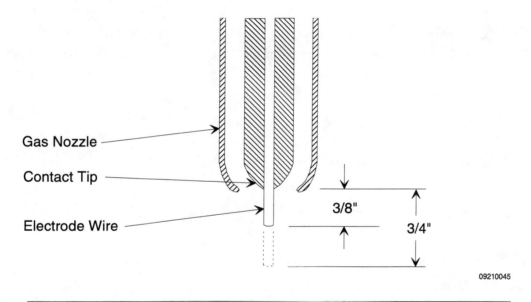

Gas Nozzle

Contact Tip

Electrode Wire

3/8"

3/4"

09210045

Figure 4-3. Stickout Range for 0.045 Inch Carbon Steel Dual-Shield Wire.

4.6.0   SHIELDING GAS

Shielding gas affects weld penetration. Welds made with self-shielding flux core wire do not penetrate as deeply as do welds made with flux core wire and with external (supplemental) gas shielding because the external gas results in a more compact and therefore hotter arc.

The two types of shielding gas generally used for FCAW and carbon steel are welding grade CO_2 and a mixture of 75 percent argon and 25 percent CO_2. The 75-25 gas mixture improves arc characteristics in out-of-position work, increases wetting action, decreases penetration and provides a finer spray metal transfer. Straight CO_2 is generally used for larger wires because it increases the heat at the arc and increases penetration.

Note   The welding voltage recommendations are for 75-25 argon-CO_2 gas mixtures. If straight CO_2 is used, increase the welding voltage 1 to 2 volts.

4.7.0   GAS NOZZLE CLEANING

With use, weld spatter accumulates on the gas nozzle and contact tip. If it is not occasionally cleaned, the gas nozzle will become blocked, restricting the shielding gas flow and causing porosity in the weld.

Clean the gas nozzle with a reamer or round file, or the tang of a standard file. After cleaning, the nozzle can be sprayed or dipped in a special anti-spatter compound. The anti-spatter helps prevent the spatter from sticking to the nozzle.

CAUTION   Use only anti-spatter material specifically designed for welding gas nozzles. Other materials may cause porosity in the welds.

4.8.0   BEAD TYPES

There are two types of bead commonly used to increase bead width:

● Stringer beads
● Weave beads

4.8.1   Stringer Beads

Stringer beads are made with little or no side-to-side motion of the electrode. The width of a stringer bead is generally no more than six times the diameter of small core (0.045 -

1/16 inch) or four times the diameter of large core (3/32 - 1/8 inch) electrode wire.

Figure 4-4 shows a stringer bead.

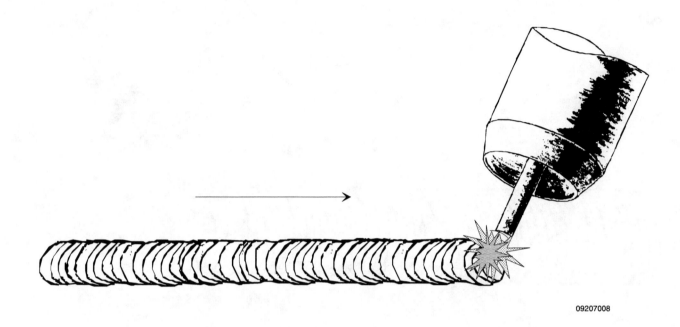

09207008

Figure 4-4. Stringer Bead.

CAUTION   The width of stringer beads is specified in the welding code or WPS being used at your site. Do not exceed the widths specified for your site.

4.8.2   Weave Beads

Weave beads are made with a side-to-side motion of the electrode. The width of the weave bead is determined by the amount of side-to-side motion.

Figure 4-5 shows a weave bead.

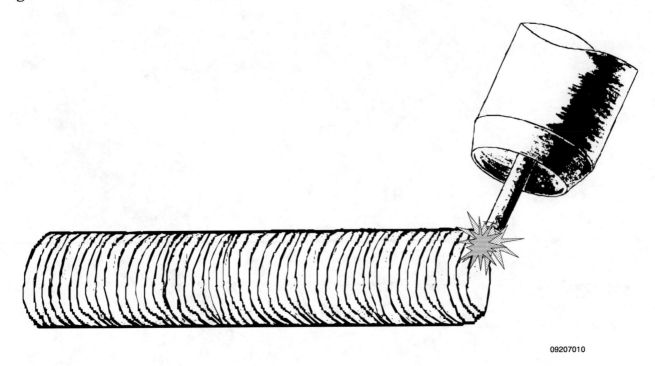

09207010

Figure 4-5. Weave Bead.

Weave beads are made with a side-to-side or up and down zigzag motion. *Figure 4-6* shows FCAW weave motions.

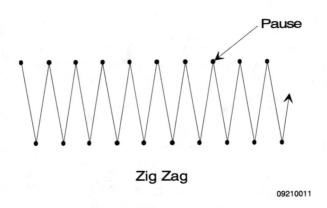

Pause

Zig Zag

09210011

Figure 4-6. FCAW Weave Motions.

When making a weave bead, care must be used at the toes to ensure proper tie-in to the base metal. To ensure proper tie-in at the toes, slow down or pause slightly at the edges. The pause at the edges will also flatten out the weld, giving it the proper profile.

CAUTION　　The width of weave beads is specified in the welding code or WPS being used at your site. Do not exceed the widths specified for your site.

4.9.0　WELD RESTARTS

A restart is the junction where a new weld connects to continue the bead of a previous weld. Restarts are important because an improperly made restart will create a weld defect. One advantage of the semi-automatic FCAW process is that restarts can be minimized. Plan the weld to eliminate restarts if possible. If a restart has to be made, it must be made so that it blends smoothly with the previous weld and does not stand out. The technique for making a FCAW restart is the same for both stringer and weave beads. Follow these steps to make a restart.

Step 1　With the weld chipped clean of slag, hold the torch at the proper angle and restart the arc directly over the center of the crater. (The welding codes do not allow arc strikes outside the area to be welded.)

Step 2　Move the electrode tip in a small circular motion over the crater to fill the crater with a molten puddle.

Step 3　As soon as the puddle fills the crater, continue the bead with the stringer or weave pattern used to make the previous weld.

Step 4　Inspect the restart. A properly made restart will blend into the bead, making it hard to detect.

If the restart has undercut, not enough time was spent in the crater to fill it. If undercut is on one side or the other, use more of a side-to-side motion as you move back into the crater. If the restart has a lump, it was overfilled. Too much time was spent in the crater before resuming the forward motion.

Continue to practice restarts until they are correct. Use the same techniques for making restarts whenever performing FCAW.

4.10.0 WELD TERMINATIONS

A weld termination is made at the end of a weld. A termination normally leaves a crater. When making a termination, the welding codes require that the crater must be filled to the full cross-section of the weld. This can be difficult, since most terminations are at the edge of a plate where welding heat tends to build up making filling the crater more difficult.

The technique for making a termination is basically the same for all FCAW. Follow these steps to make a termination.

Step 1   As you approach the end of the weld, start to bring the gun up to a 0-degree longitudinal angle (gun normal to the bead) and slow forward travel.

Step 2   Stop forward movement about 1/8 inch from the end of the plate and slowly angle the gun to about 10 degrees toward the start of the weld (drag or backhand weld).

Step 3   Move about 1/8 inch toward the start of the weld and release the trigger when the crater is filled.

Step 4   Inspect the termination. The crater should be filled to the full cross-section of the weld.

Figure 4-7 shows a terminating a weld.

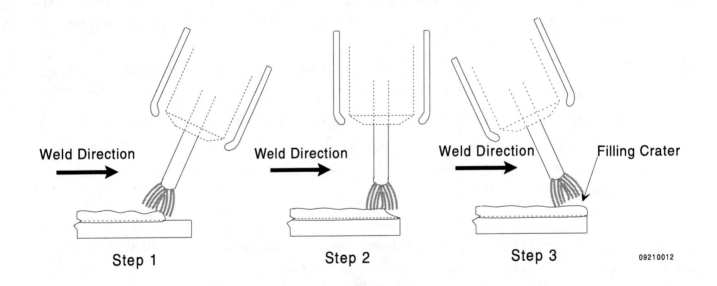

Figure 4-7. Terminating a Weld.

5.0.0   PRACTICE STRINGER BEADS

Practice running stringer beads in the flat position. Experiment with different voltages, gun angles and stickouts. Change the wire feed speed and observe the bead size. Try different gun angles and observe the bead shape and size.

Follow these steps to run stringer beads.

Step 1 Hold the gun at the desired angle with the electrode tip directly over the point where the weld is to begin and pull the gun trigger.

Step 2 Hold the arc in place until the weld puddle begins to form.

Step 3 Slowly advance the arc, while maintaining the gun angle.

Step 4 Continue to weld until a bead about 2 to 3 inches long is formed and then break the arc by releasing the trigger. A small crater will be left at the point where the arc was broken.

Step 5 Chip the slag and inspect the bead for:

- Straightness of the bead
- Uniform rippled appearance of the bead face
- Smooth flat transition with complete fusion at the toes of weld
- No porosity
- No undercut
- No cold lap
- No lack of fusion
- No incomplete penetration
- No pinholes (fisheyes)
- No slag inclusions

Step 6 Continue practicing stringer beads until you can make acceptable welds every time.

6.0.0 PRACTICE WEAVE BEADS

Practice running weave beads about 3/4 inch wide in the flat position. Experiment with different weave motions, lead angles and stickouts. Change the wire feed speed and observe the bead size. Try different gun angles and observe the bead shape and size.

Follow these steps to run weave beads.

Step 1 Hold the gun at the desired angle with the electrode tip directly over the point where the weld is to begin and pull the gun trigger.

Step 2 Hold the arc in place until the weld puddle begins to form.

Step 3 Slowly advance the arc using the zigzag weave motion, while maintaining the gun angle.

Step 4 Continue to weld until a bead about 2 to 3 inches long is formed and then break the arc by releasing the trigger. A small crater will be left at the point where the arc was broken.

Step 5 Chip the slag and inspect the bead for:

- Straightness of the bead
- Uniform rippled appearance of the bead face
- Smooth flat transition with complete fusion at the toes of weld
- No porosity
- No undercut
- No cold lap
- No lack of fusion
- No Incomplete penetration
- No Pinholes (fisheyes)
- No slag inclusions

Step 6 Continue practicing weave beads until you can make acceptable welds every time.

SELF CHECK REVIEW 1

1. Before welding, how do you check for a fire extinguisher?
2. Why is it important to conserve steel used for practice welding?
3. How is filler wire specified?
4. What type of DC welding machine is best for FCAW?
5. When setting up a FCAW gun, what should be checked for?
6. What are ways to control arc blow?
7. What is the maximum practical longitudinal gun angle for FCAW?
8. What longitudinal gun angle gives the maximum penetration with FCAW?
9. What happens to penetration when stickout is increased?
10. Is weld penetration deeper or shallower when running FCAW with external gas shielding?
11. What can happen to the weld if the wrong type anti-spatter material is used on the gas nozzle?
12. What specifically defines the maximum width of a stringer bead?

ANSWERS TO SELF CHECK REVIEW 1

1. Check the location of the nearest fire extinguisher. Do not proceed unless the extinguisher is charged and you know how to use it. (2.1.0)
2. Steel for practice welding is expensive and difficult to obtain. (2.2.0)
3. Filler wire is specified by size (diameter) and metal composition (alloy). (2.3.0)
4. A constant voltage (constant potential) machine is best for FCAW. (2.4.0)
5. The gun adapter (contact tip) and wire drive rollers should be checked for correct wire size and tension adjustment. (2.4.0)
6. Change the position of the ground clamp. Change the angle of the electrode. (3.0.0)
7. The maximum gun angle is 55 degrees from a normal to the weld bead. (4.4.0)
8. Maximum penetration is achieved with a 25 degree backhand (drag) angle. (4.4.0)
9. Penetration decreases. (4.5.0)
10. Penetration is deeper. (4.6.0)
11. Porosity can develop in the weld. (4.7.0)
12. The WPS or site welding quality standards define the maximum width of a stringer bead. (4.8.1)

7.0.0 OVERLAPPING BEADS

Overlapping beads are made by depositing connective weld beads parallel to one another. The parallel beads overlap forming a flat surface. This is also called padding. Overlapping beads are used to build up a surface and for making multipass welds. Both stringer and weave beads can be overlapped.

Properly overlapped beads will form a relatively flat surface when viewed from the end.

Figure 7-1 shows proper and improper overlapping beads.

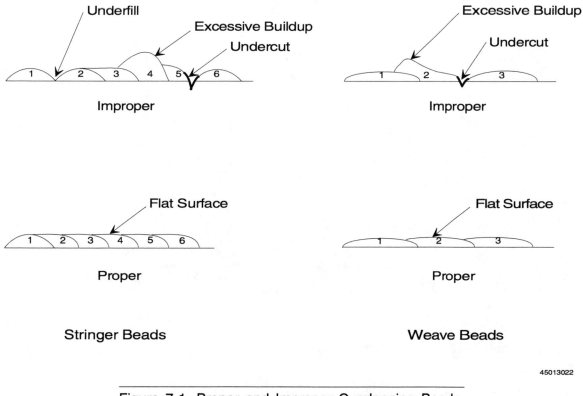

45013022

Figure 7-1. Proper and Improper Overlapping Beads.

7.1.0 PRACTICE OVERLAPPING STRINGER BEADS

Practice welding overlapping stringer beads using 0.045 inch flux core dual shield carbon steel electrode wire and 75/25 argon-carbon dioxide mixture shielding gas.

Follow these steps to make overlapping stringer beads.

Step 1 Mark out a 4-inch square on a piece of steel.

Step 2 Weld a stringer bead along one edge.

Step 3 Chip the slag from the bead.

Step 4 Position the gun at a transverse angle of 10 to 15 degrees toward the side of the previous bead to get proper tie-in and pull the trigger.

Figure 7-2 shows the transverse gun angle for overlapping stringer beads.

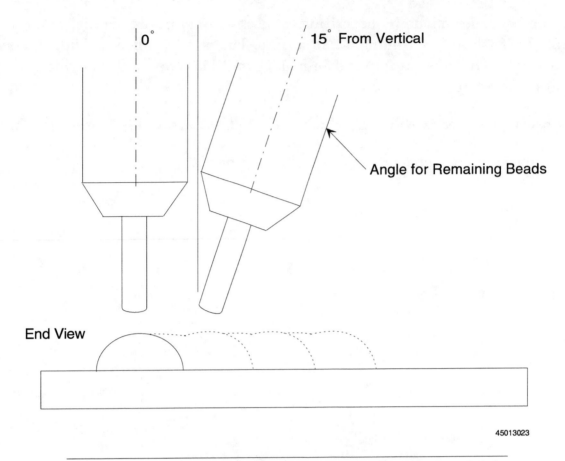

Figure 7-2. Transverse Gun Angle For Overlapping Stringer Beads.

Step 5 Continue running overlapping stringer beads until the 4-inch square is covered. Chip each bead before running the next bead.

Note The base metal will get very hot as it is being built up. If necessary, hold it with pliers and cool it with water.

Step 6 Continue building layers of stringer beads, one on top of the other, until the technique is perfected.

8.0.0 FILLET WELDS

Fillet welds are commonly used to weld lap and tee joints. The weld position is determined by the axis of the weld. The standard positions for fillet welding are flat or 1F (the "F" indicates fillet), horizontal or 2F, vertical or 3F, and overhead or 4F. The axis of the weld can rotate plus or minus 15 degrees and the weld can be inclined up to 15 degrees within a given position.

WELDING TWO TRAINEE TASK MODULE 09210

Figure 8-1 shows the standard fillet weld positions.

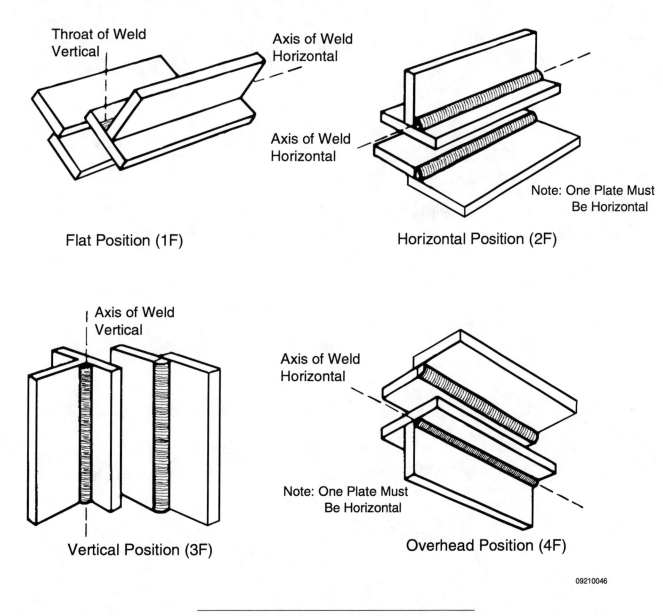

Throat of Weld
Vertical

Axis of Weld
Horizontal

Axis of Weld
Horizontal

Note: One Plate Must
Be Horizontal

Flat Position (1F)

Horizontal Position (2F)

Axis of Weld
Vertical

Axis of Weld
Horizontal

Note: One Plate Must
Be Horizontal

Vertical Position (3F)

Overhead Position (4F)

09210046

Figure 8-1. Standard Fillet Weld Positions.

Fillet welds can be concave or convex, depending on the WPS or site quality standards. Welding codes require that a fillet weld have a uniform concave or convex face, although a slightly nonuniform face is acceptable. Convexity is the distance the weld extends above a line drawn between the roots of the weld. The convexity of a fillet weld or individual surface bead must not exceed 0.07 times the actual face width or individual surface bead, plus 0.06-inch.

A fillet weld is unacceptable and must be repaired if the profile has insufficient throat, excess convexity, excessive undercut, overlap, insufficient leg or inadequate penetration.

Figure 8-2 shows acceptable and unacceptable fillet weld profiles.

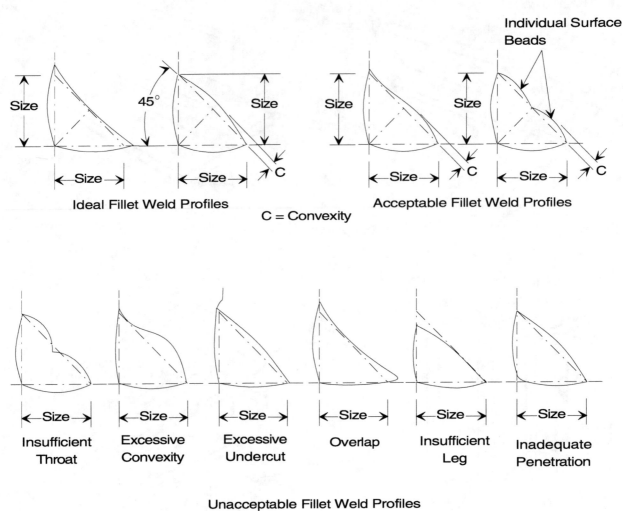

Figure 8-2. Acceptable and Unacceptable Fillet Weld Profiles.

8.1.0   PRACTICE FLAT (1F) FILLET WELDS

Practice flat fillet welds by welding stringer beads in a tee joint using 0.045 inch flux core dual shield carbon steel electrode wire and shielding gas. Set the welding parameters (voltage and wire feed speed) at the high end of the ranges.

When making flat fillet welds, pay close attention to the gun angle and travel speed. For the first bead, the transverse gun angle is 45 degrees to both plate surfaces. The angle is adjusted for all subsequent beads. Increase or decrease the travel speed to control the amount of weld metal build-up.

Follow these steps to make a flat fillet weld.

Step 1   Tack two plates together to form a tee joint for the fillet weld coupon. Chip the slag from the tack welds.

Figure 8-3 shows the fillet weld coupon.

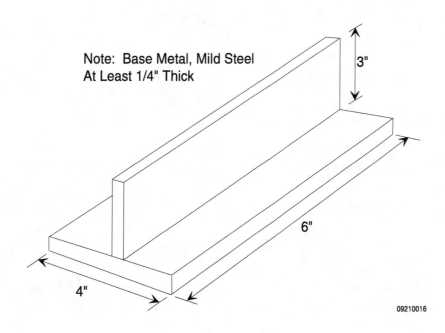

Note: Base Metal, Mild Steel
At Least 1/4" Thick

09210016

Figure 8-3. Fillet Weld Coupon.

Step 2   Tack the coupon to the welding table or positioning arm in the 1F (flat) position.

Step 3   Run the root bead along the root of the joint using a transverse gun angle of 45 degrees with a 5- to 10-degree longitudinal gun angle. Thoroughly chip the slag from the bead.

Step 4   Run the second bead along one toe of the first weld, overlapping about 75 percent of the first bead. Use a transverse gun angle of 70 degrees with a 5- to 10-degree longitudinal gun angle and a slight oscillation. Chip the bead.

Step 5   Run the third bead along the opposite toe of the first weld, filling the groove created when the second bead was run. Use a transverse gun angle of 30 degrees with a 5- to 10-degree longitudinal gun angle and a slight oscillation. Chip the bead.

Figure 8-4 shows weld bead sequences 1, 2 and 3, and gun angles.

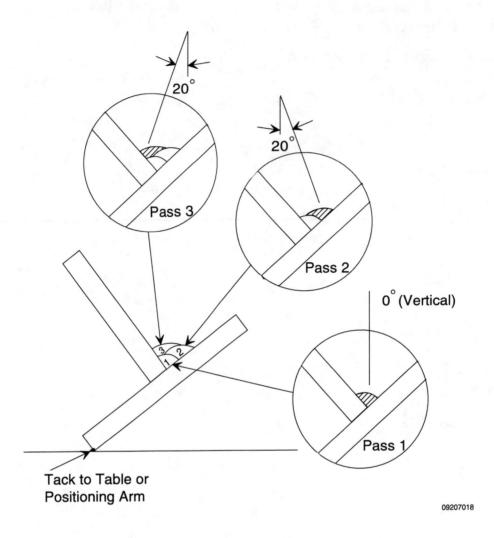

Figure 8-4. Bead Sequences 1, 2 and 3 and Gun Angles.

Step 6 Run the fourth bead along the outside toe of the second weld, overlapping about half the second bead. Use a transverse gun angle of 70 degrees with a 5- to 10-degree longitudinal gun angle and a slight oscillation. Chip the bead.

Step 7 Run the fifth bead along the inside toe of the fourth weld, overlapping about half the fourth bead that was run. Use a transverse gun angle of 70 degrees with a 5- to 10-degree longitudinal gun angle and a slight oscillation. Chip the bead.

Step 8 Run the sixth bead along the toe of the fifth weld, filling the groove created when the fifth bead was run. Use a transverse gun angle of 30 degrees with a 5- to 10-degree longitudinal gun angle and a slight oscillation. Chip the bead.

Figure 8-5 shows weld bead sequences 4, 5 and 6 and the gun angles.

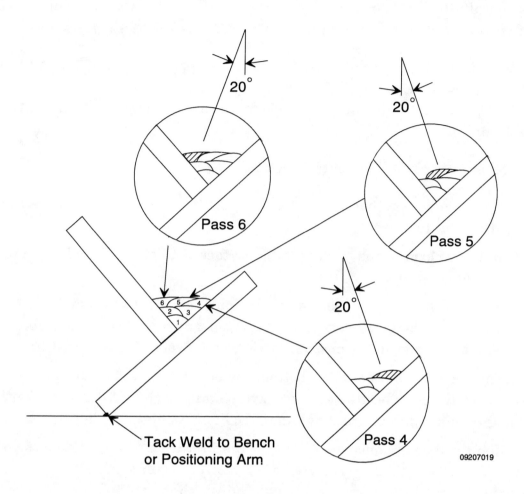

Figure 8-5. Bead Sequences 4, 5 and 6 and Gun Angles.

Step 9 Inspect the weld. The weld is acceptable if it has:

- Uniform rippled appearance on the bead face
- Craters and restarts filled to the full cross-section of the weld
- Uniform weld size, plus or minus 1/16 inch
- Acceptable weld profile in accordance with AWS D1.1
- Smooth transition with complete fusion at the toes of the weld
- No porosity
- No undercut
- No cold lap
- No lack of fusion
- No Incomplete penetration
- No slag inclusions

8.2.0 PRACTICE HORIZONTAL (2F) FILLET WELDS

Practice horizontal fillet welds by welding multi-pass convex fillet welds in a tee joint using 0.045 inch flux core dual shield carbon steel electrode wire and shielding gas.

When making horizontal fillet welds, pay close attention to the gun angle and travel speed. For the root bead, the gun angle is 45 degrees to the plate surface. The angle is adjusted slightly for all subsequent beads. Increase or decrease the travel speed to control the amount of weld metal build-up.

Follow these steps to make a horizontal fillet weld.

Step 1 Tack two plates together to form a tee joint for the fillet weld coupon. Chip the slag from the tack welds.

Step 2 Clamp the coupon to the welding table or tack it to the positioning arm in the 2F (horizontal) position.

Step 3 Run the first bead along the root of the joint using a transverse gun angle of 45 degrees with a 5- to 10-degree longitudinal gun angle. Use a slight oscillation (circular or up and down motion) to tie-in the weld at the toes. Chip the bead.

Step 4 Run the second bead along the bottom toe of the first weld, overlapping about 75 percent of the first bead. Use a transverse gun angle of 70 degrees with a 5- to 10-degree longitudinal gun angle and a slight oscillation. Chip the bead.

Step 5 Run the third bead along the top toe of the first weld, filling the platform created when the second bead was run. Use a transverse gun angle of 30 degrees with a 5- to 10-degree longitudinal gun angle and a slight oscillation. Chip the bead.

Figure 8-6 shows weld bead sequences 1, 2 and 3, and the gun angles.

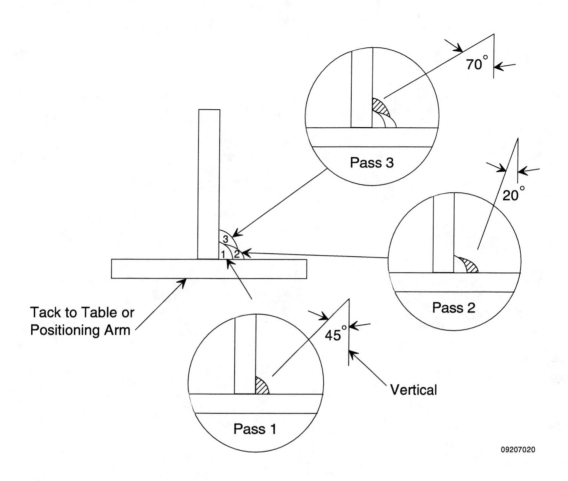

Figure 8-6. Bead Sequences 1, 2 and 3 and Gun Angles.

Step 6 Run the fourth bead along the bottom toe of the second weld, overlapping about half the second bead. Use a transverse gun angle of 70 degrees with a 5- to 10-degree longitudinal gun angle and a slight oscillation. Chip the bead.

Step 7 Run the fifth bead along the top toe of the fourth weld, overlapping about half the fourth bead that was run. Use a transverse gun angle of 70 degrees with a 5- to 10-degree longitudinal gun angle and a slight oscillation. Chip the bead.

Step 8 Run the sixth bead along the top toe of the third weld, filling the platform created when the fifth bead was run. Use a transverse gun angle of 30 degrees with a 5- to 10-degree longitudinal gun angle and a slight oscillation. Chip the bead.

Figure 8-7 shows weld bead sequences 4, 5 and 6 and the gun angles.

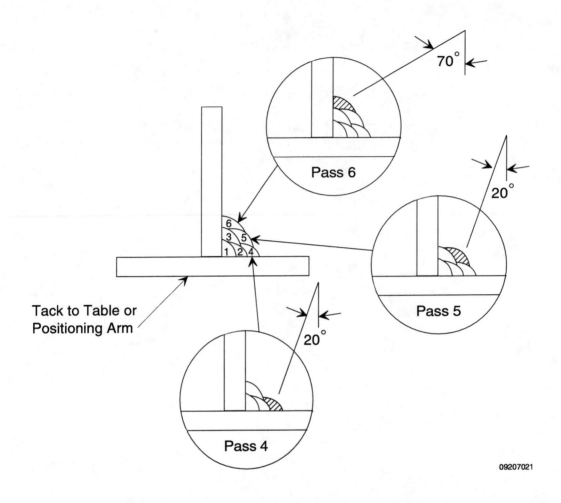

Figure 8-7. Bead Sequences 4,5 and 6 and Gun Angles.

Step 9   Inspect the weld. The weld is acceptable if it has:

- Uniform rippled appearance on the bead face
- Craters and restarts filled to the full cross-section of the weld
- Uniform weld size, plus or minus 1/16 inch
- Acceptable weld profile in accordance with AWS D1.1
- Smooth transition with complete fusion at the toes of the weld
- No porosity
- No undercut
- No cold lap
- No lack of fusion
- No Incomplete penetration
- No slag inclusions

WELDING TWO TRAINEE TASK MODULE 09210

8.3.0   PRACTICE VERTICAL UP (3F) FILLET WELDS

Practice vertical up fillet welds by welding multi-pass convex fillet welds in a tee joint using 0.045 inch flux core dual shield carbon steel electrode wire and shielding gas. When vertical welding, either stringer or weave beads can be used. On the job, the site WPS or Quality Standard will specify which technique to use.

Note   Check with your instructor to see if you should run stringer beads or weave beads or practice both techniques.

When making vertical fillet welds pay close attention to the gun angles and travel speed. For the first bead the transverse gun angle is 45 degrees. The angle is adjusted for all other welds. The angle is increased to deposit beads on one side and the angle is decreased to deposit beads on the other side. Increase or decrease the travel speed to control the amount of weld metal build-up.

8.3.1   Vertical Fillet Weld With Stringer Beads

Follow these steps to make a vertical fillet weld with stringer beads.

Step 1   Tack two plates together to form a tee joint for the fillet weld coupon. Chip the slag from the tack welds.

Step 2   Tack-weld the coupon to the positioning arm in the vertical position.

Step 3   Run the root bead (first bead) up the root of the joint (starting from the bottom) using a transverse gun angle of 45 degrees with a 10- to 15-degree upward angle. Pause in the weld puddle to fill the crater. Chip the bead.

Figure 8-8 shows stringer bead sequences 1 through 6 and the transverse gun angles.

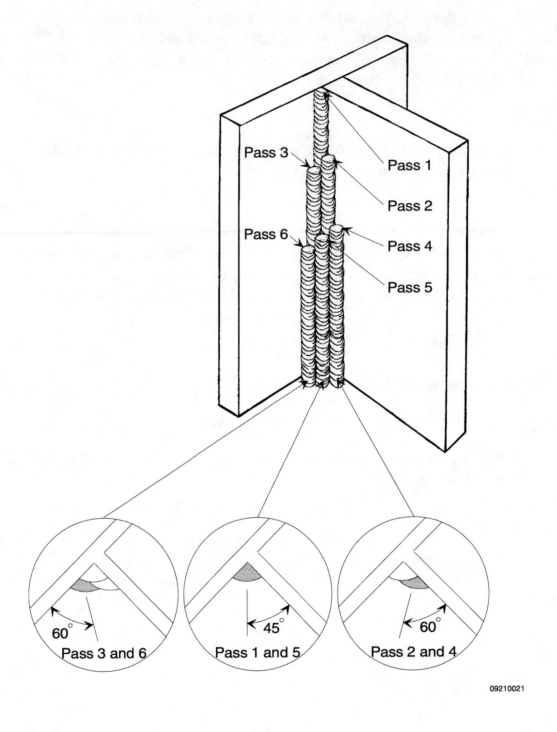

09210021

Figure 8-8. Stringer Bead Sequences 1 Through 6 and Transverse Gun Angles.

38

WELDING TWO TRAINEE TASK MODULE 09210

Step 4   Run the second bead along one toe of the first weld, overlapping about 75 percent of the first bead. Use a transverse gun angle of 60 degrees (from the plate face being welded) with a 5- to 10-degree longitudinal gun angle and a slight oscillation. Chip the bead.

Step 5   Run the third bead along the other toe of the first weld, filling the groove created when the second bead was run. Use a transverse gun angle of 60 degrees (from the plate being welded) with a 5- to 10-degree longitudinal gun angle and a slight oscillation. Chip the bead.

Step 6   Run the fourth bead along the plate toe of the second weld, overlapping about half the second bead. Use a transverse gun angle of 60 degrees (from the plate being welded) with a 5- to 10-degree longitudinal gun angle and a slight oscillation. Chip the bead.

Step 7   Run the fifth bead along the inner toe (on bead 3) of the fourth weld, overlapping about half the fourth bead. Use a transverse gun angle of 60 degrees (from the plate with the fourth bead) with a 5- to 10-degree longitudinal gun angle and a slight oscillation. Chip the bead.

Step 8   Run the sixth bead along the inner toe of the fifth weld, filling the groove created when the fifth bead was run. Use a transverse gun angle of 60 degrees (from the plate with the third bead) with a 5- to 10-degree longitudinal gun angle and a slight oscillation. Chip the bead.

Step 9   Inspect the weld. The weld is acceptable if it has:

- Uniform rippled appearance on the bead face
- Craters and restarts filled to the full cross-section of the weld
- Uniform weld size, plus or minus 1/16 inch
- Acceptable weld profile in accordance with AWS D1.1
- Smooth transition with complete fusion at the toes of the weld
- No porosity
- No undercut
- No cold lap
- No lack of fusion
- No Incomplete penetration
- No slag inclusions

8.3.2   Vertical Fillet Weld With Weave Beads

Follow these steps to make a vertical fillet weld with weave beads.

Step 1   Tack two plates together to form a tee joint for the fillet weld coupon. Chip the slag from the tack welds.

Step 2   Tack-weld the coupon to the positioning arm in the vertical position.

Step 3   Run the root bead (first bead) up the root of the joint (starting from the bottom) using a transverse gun angle of 45 degrees with a 10- to 15-degree upward angle. Pause in the weld puddle to fill the crater. Chip the bead.

Step 4   Run the second bead using a weave technique. Use a slow motion across the face of the weld, pausing at each toe for penetration and to fill the crater. Adjust the travel speed across the face of the weld to control the build-up. Chip the bead.

Figure 8-9 shows the bead sequence for vertical weave beads.

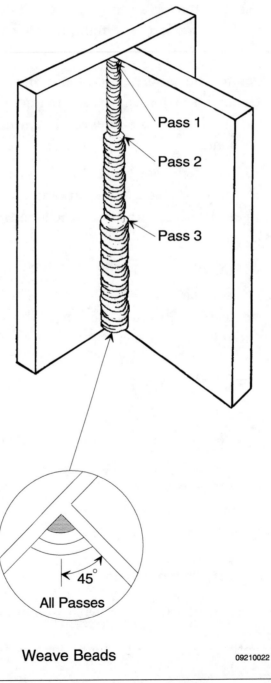

Weave Beads   09210022

Figure 8-9. Bead Sequence For Vertical Weave Beads.

Step 5   Continue to run weld beads as shown in *Figure 8-9*. Chip each bead before running the next bead.

Step 6   Inspect the weld. The weld is acceptable if it has:

- Uniform rippled appearance on the bead face
- Craters and restarts filled to the full cross-section of the weld
- Uniform weld size, plus or minus 1/16 inch
- Acceptable weld profile in accordance with AWS D1.1
- Smooth transition with complete fusion at the toes of the weld
- No porosity
- No undercut
- No cold lap
- No lack of fusion
- No Incomplete penetration
- No slag inclusions

8.4.0   PRACTICE OVERHEAD (4F) FILLET WELDS

Practice overhead fillet welds in a tee joint using stringer beads to make a six-pass convex fillet weld. Use 0.045 inch flux core dual shield carbon steel electrode wire and shielding gas. Pay close attention to gun angles and travel speed. The gun angle is adjusted for all subsequent beads. Stickout length is critical. Too little and the bead will be porous with excessive spatter. Too much and the bead will have poor fusion.

Follow these steps to make an overhead fillet weld with stringer beads.

Step 1   Tack two plates together to form a tee joint for the fillet weld coupon. Chip the slag from the tack welds.

Step 2   Tack weld the coupon so it is in the overhead position.

Step 3   Run the first bead along the root of the joint using a transverse gun angle of 45 degrees with a 5- to 10-degree longitudinal gun angle. Use a slight oscillation (side-to-side motion) to tie-in the weld at the toes. Chip the bead.

Step 4   Run the second bead along the bottom toe of the first weld, overlapping about 75 percent of the first bead. Use a transverse gun angle of 30 degrees with a 5- to 10-degree longitudinal gun angle and a slight oscillation. Chip the bead.

Step 5   Run the third bead along the top toe of the first weld, filling the groove created when the second bead was run. Use a transverse gun angle of 70 degrees with a 5- to 10-degree longitudinal gun angle and a slight oscillation. Chip the bead.

Figure 8-10 shows stringer bead sequences 1, 2 and 3 and transverse gun angles.

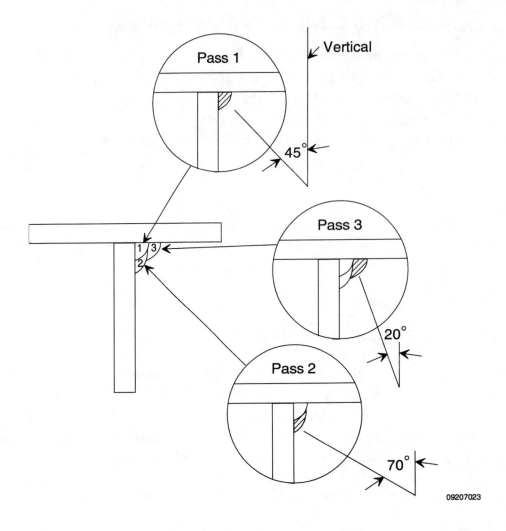

Figure 8-10. Stringer Bead Sequences 1, 2 and 3 and Transverse Gun Angles.

Step 6 Run the fourth bead along the bottom toe of the second weld, overlapping about half the second bead. Use a transverse gun angle of 40 degrees with a 5- to 10-degree longitudinal gun angle and a slight oscillation. Chip the bead.

Step 7 Run the fifth bead along the top toe of the fourth weld, overlapping about half the fourth bead that was run. Use a transverse gun angle of 70 degrees with a 5- to 10-degree longitudinal gun angle and a slight oscillation. Chip the bead.

Step 8 Run the sixth bead along the top toe of the third weld, filling the groove created when the fifth bead was run. Use a transverse gun angle of 70 degrees with a 5- to 10-degree longitudinal gun angle and a slight oscillation. Chip the bead.

Figure 8-11 shows stringer bead sequences 4, 5 and 6 and transverse gun angles.

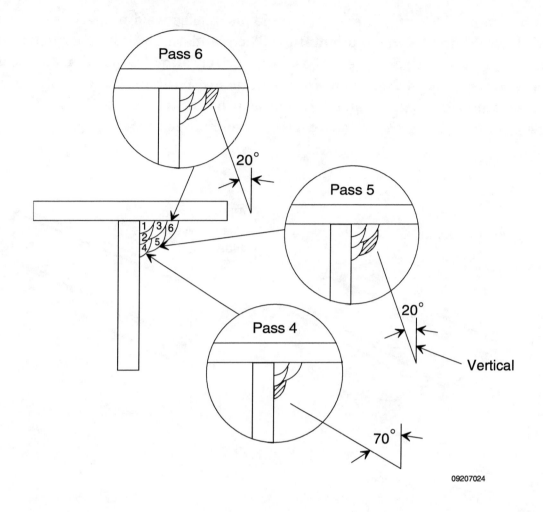

Figure 8-11. Stringer Bead Sequences 4, 5 and 6 and Transverse Gun Angles.

Step 9   Inspect the weld. The weld is acceptable if it has:

- Uniform rippled appearance on the bead face
- Craters and restarts filled to the full cross-section of the weld
- Uniform weld size, plus or minus 1/16 inch
- Acceptable weld profile in accordance with AWS D1.1
- Smooth transition with complete fusion at the toes of the weld
- No porosity
- No undercut
- No cold lap
- No lack of fusion
- No Incomplete penetration
- No slag inclusions

9.0.0 GROOVE WELDS

Groove welds are used to join plates in the same plane. The weld position is determined by the axis of the weld and the plate orientation. The standard positions for groove welds are the flat or 1G position (the "G" means groove), horizontal or 2G position, vertical or 3G-position, and overhead or 4G position. In any standard position, the axis of the weld can rotate plus or minus 15 degrees and the weld can be inclined up to 15 degrees within a position. *Figure 9-1* shows the standard groove welding positions.

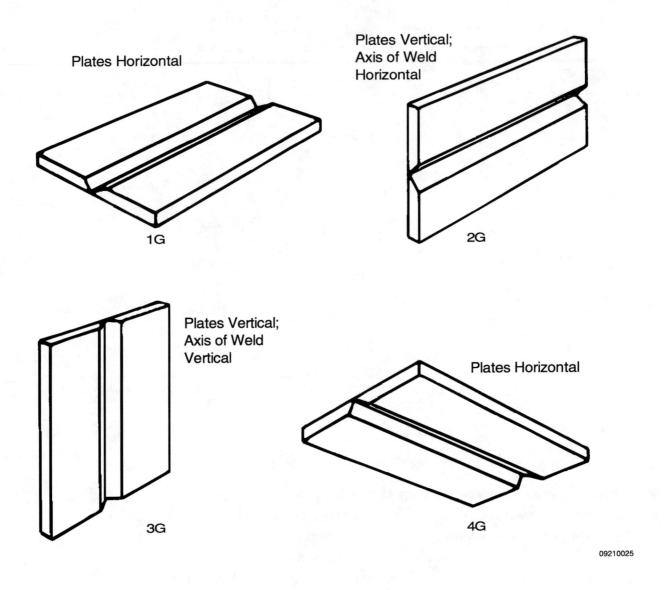

Figure 9-1. Standard Groove Welding Positions.

The groove welds in this module are V-groove open butt welds. Each coupon is beveled at 30-degrees and has a 3/32-inch land and a 3/32-inch root opening.

Note   Most codes allow a land of 1/16 to 1/8 inch and a root opening of 1/16 to 1/8 inch. Adjust the root opening and/or land within the limits as needed to obtain the proper root penetration.

Figure 9-2 shows a prepared groove weld coupon.

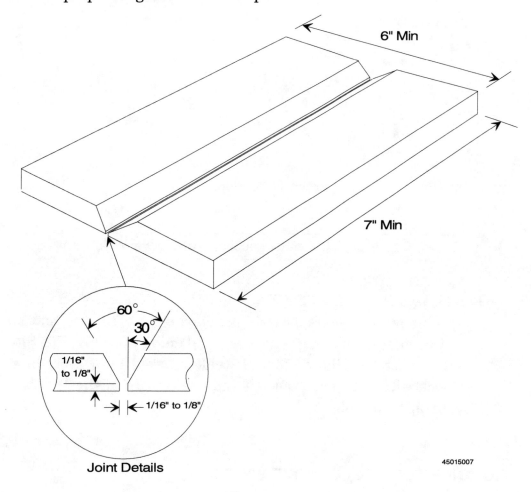

Figure 9-2. Prepared Groove Weld Coupon.

Groove welds should be made with slight reinforcement (not exceeding 1/8 inch) and a gradual transition to the base metal at each toe. Groove welds must not have excess convexity, insufficient throat, excessive undercut or overlap. If a groove weld has any of these defects, it must be repaired.

Figure 9-3 shows acceptable and unacceptable groove weld profiles.

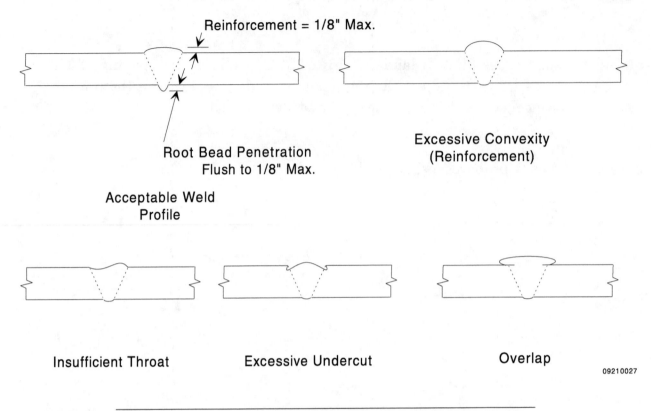

Figure 9-3. Acceptable and Unacceptable Groove Weld Profiles.

CAUTION   Refer to your site's WPS (Welding Procedure Specification) for specific requirements on weld reinforcement. The information in this manual is provided as a general guide line only. The Site WPS must be followed for all welds covered by a procedure. Check with your supervisor if you are unsure of the WPS for your application.

9.1.0   PRACTICE FLAT (1G) GROOVE WELDS

Practice open-root V-groove welds in the flat (1G) position with multi-pass stringer beads using 0.045 inch flux core dual shield carbon steel electrode wire and shielding gas.

For the root pass of a flat position groove weld, keep the gun (nozzle) transverse angle at 90 degrees (perpendicular) to the plate surface. Use a 45- to 55-degree (from the vertical) longitudinal drag (also called trailing or pulling) angle to prevent the slag from flowing ahead of the weld puddle. (The longitudinal angle is measured from a perpendicular to the weld axis, and for a flat position groove weld the perpendicular is vertical.) In a "drag" longitudinal angle, the electrode tip trails the gun nozzle, pointing away from the travel direction and toward the weld bead.

Figure 9-4 shows the root pass gun angle for the 1G position.

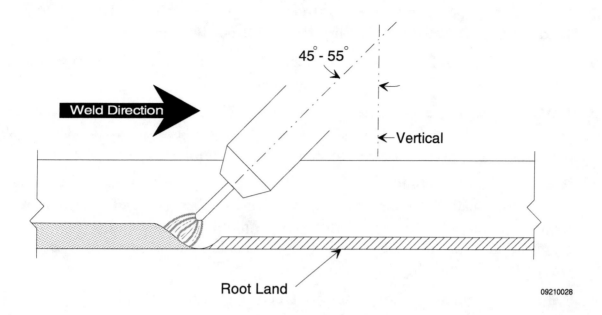

Figure 9-4. Root Pass Gun Angle for the 1G Position.

The remaining passes will be made with a 5- to 10-degree longitudinal drag angle.

Figure 9-5 shows the longitudinal gun angle for the remaining passes.

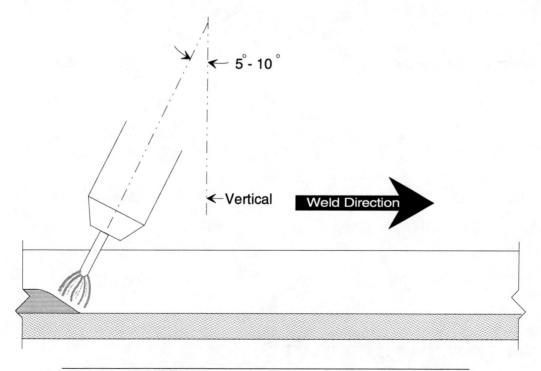

Figure 9-5. Longitudinal Gun Angle for the Remaining Passes.

Follow these steps to make a flat (1G) groove weld.

Step 1 Tack-weld the practice coupon together with a root opening of 3/32 inch and a root land (root face) of 3/32 inch. Chip the slag from the tack welds.

Step 2 Clamp or tack-weld the coupon to the positioning arm in the 1G (flat) position, and adjust the arm to a comfortable weld position.

Step 3 Run the root bead using a transverse gun angle of 90 degrees (perpendicular to the plate surface) and a 45-55 degree longitudinal drag angle. Use a slight oscillation to control penetration at the root opening. Chip the bead.

Step 4 Run the second bead along one toe of the first weld, overlapping about 75 percent of the first bead. All remaining stringer beads will be welded with a transverse angle of 20 degrees (from a normal to the plate surface), and a 5- to 10-degree longitudinal drag angle. Use a slight side-to-side (zigzag) oscillation to tie-in the welds at the toes. Chip the bead.

Figure 9-6 shows the transverse gun angles for beads 2 through 6.

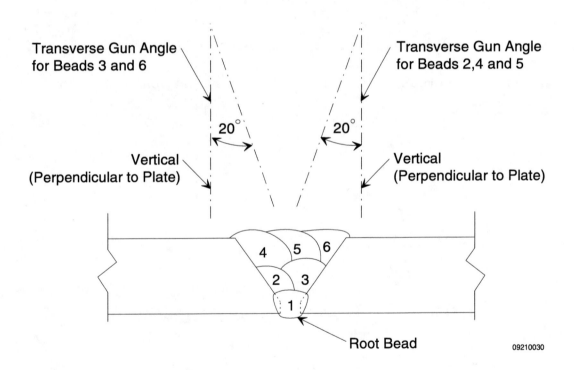

Figure 9-6. Transverse Gun Angles for Beads 2 Through 6.

WELDING TWO TRAINEE TASK MODULE 09210

Step 5 Run the third bead along the other toe of the first weld, and fill the groove formed by the second bead. Use a transverse angle of 20 degrees, and a 5- to 10-degree longitudinal drag angle. Use a slight side-to-side oscillation to tie-in the welds at the toes. Chip the bead.

Step 6 Run the fourth bead along the outside toe of the second weld, overlapping about half the second bead. Use a transverse angle of 20 degrees, and a 5- to 10-degree longitudinal drag angle. Use a slight side-to-side oscillation to tie-in the welds at the toes. Chip the bead.

Step 7 Run the fifth bead along the inside toe of the fourth weld, overlapping about half the fourth bead. Use a transverse angle of 20 degrees, and a 5- to 10-degree longitudinal drag angle. Use a slight side-to-side oscillation to tie-in the welds at the toes. Chip the bead.

Step 8 Run the sixth bead along the outside toe of the fifth weld, filling the groove created when the fifth bead was run. Use a transverse angle of 20 degrees, and a 5- to 10-degree longitudinal drag angle. Use a slight side-to-side oscillation to tie-in the welds at the toes. Chip the bead.

Step 9 Inspect the weld. The weld is acceptable if it has:

- Uniform rippled appearance on the bead face
- Craters and restarts filled to the full cross-section of the weld
- Uniform weld size plus or minus 1/16 inch
- Acceptable weld profile in accordance with the *ASME Boiler and Pressure Vessel Code*
- Complete uniform root penetration at least flush with the base metal to a maximum buildup of 3/32 inch
- Smooth flat transition with complete fusion at the toes of the weld
- No porosity
- No undercut
- No slag inclusions
- No cracks
- Acceptable guided bend test results

9.2.0   PRACTICE HORIZONTAL (2G) GROOVE WELDS

Practice open-root V-groove welds in the horizontal (2G) position with multi-pass stringer beads using 0.045 inch flux core dual shield carbon steel electrode wire and shielding gas.

When making horizontal groove welds, pay close attention to the gun angle and travel speed. For the root pass, keep the gun (nozzle) transverse angle at 90 degrees (perpendicular) to the plate surface. Use a 45- to 55-degree longitudinal drag (or trailing or pulling) angle to prevent the slag from flowing ahead of the weld puddle.

The transverse angle is adjusted for all subsequent beads. The nozzle is raised about 20 degrees (above the horizontal) to deposit beads on the upper bevel face, and lowered about 20 degrees (below the horizontal) to deposit beads on the lower bevel face. Use a slight side-to-side movement, and increase or decrease the travel speed to control the amount of weld metal build-up.

Follow these steps to make a horizontal groove weld.

Step 1 Tack-weld the practice coupon together with a root opening of 3/32 inch and a root land of 3/32 inch. Chip the slag from the tack welds.

Step 2 Clamp or tack-weld the coupon to the positioning arm in the 2G (horizontal) position, and then adjust the arm to a comfortable welding position.

Step 3 Run the root bead (bead 1) using a transverse gun angle of 90 degrees (nozzle is horizontal), and a 45- to 55-degree longitudinal drag angle. Use a slight oscillation to control penetration at the root opening. Chip the bead.

Step 4 Run the second bead along the bottom toe of the first weld, overlapping about 75 percent of the first bead. All remaining beads will be welded with a transverse angle of 20 degrees (from the horizontal), and a 5- to 10-degree longitudinal drag angle. Use a slight side-to-side oscillation to tie-in the welds at the toes. Chip the bead.

Figure 9-7 shows the transverse gun angles for 2G position welds.

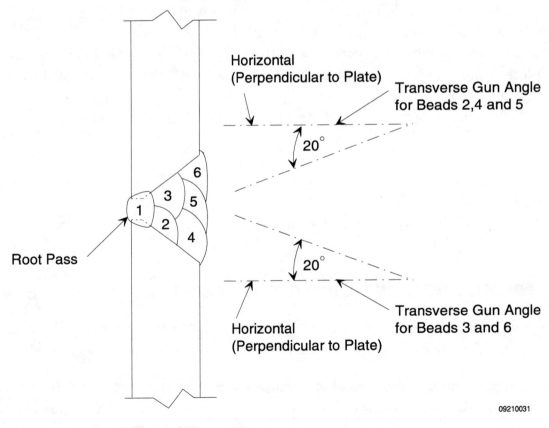

Figure 9-7. Transverse Gun Angles for 2G Position Welds.

WELDING TWO TRAINEE TASK MODULE 09210

Step 5 Run the third bead along the upper toe of the first weld, and fill the groove formed by the second bead. Use a transverse angle of 20 degrees (nozzle points 20 degrees above the horizontal), and a 5- to 10-degree longitudinal drag angle. Use a slight side-to-side oscillation to tie-in the welds at the toes. Chip the bead.

Step 6 Run the fourth bead along the bottom toe of the second weld, overlapping about half the second bead. Use a transverse angle of 20 degrees, and a 5- to 10-degree longitudinal drag angle. Use a slight side-to-side oscillation to tie-in the welds at the toes. Chip the bead.

Step 7 Run the fifth bead along the top toe of the fourth weld, overlapping about half the fourth bead that was run. Use a transverse angle of 20 degrees (nozzle points 20 degrees below horizontal), and a 5- to 10-degree longitudinal drag angle. Use a slight side-to-side oscillation to tie-in the welds at the toes. Chip the bead.

Step 8 Run the sixth bead along the top toe of the fifth weld, filling the groove created when the fifth bead was run. Use a transverse angle of 20 degrees (nozzle points 20 degrees above the horizontal), and a 5- to 10-degree longitudinal drag angle. Use a slight side-to-side oscillation to tie-in the welds at the toes. Chip the bead.

Step 9 Inspect the weld. The weld is acceptable if it has:

- Uniform rippled appearance on the bead face
- Craters and restarts filled to the full cross-section of the weld
- Uniform weld size plus or minus 1/16 inch
- Acceptable weld profile in accordance with the *ASME Boiler and Pressure Vessel Code*
- Complete uniform root penetration at least flush with the base metal to a maximum buildup of 3/32 inch
- Smooth flat transition with complete fusion at the toes of the weld
- No porosity
- No undercut
- No slag inclusions
- No cracks
- Acceptable guided bend test results

9.3.0 PRACTICE VERTICAL UP (3G) GROOVE WELDS

Practice multi-pass groove welds on mild steel plate using stringer beads with 0.045 inch flux core dual shield carbon steel electrode wire and shielding gas. Set the welding parameters at the lower ends of the suggested ranges. Out-of-position welds are made at lower welding voltages and currents than flat or horizontal welds to prevent the molten weld metal from sagging or running.

Note   When vertical welding, either stringer or weave beads can be used. On the job, the WPS or site quality standards will specify which technique to use. Check with your instructor to see if you should run stringer beads or weave beads, or practice both techniques.

Note   The root bead can be run either vertical up or vertical down. Use the direction preferred at your site or as specified in the WPS or site quality standards. The filler and cover beads are always run vertical up.

Run the root bead using a transverse gun angle of 90 degrees (nozzle centerline bisects the V-groove) and a 45-55 degree longitudinal drag angle. The transverse angle is adjusted for all other welds.

9.3.1   Vertical Groove Weld With Stringer Beads

Follow these steps to make a vertical groove weld with stringer beads.

Step 1   Tack-weld the practice coupon together with a root opening of 3/32 inch and a root land of 1/16 inch for vertical down or 3/32 inch for vertical up. Chip the slag from the tack welds.

Step 2   Clamp or tack-weld the coupon to the positioning arm in the 3G (vertical) position, and then adjust the arm for a comfortable welding position.

Step 3   Run the root bead with a transverse gun angle of 0 degrees (nozzle centerline bisects the V-groove) and a 45- to 55-degree drag angle. Chip the bead.

Step 4   Run the second bead (vertical up) along one toe of the first weld, overlapping about 75 percent of the first bead. All remaining beads will be welded vertical up with a transverse angle of 20 degrees (from a plane bisecting the V-groove), and a 5- to 10-degree longitudinal drag angle. Use a slight side-to-side oscillation to tie-in the welds at the toes. Chip the bead.

Figure 9-8 shows the stringer bead sequence and transverse gun angles for vertical up welds.

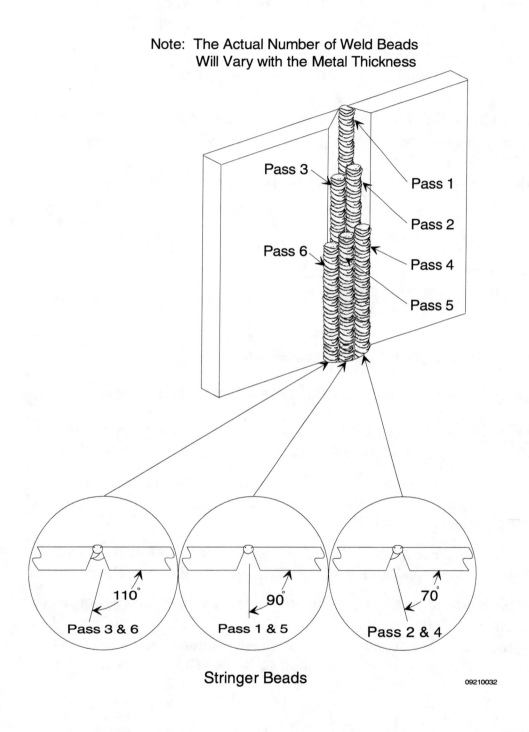

Note: The Actual Number of Weld Beads
Will Vary with the Metal Thickness

Pass 3

Pass 1

Pass 2

Pass 6

Pass 4

Pass 5

110°
Pass 3 & 6

90°
Pass 1 & 5

70°
Pass 2 & 4

Stringer Beads

09210032

Figure 9-8. Stringer Bead Sequence and Transverse Gun Angles for Vertical Up Welds.

Step 5   Continue to run vertical up stringer beads as shown in *Figure 9-8*. Chip each bead before running the next bead.

Step 6   Inspect the weld. The weld is acceptable if it has:

- Uniform rippled appearance on the bead face
- Craters and restarts filled to the full cross-section of the weld
- Uniform weld size plus or minus 1/16 inch
- Acceptable weld profile in accordance with the *ASME Boiler and Pressure Vessel Code*
- Complete uniform root penetration at least flush with the base metal to a maximum buildup of 3/32 inch
- Smooth flat transition with complete fusion at the toes of the weld
- No porosity
- No undercut
- No slag inclusions
- No cracks
- Acceptable guided bend test results

9.3.2   Vertical Groove Weld With Weave Beads

Follow these steps to make a vertical groove weld with weave beads.

Step 1   Tack-weld the practice coupon together with a root opening of 3/32 inch and a root land of 1/16 inch for vertical down or 3/32 inch for vertical up. Chip the slag from the tack welds.

Step 2   Clamp or tack-weld the coupon to the positioning arm in the 3G (vertical) position, and then adjust the arm for a comfortable welding position.

Step 3   Run the root bead with a transverse gun angle of 90 degrees (nozzle centerline bisects the V-groove) and a 45- to 55-degree drag angle. Chip the bead.

Step 4   The second bead is a vertical up weave bead across the root bead. Use a zigzag pattern across the root bead, and pause at each toe for good fusion. Use a longitudinal drag angle of 5- to 10-degrees. Increase or decrease the travel speed to control the amount of weld metal build-up. Chip the bead.

Figure 9-9 shows the zigzag pattern for weave beads.

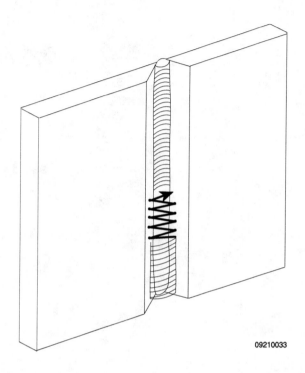

09210033

Figure 9-9. Zigzag Pattern for Weave Beads.

Figure 9-10 shows the sequence and transverse gun angle for vertical up weave beads.

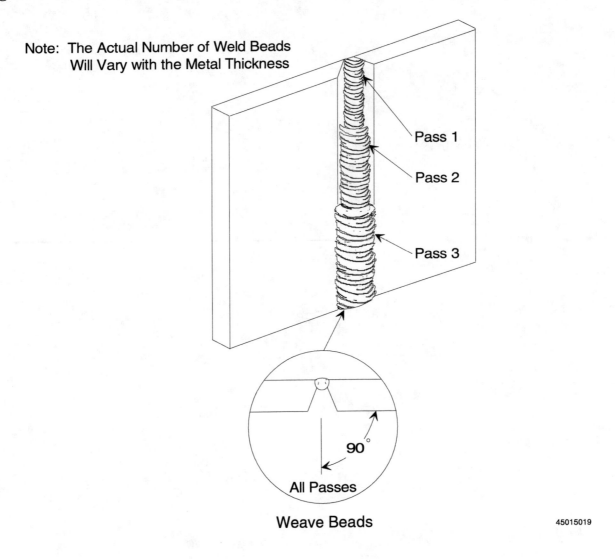

Note: The Actual Number of Weld Beads
Will Vary with the Metal Thickness

Pass 1

Pass 2

Pass 3

90°

All Passes

Weave Beads

45015019

Figure 9-10. Sequence and Transverse Gun Angle for Vertical Up Weave Beads.

Step 5 Continue to run vertical up weave beads until the weld is completed. Continue to use the zigzag pattern and the 5- to 10-degree longitudinal drag angle. Chip each bead before running the next bead.

Step 6 Inspect the weld. The weld is acceptable if it has:

- Uniform rippled appearance on the bead face
- Craters and restarts filled to the full cross-section of the weld
- Uniform weld size plus or minus 1/16 inch
- Acceptable weld profile in accordance with the *ASME Boiler and Pressure Vessel Code*
- Complete uniform root penetration at least flush with the base metal to a maximum buildup of 3/32 inch
- Smooth flat transition with complete fusion at the toes of the weld

- No porosity
- No undercut
- No slag inclusions
- No cracks
- Acceptable guided bend test results

9.4.0   PRACTICE OVERHEAD (4G) GROOVE WELDS

Practice overhead groove welds on mild steel plate using stringer beads with 0.045 inch flux core dual shield carbon steel electrode wire and shielding gas.

Note   FCAW overhead welds are made with lower welding voltages and currents than flat or horizontal welds, to reduce heat and prevent the molten weld metal from sagging or running.

Use a transverse gun angle of 0 degrees (bisects the V-groove) and a 45- to 55-degree longitudinal drag angle for the root bead.

Figure 9-11 shows the longitudinal gun angle for an overhead root bead.

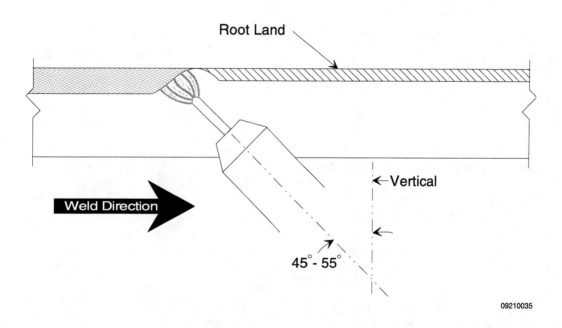

Figure 9-11. Longitudinal Gun Angle for an Overhead Root Bead.

Follow these steps to make an overhead groove weld with stringer beads.

Step 1   Tack-weld the practice coupon together with a root opening of 3/32 inch and a root land of 3/32 inch. Chip the slag from the tack welds.

Step 2   Clamp or tack-weld the coupon to the positioning arm in the 4G (overhead) position, and then adjust the arm for a comfortable welding position.

Step 3   Run the root bead using a transverse gun angle of 0 degrees (nozzle centerline bisects the V-groove) and a 45- to 55-degree drag angle. Chip the bead.

Step 4   Run the second bead along one toe of the first weld, overlapping about 75 percent of the first bead. All remaining beads will be welded with a transverse angle of 20 degrees (from a normal to the plate surface), and a 5- to 10-degree longitudinal drag angle. Use a slight side-to-side (zigzag) oscillation to tie-in the welds at the toes. Chip the bead.

Figure 9-12 shows overhead transverse gun angles for beads 2 through 6.

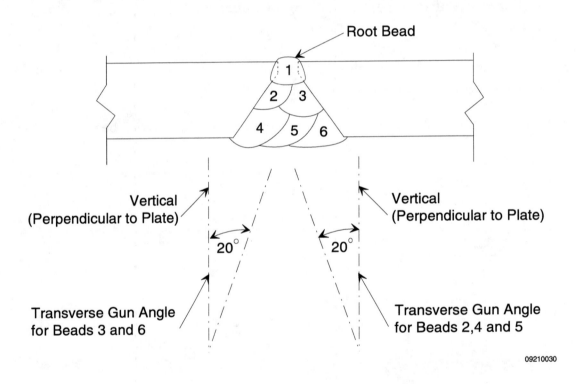

Figure 9-12. Overhead Transverse Gun Angles for Beads 2 Through 6.

Step 5   Run the third bead along the other toe of the first weld, and fill the groove formed by the second bead. Use a transverse angle of 20 degrees, and a 5- to 10-degree longitudinal drag angle. Use a slight side-to-side oscillation to tie-in the welds at the toes. Chip the bead.

Step 6 Run the fourth bead along the outside toe of the second weld, overlapping about half the second bead. Use a transverse angle of 20 degrees, and a 5- to 10-degree longitudinal drag angle. Use a slight side-to-side oscillation to tie-in the welds at the toes. Chip the bead.

Step 7 Run the fifth bead along the inside toe of the fourth weld, overlapping about half the fourth bead that was run. Use a transverse angle of 20 degrees, and a 5- to 10-degree longitudinal drag angle. Use a slight side-to-side oscillation to tie-in the welds at the toes. Chip the bead.

Step 8 Run the sixth bead along the outside toe of the fifth weld, filling the groove created when the fifth bead was run. Use a transverse angle of 20 degrees, and a 5- to 10-degree longitudinal drag angle. Use a slight side-to-side oscillation to tie-in the welds at the toes. Chip the bead.

Step 6 Inspect the weld. The weld is acceptable if it has:

- Uniform rippled appearance on the bead face
- Craters and restarts filled to the full cross-section of the weld
- Uniform weld size plus or minus 1/16 inch
- Acceptable weld profile in accordance with the *ASME Boiler and Pressure Vessel Code*
- Complete uniform root penetration at least flush with the base metal to a maximum buildup of 3/32 inch
- Smooth flat transition with complete fusion at the toes of the weld
- No porosity
- No undercut
- No slag inclusions
- No cracks
- Acceptable guided bend test results

SELF CHECK REVIEW 2

1. How do you angle the gun to get tie-in when running overlapping stringer beads?
2. What is another name for the horizontal fillet weld position?
3. When making a horizontal fillet weld, what are the gun angles for the first pass?
4. When welding vertically, what determines whether you run stringers or weave beads?
5. What are the gun angles for the second pass of an overhead fillet weld?

ANSWERS TO SELF CHECK REVIEW 2

1. Position the gun at a transverse angle of 10 to 15 degrees toward the side of the previous bead. (7.1.0)
2. The horizontal position is also called the 2F position. (8.0.0)
3. The gun is positioned at a 45 degree transverse angle and a 5- to 10-degree longitudinal angle. (8.2.0)
4. WPS or site quality standards will specify the bead type to use. (8.3.0)
5. Use a 70 degree transverse angle and a 5- to 10-degree longitudinal angle. (8.4.0)

10.0.0 PERFORMANCE QUALIFICATIONS TASKS

The following tasks are designed to evaluate your ability to run beads and fillet welds with FCAW equipment. Perform each task when you are instructed to do so by your instructor. As you complete each task, show it to your instructor for evaluation. Do not proceed to the next task until instructed to do so by your instructor.

10.1.0 MAKE A FILLET WELD IN THE (1F) FLAT POSITION

Using 0.045 inch flux core dual shield carbon steel electrode wire and shielding gas, make a six-pass fillet weld using stringer beads on mild steel plate as shown in *Figure 10-1*.

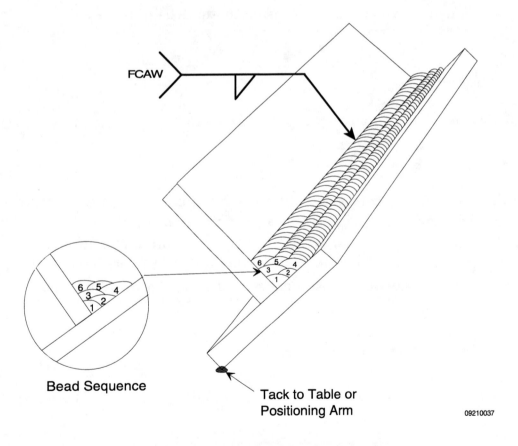

Figure 10-1. Fillet Weld in the 1F Position.

WELDING TWO TRAINEE TASK MODULE 09210

Criteria For Acceptance:

- Uniform rippled appearance on the bead face
- Craters and restarts filled to the full cross-section of the weld
- Smooth flat transition with complete fusion at the toes of one bead into the face of the previous bead
- No porosity
- No undercut
- No cold lap
- No lack of fusion
- No slag inclusions

10.2.0 MAKE A FILLET WELD IN THE (2F) HORIZONTAL POSITION

Using 0.045 inch flux core dual shield carbon steel electrode wire and shielding gas, make a six-pass fillet weld using stringer beads on mild steel plate as indicated in *Figure 10-2*.

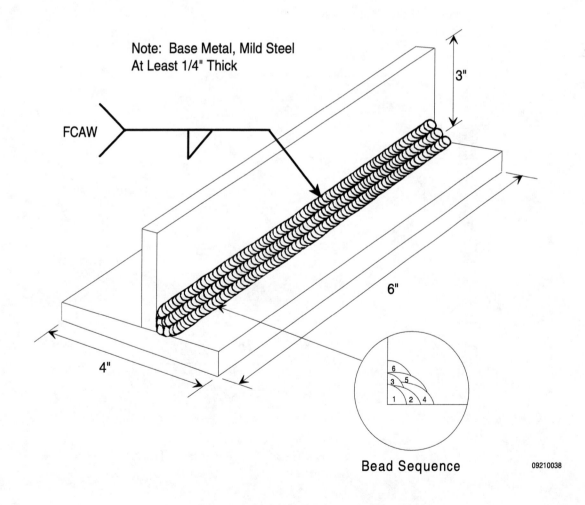

Bead Sequence

09210038

Figure 10-2. Fillet Weld in the 2F Position.

Criteria For Acceptance:

- Uniform rippled appearance on the bead face
- Craters and restarts filled to the full cross-section of the weld
- Smooth flat transition with complete fusion at the toes of one bead into the face of the previous bead
- No porosity
- No undercut
- No cold lap
- No lack of fusion
- No slag inclusions

10.3.0 MAKE A FILLET WELD IN THE (3F) VERTICAL POSITION

Using 0.045 inch flux core dual shield carbon steel electrode wire and shielding gas, make a vertical fillet weld on mild steel plate as indicated in *Figure 10-3*.

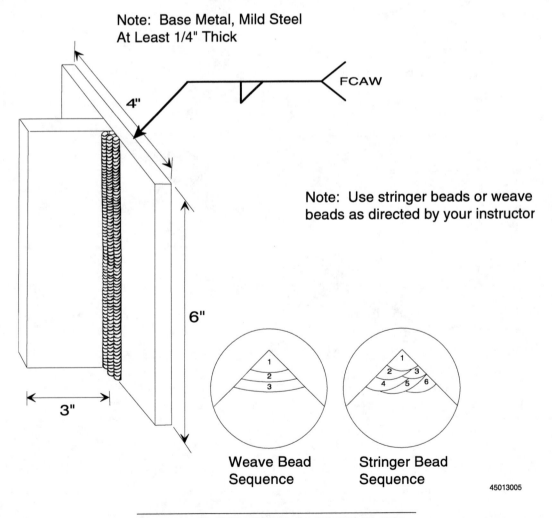

Note: Base Metal, Mild Steel At Least 1/4" Thick

FCAW

Note: Use stringer beads or weave beads as directed by your instructor

Weave Bead Sequence

Stringer Bead Sequence

45013005

Figure 10-3. Fillet Weld in the 3F Position.

Criteria For Acceptance:

- Uniform rippled appearance on the bead face
- Craters and restarts filled to the full cross-section of the weld
- Smooth flat transition with complete fusion at the toes of one bead into the face of the previous bead
- No porosity
- No undercut
- No cold lap
- No lack of fusion
- No slag inclusions

10.4.0   MAKE A FILLET WELD IN THE (4F) OVERHEAD POSITION

Using 0.045 inch flux core dual shield carbon steel electrode wire and shielding gas, make a six-pass fillet weld using stringer beads on mild steel plate as indicated in *Figure 10-4*.

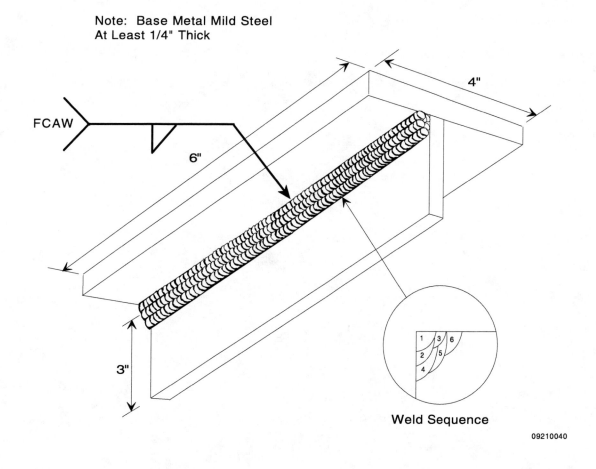

Figure 10-4. Fillet Weld in the 4F Position.

Criteria For Acceptance:

- Uniform rippled appearance on the bead face
- Craters and restarts filled to the full cross-section of the weld
- Smooth flat transition with complete fusion at the toes of one bead into the face of the previous bead
- No porosity
- No undercut
- No cold lap
- No lack of fusion
- No slag inclusions

10.5.0   MAKE A GROOVE WELD IN THE (1G) FLAT POSITION

Using 0.045 inch flux core dual shield carbon steel electrode wire and shielding gas, make a multi-pass groove weld using stringer beads on mild steel plate as shown in *Figure 10-5.*

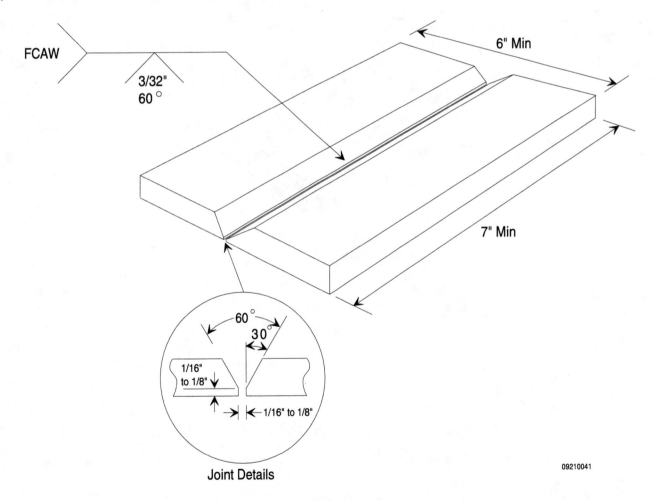

Figure 10-5. Groove Weld in the 1G Position.

Criteria For Acceptance:

- Uniform rippled appearance on the bead face
- Craters and restarts filled to the full cross-section of the weld
- Uniform weld size plus or minus 1/16 inch
- Acceptable weld profile in accordance with the *ASME Boiler and Pressure Vessel Code*
- Complete uniform root penetration at least flush with the base metal to a maximum buildup of 3/32 inch
- Smooth flat transition with complete fusion at the toes of the weld
- No porosity
- No undercut
- No slag inclusions
- No cracks
- Acceptable guided bend test results

10.6.0   MAKE A GROOVE WELD IN THE (2G) HORIZONTAL POSITION

Using 0.045 inch flux core dual shield carbon steel electrode wire and shielding gas, make a multi-pass groove weld on mild steel plate as indicated in *Figure 10-6*.

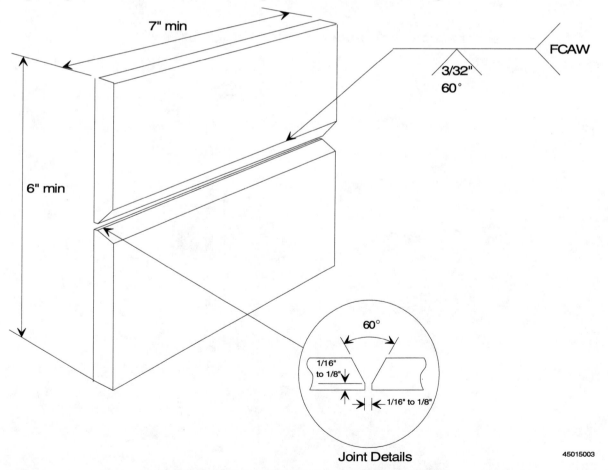

Figure 10-6. Groove Weld in the 2G Position.

Criteria For Acceptance:

- Uniform rippled appearance on the bead face
- Craters and restarts filled to the full cross-section of the weld
- Uniform weld size plus or minus 1/16 inch
- Acceptable weld profile in accordance with the *ASME Boiler and Pressure Vessel Code*
- Complete uniform root penetration at least flush with the base metal to a maximum buildup of 3/32 inch
- Smooth flat transition with complete fusion at the toes of the weld
- No porosity
- No undercut
- No slag inclusions
- No cracks
- Acceptable guided bend test results

10.7.0   MAKE A GROOVE WELD IN THE (3G) VERTICAL POSITION

Using 0.045 inch flux core dual shield carbon steel electrode wire and shielding gas, make a multi-pass groove weld on mild steel plate as indicated in *Figure 10-7*.

Note   Run the root vertical up or vertical down as specified by your instructor.

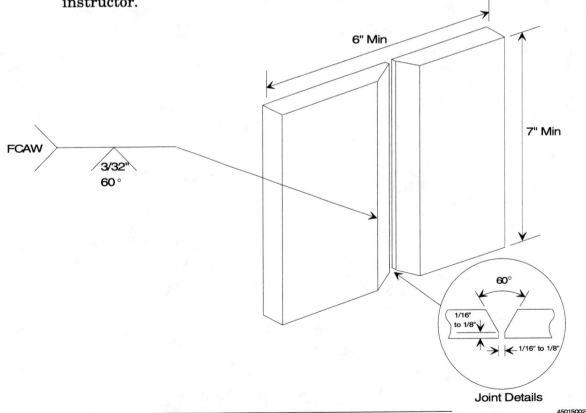

Figure 10-7. Groove Weld in the 3G Position.

WELDING TWO TRAINEE TASK MODULE 09210

Criteria For Acceptance:

- Uniform rippled appearance on the bead face
- Craters and restarts filled to the full cross-section of the weld
- Uniform weld size plus or minus 1/16 inch
- Acceptable weld profile in accordance with the *ASME Boiler and Pressure Vessel Code*
- Complete uniform root penetration at least flush with the base metal to a maximum buildup of 3/32 inch
- Smooth flat transition with complete fusion at the toes of the weld
- No porosity
- No undercut
- No slag inclusions
- No cracks
- Acceptable guided bend test results

10.8.0   MAKE A GROOVE WELD IN THE (4G) OVERHEAD POSITION

Using 0.045 inch flux core dual shield carbon steel electrode wire and shielding gas, make a multi-pass groove weld on mild steel plate as indicated in *Figure 10-8*.

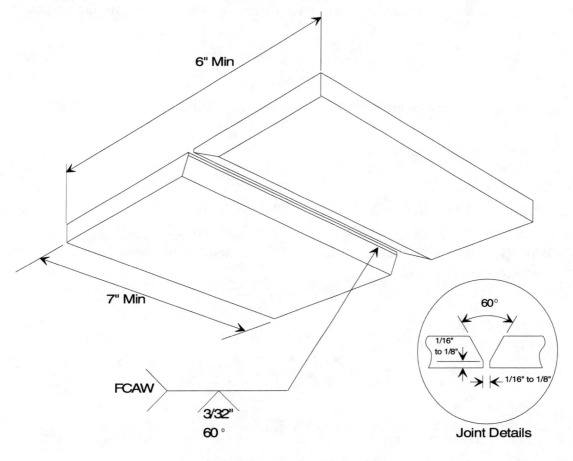

45015004

Figure 10-8. Groove Weld in the 4G Position.

Criteria For Acceptance:

- Uniform rippled appearance on the bead face
- Craters and restarts filled to the full cross-section of the weld
- Uniform weld size plus or minus 1/16 inch
- Acceptable weld profile in accordance with the *ASME Boiler and Pressure Vessel Code*
- Complete uniform root penetration at least flush with the base metal to a maximum buildup of 3/32 inch
- Smooth flat transition with complete fusion at the toes of the weld
- No porosity
- No undercut
- No slag inclusions
- No cracks
- Acceptable guided bend test results

SUMMARY

Setting up FCAW equipment, preparing the welding work area and running stringer and weave beads to make acceptable fillet and groove welds in all positions are essential skills a welder must have to perform difficult welding procedures. Practice these welds until you can consistently produce acceptable welds as defined in the criteria for acceptance.

References

For advanced study of topics covered in this Task Module, the following works are suggested:

Welding Handbook, Volume 5, Seventh Edition, The American Welding Society, Miami, FL, 1984, 1-800-334-9353.

Welding Technology, (1982) Gower A. Kennedy, The Bobbs-Merrill Company, Inc., 4300 West 62nd Street, Indianapolis, IN 46268, Phone 317-298-5686.

Welding Skills, Giachino and Weeks, American Technical Publishers Inc., Homewood, IL, 1985, 1-800-323-3471.

PERFORMANCE / LABORATORY EXERCISES

1. Practice stringer beads using carbon steel wire and 75/25 argon-carbon dioxide gas.
2. Practice weave beads using carbon steel wire and 75/25 argon-carbon dioxide gas.
3. Practice overlapping stringer beads using carbon steel wire and 75/25 argon-carbon dioxide gas.
4. Practice overlapping weave beads using carbon steel wire and 75/25 argon-carbon dioxide gas.
5. Practice overlapping stringer beads using carbon steel wire and 75/25 argon-carbon dioxide gas.
6. Practice flat (1F) position fillet welds on steel plate using carbon steel wire and 75/25 argon-carbon dioxide gas.
7. Practice horizontal (2F) position fillet welds on steel plate using carbon steel wire and 75/25 argon-carbon dioxide gas.
8. Practice vertical up (3F) position fillet welds on steel plate using carbon steel wire and 75/25 argon-carbon dioxide gas.
9. Practice overhead (4F) position fillet welds on steel plate using carbon steel wire and 75/25 argon-carbon dioxide gas.
10. Practice flat (1G) position groove welds on steel plate using carbon steel wire and 75/25 argon-carbon dioxide gas.
11. Practice horizontal (2G) position groove welds on steel plate using carbon steel wire and 75/25 argon-carbon dioxide gas.
12. Practice vertical up (3G) position groove welds on steel plate using carbon steel wire and 75/25 argon-carbon dioxide gas.
13. Practice overhead (4G) position groove welds on steel plate using carbon steel wire and 75/25 argon-carbon dioxide gas.

The NCCER makes every effort to keep these manuals up-to-date and free of technical errors. We appreciate your help in this process. If you have an idea for improving this manual, or if you find an error, a typographical mistake, or an inaccuracy in the *Wheels of Learning*, please write us, using this form or a photocopy. Be sure to include the exact module number, page number, a description of the problem, and the correction, if possible. We'll do our best to correct it in later editions. Thank you for your assistance.

Write: *Wheels of Learning*
National Center for Construction Education and Research
P.O. Box 141104
Gainesville, FL 32614-1104
Fax: 352-334-0932

WHEELS OF LEARNING USER UPDATE

Please let us know if you have found an inaccuracy, error, or other problem in a *Wheels of Learning* manual. Use this form or write us a letter. Please be sure to tell us the exact module name and module number, the page number, and the problem. Thanks for your help.

Craft _____ Module Name _____

Module Number _____ Page Number(s) _____

Description of Problem _____

_____

_____

(Optional) Correction of Problem _____

_____

_____

(Optional) Your Name and Address _____

_____

GTAW -- Equipment and Filler Metals

Module 09304

Welder Trainee Task Module 09304

NATIONAL
CENTER FOR
CONSTRUCTION
EDUCATION AND
RESEARCH

GAS TUNGSTEN ARC WELDING - EQUIPMENT AND FILLER METALS

OBJECTIVES

Upon completion of this module, the trainee will be able to:

1. Explain Gas Tungsten Arc Welding (GTAW) safety.
2. Identify and explain GTAW equipment.
3. Identify and explain GTAW filler metals.
4. Identify and explain GTAW shielding gasses.
5. Set up GTAW welding equipment.

Prerequisites

Successful completion of the following Task Module(s) is required before beginning study of this Task Module:

- Safety (Common Core)
- Rigging/Material Handling (Common Core)

Required Student Materials

None

Course Map Information

This course map shows all of the *Wheels of Learning* task modules in the third level of the Welding curricula. The suggested training order begins at the bottom and proceeds up. Skill levels increase as a trainee advances on the course map. The training order may be adjusted by the local Training Program Sponsor.

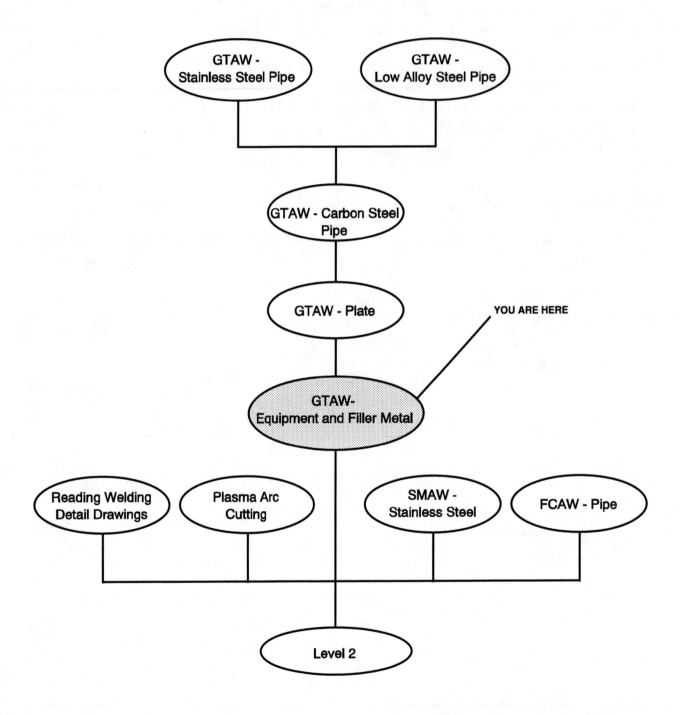

TABLE OF CONTENTS

Trade Terms Introduced in This Module

Alternating Current (AC): Electrical current that reverses its flow at set intervals.

Amperage: A measurement of the rate of flow of electric current.

Arc: Flow of electricity through an air gap.

Arc Flash: Burns to the eyes or skin caused by the arc.

Direct Current (DC): Electrical current that flows in one direction only.

Direct Current Reverse Polarity (DCRP): A welding setup where the torch electrode lead is connected to the welding machine's positive (+) output terminal, and the work lead (ground lead) is connected to the welding machine's negative (-) output terminal.

Direct Current Straight Polarity (DCSP): A welding setup where the torch electrode lead is connected to the welding machine's negative (-) output terminal, and the work lead (ground lead) is connected to the welding machine's positive (+) output terminal.

Duty Cycle: The percentage of a ten minute period in which a welding machine can deliver its rated output.

Ground: An object that makes an electrical connection with the earth. Also, the welding cable (work lead) attached to the base metal.

High Frequency: A very high voltage (several thousand), extremely low amperage current that alternates at high frequency (several million cycles per second). It is superimposed on the welding current to help establish and maintain the arc.

Hot Work Permit: Official authorization from a site manager to perform work which may pose a fire hazard.

Out-Of-Position Welding: Any welding position other than the flat position; includes vertical, horizontal, and overhead welding.

Polarity: The direction of flow of electrical current in a direct current welding circuit.

Primary Current: Electrical current received from conventional power lines used to energize welding machines.

Tungsten: A common name for the virtually nonconsumable GTAW electrode and the principal metal from which most are constructed.

Voltage: A measurement of electromotive force or pressure that causes current to flow in a circuit.

1.0.0 INTRODUCTION

Gas Tungsten Arc Welding (GTAW) is an arc welding process. It uses a welding power supply to produce an electric arc between a virtually nonconsumable tungsten electrode and the base metal. The arc melts and fuses the base metal forming the weld. If required, a filler metal can be added. In manual welding the filler metal is usually a hand-held metal rod or wire that is manually fed into the leading edge of the weld metal pool. An inert gas or gas mixture is used to shield the electrode and molten weld metals to prevent oxidation and contamination by the atmosphere. The finished weld has no slag because no flux is used. Also, since no filler metal is transported within the arc, the process produces little or no spatter. *Figure 1-1* shows the basic GTAW process.

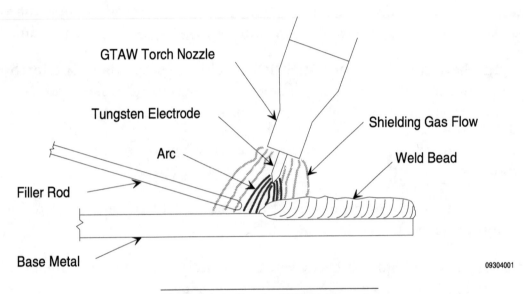

Figure 1-1. Basic GTAW Process

This module provides an overview of the different items of equipment used for GTAW. Topics include safety and safety equipment, GTAW equipment, electrodes types and preparation, filler metal types and classifications, shielding gas characteristics and gas systems, and equipment set-up. Upon completion of this unit, trainees will be aware of the safety concerns associated with GTAW and will be able to select and set up GTAW equipment safely and efficiently.

2.0.0 GTAW SAFETY

Intense heat and arc flash are the two principal hazards associated with gas tungsten arc welding. The intense heat of the process can cause severe burns. The arc used in the process gives off intense ultraviolet and infrared rays which will burn unprotected eyes and skin much as a sunburn does, only much more severely. Severe arc burn to the eyes can cause permanent eye damage. The following sections will explain how gas tungsten arc welding can be safely performed.

2.1.0 PERSONAL PROTECTIVE EQUIPMENT

Because of the potential dangers of GTAW, welders must use personal protective equipment. This equipment includes the proper work clothing, boots, gloves, safety glasses and special welding helmets. This personal protective equipment is explained in the following sections.

2.1.1 Work Clothing

To avoid burns from ultraviolet rays, welders performing GTAW must wear appropriate clothing. Never wear polyester or other synthetic fibers. Sparks or intense heat will melt these materials, causing severe burns. Wool or cotton is more resistant to sparks and so should be worn. Dark clothing is also preferred because dark clothes minimize the reflection of arc rays which could be deflected under the welding helmet.

To prevent sparks from lodging in clothing and causing a fire or burns, collars should be kept buttoned and pockets should have flaps that can be buttoned to keep out the sparks. A soft cotton cap worn with the visor reversed will protect the top of your head and keep sparks from going down the back of your collar. Pants should be cuffless and hang straight down the leg. Pant cuffs will catch sparks and catch on fire. Never wear frayed or fuzzy materials. These materials will trap sparks which will catch the clothing on fire. Low-top shoes should never be worn while welding. Sparks will fall into the shoes, causing severe burns. Leather work boots at least eight inches high should be used to protect against sparks and arc flash.

Figure 2-1 shows proper work clothing.

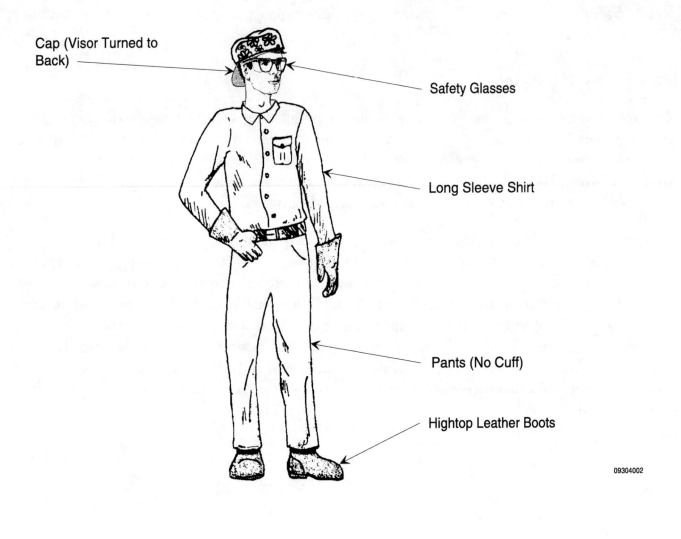

Figure 2-1. Proper Work Clothing

2.1.2 Leathers

For additional protection, welders often wear leathers over their work clothing. Leather aprons, split leg aprons, sleeves and jackets are flexible enough to allow ease of movement while providing added protection from sparks and heat. Because there is very little spatter or sparks associated with GTAW, leathers are usually not necessary.

WELDING THREE TRAINEE TASK MODULE 09304

2.1.3 Welding Gloves

Leather gauntlet-type gloves are designed specifically for welding. They must be worn when performing any type of arc welding to protect against spattering hot metal and the ultraviolet and infrared arc rays. When performing GTAW at lower amperages, special thin flexible gauntlet-type leather gloves are often worn to allow better feel and control of the filler rod and torch.

Leather welding gloves are not designed to handle hot metal. Pliers, tongs or some other means should be used to handle hot metal.

CAUTION Picking up hot metal with leather welding gloves will burn the leather, causing it to shrivel and become hard. Never handle hot metal with leather welding gloves.

Figure 2-2 shows typical leather gauntlet-type welding gloves.

09304003

Figure 2-2. Leather Welding Gloves

2.2.0 EYE PROTECTION

Eyes are delicate and can be easily damaged. The welding arc gives off ultraviolet and infrared rays which will burn unprotected eyes. The burns caused by the arc are called arc flash. Arc flash causes blistering of the outer eye. This blistering feels like sand or grit in the eyes. It can be caused from looking directly at the arc or from receiving reflected glare from the arc.

2.2.1 Safety Glasses and Goggles

Safety glasses and goggles protect the eyes from flying debris. Safety glasses have impact-resistant lenses. Side shields should be fitted to safety glasses to prevent flying debris from entering from the side. Some safety glasses are equipped with shaded lenses to protect against glare.

WARNING!	Shaded safety glasses will not provide sufficient protection from the arc. Looking directly at the arc with shaded safety glasses will result in severe burns to the eyes and face.

Safety goggles are contoured to fit the wearer's face. Goggles are a very efficient means of protecting the eyes from flying debris. They are often worn over safety glasses to provide extra protection when grinding or performing surface cleaning. *Figure 2-3* shows safety glasses and goggles.

Side Shields

Safety Glasses

Clear Safety Goggles

09304004

Figure 2-3. Safety Glasses and Goggles

2.2.2 Welding Shields

Welding shields (also called welding helmets) provide eye and face protection for welders. Some shields are equipped with handles, but most are designed to be worn on the head. They either connect to helmet-like headgear or attach to a hardhat. Shields designed to be worn on the head have pivot points where they attach to the headgear. They can be raised when not needed. Welding shields are made of dark, non-flammable material. The welder observes the arc through a window that is either 2- by 4-1/4 inches or 4-1/2 by 5-1/4 inches. The window contains a glass filter plate and a clear glass or plastic safety lens. The safety lens is on the outside to protect the more costly filter plate from damage by spatter and debris. For additional filter plate protection a clear safety lens is sometimes also placed on the inside of the filter plate. On most welding shields the window is fixed in the shield. However some welding shields have a hinge on the 2- by 4-1/4 inch window. The hinged window containing the filter plate can be raised, leaving a separate clear safety lens. This protects the welder's face from flying debris during surface cleaning. *Figure 2-4* shows typical welding shields.

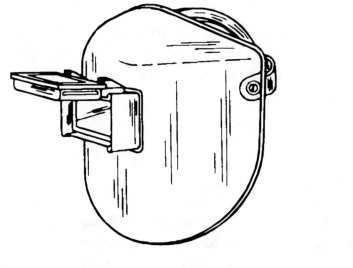 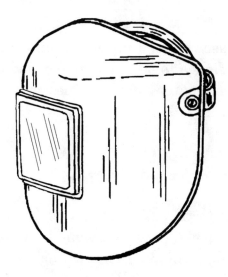

Hinged 2" X 4-1/4" Window   4-1/2" X 5-1/4" Window

09304005

Figure 2-4. Welding Shields

Filter plates come in varying shades. The shade required depends on the maximum amount of amperage to be used. The higher the amperage, the darker the filter plate must be. Filter plates are graded by numbers. The larger the number, the darker the filter plate. The American National Standards Institute (ANSI) publication *Z87.1, Practices for Occupational and Educational Eye and Face Protection,* provides guidelines for selecting filter plates. The following recommendations are based on these guidelines.

	Welding Current (Amps)	Lowest Shade Number	Comfort Shade Number
Gas Tungsten Arc	less than 50	8	10
Welding (GTAW)	50 - 150	8	12
	150 - 500	10	14

To select the best shade lens, first start with the darkest lens recommended. If it is difficult to see, try a lighter shade lens until there is good visibility. However, do not go below the lowest recommended number.

WARNING!   Using a filter plate with a lower shade number than is recommended for the welding current (amps) being used can result in severe arc flash (burns) to the eyes.

With use the lens in the welding shield will become dirty. The safety lens and filter plate are designed to be removed and cleaned in the same manner that safety glasses are cleaned. When the outer safety lens becomes too heavily impregnated with weld spatter or scratched, replace it.

2.3.0   EAR PROTECTION

Ear protection is necessary to prevent ear injury from flying debris or loud noise during grinding, chipping, and some types of cutting procedures. One source of damaging noise is pneumatic chipping and scaling hammers. Welders must also protect their ears from flying sparks and weld spatter that may enter the ear canal, causing painful burns. It is common for welders who do not protect their ears to suffer from perforated ear drums caused by sparks. *Figure 2-5* shows earmuffs and earplugs.

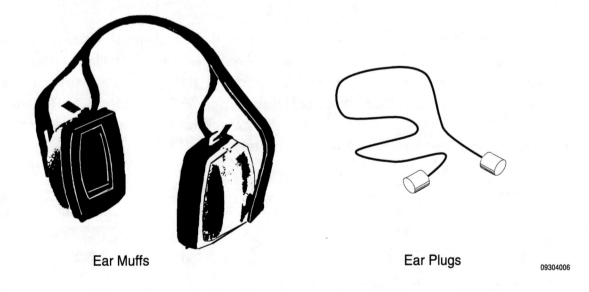

Ear Muffs   Ear Plugs

09304006

Figure 2-5. Earmuffs and Earplugs

2.4.0 VENTILATION

The fumes and gases associated with gas tungsten arc welding can be hazardous if the appropriate safety precautions are not observed. The following section will define these hazards and describe the appropriate safety precautions.

2.4.1 Fume Hazard

Metals heated during GTAW may give off toxic fumes. These fumes are not considered dangerous as long as there is adequate ventilation. Adequate ventilation can be a problem in tight or cramped working quarters. The following general rules can be used to determine if there is adequate ventilation:

- The welding area should contain at least 10,000 cubic feet of air for each welder
- There should be positive air circulation
- Air circulation should not be blocked by partitions, structural barriers, or equipment

Even when there is adequate ventilation, the welder should try to avoid inhaling welding fumes. The heated fumes generally rise straight up from the welding arc. Position yourself to avoid it. A small fan may also be used to divert the fumes, but care must be taken to keep the fan from blowing directly on the arc. The fan could blow away the shielding gas which must be present at the arc in order to protect the molten metal from the atmosphere.

CAUTION   Disturbance of the inert gas shield around the arc will cause weld defects. Never point a fan or compressed air directly at the arc.

If adequate ventilation cannot be ensured or if the welder cannot avoid the welding fumes, a respirator should be used.

2.4.2 Respirator

For most GTAW applications, a filter respirator offers adequate protection from fumes. The filter respirator is a mask which covers the nose and mouth. The air intake contains a cloth or paper filter to remove impurities. When additional protection is required, an air-line respirator should be used. It is similar to the filter respirator, except that it provides breathing air through a hose from an external source. The breathing air can be supplied from a cylinder of compressed breathing air, from special compressors that furnish breathing air or from special breathing air filters attached to compressed air lines.

WARNING! Air line respirators must only be used with pure breathing air sources. Standard compressed air contains oil, which is toxic. Standard compressed air can only be used for breathing if it is cleaned by a special breathing air filter placed in the line just before the respirator.

Figure 2-6 shows respirators.

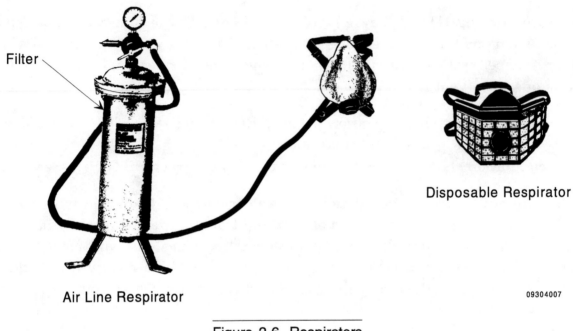

Air Line Respirator

Disposable Respirator

09304007

Figure 2-6. Respirators

| **WARNING!** | A filter type respirator will not supply oxygen in a low-oxygen environment. If normal oxygen concentration has been reduced because of replacement by shielding gasses, an air-line respirator must be used or suffocation and death could result. |

2.4.3 Suffocation

In confined spaces, the shielding gas used with the GTAW process can accumulate and displace the oxygen atmosphere, making suffocation possible. When using the GTAW process in a confined space, the welder must have positive ventilation to remove the shielding gas.

| **WARNING!** | When using GTAW in a confined space, positive ventilation must be provided. Failure to provide adequate ventilation could result in suffocation and death. |

2.5.0   ELECTRICAL SHOCK HAZARD

Most GTAW welding machines are connected to alternating current (AC) voltages of 208-460 volts. Contact with these voltages can cause extreme shock and possibly death. For this reason never open a welding machine with power on the primary circuit. To prevent electrical shock hazard, the welding machine must always be grounded.

A ground is an object that makes an electrical connection with the earth. Electrical grounding is necessary to prevent an electrical shock which can result from contact with a defective welding machine or other electrical device. The ground provides protection by providing a path from the equipment to ground for stray electrical current created by a short or other defect. The short or other defect will cause the fuse or breaker to open (blow), stopping the flow of current. Without a ground the stray electricity would be present in the frame of the equipment. When someone touched the frame the stray current would go to ground through the person.

| WARNING! | Death by electrocution can occur if equipment is not properly grounded. |

To ensure proper electrical grounding, a welding machine which is using single-phase power must have a three-prong outlet plug. Machines which use three-phase power must have a four-prong outlet plug. Welding machines must only be plugged into electrical outlets which have been properly installed by licensed electricians to ensure the ground has been properly connected. *Figure 2-7* shows grounded outlet plugs.

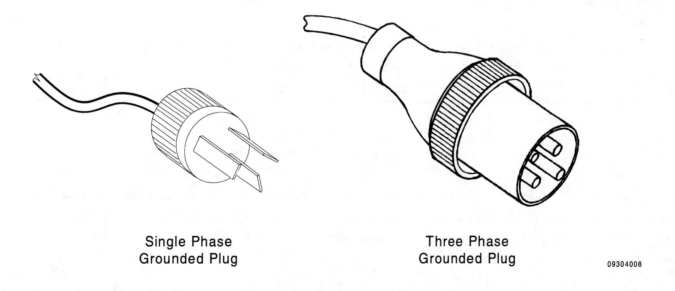

Single Phase
Grounded Plug

Three Phase
Grounded Plug

09304008

Figure 2-7. Grounded Outlet Plugs

2.6.0 AREA SAFETY

Before beginning a welding job, the area must be checked for safety hazards.

Check the area for water. If possible, remove standing water from floors and work surfaces where welding cables or machines are in use in order to avoid injury from slipping or electric shock. When it is necessary to work in a damp environment, make sure that you are well-insulated by wearing rubber boots and gloves. Protect equipment by placing it on pallets above the water and shielding it from overhead leaks. To avoid electric shock, never weld with wet gloves.

Check the area for fire hazards. Welding generates sparks that can fly ten feet or more and can drop several floors. Any flammable materials in the area must be moved or covered. Always have an approved fire extinguisher on hand before starting any welding job.

Other workers in the area must be protected from arc flash. If possible, enclose the welding area with screens to avoid exposing others to harmful rays or flying debris. Inform everyone in the area that you are going to be welding so that they can exercise appropriate caution.

2.7.0
HOT WORK PERMITS AND FIRE WATCH

Most sites require the use of hot work permits and fire watches. Severe penalties are imposed for violation of the hot work permit and fire watch standards.

WARNING!   Never perform any type of heating, cutting or welding until you have obtained a hot work permit and established a fire watch. If you are unsure of the procedure, check with your supervisor. Violation of the hot work permit and fire watch can result in fires and injury.

A hot work permit is an official authorization from the site manager to perform work which may pose a fire hazard. The permit will include information such as the time, location and type of work being done. The hot work permit system promotes the development of standard fire safety guidelines. Permits also help managers to keep records of who is working where and at what time. This information is essential in the event of an emergency or when personnel need to be evacuated.

Welders should always have a fire watch. During a fire watch, one person (other than the welder) constantly scans the work area for fires. Fire watch personnel should have ready access to fire extinguishers and alarms and know how to use them.

WELDING THREE TRAINEE TASK MODULE 09304

2.8.0 WELDING CONTAINERS

Before welding or cutting containers such as tanks, barrels and other vessels, check to see if they contain combustible and/or hazardous materials or residues of these materials. Such materials include:

- Petroleum products
- Chemicals that give off toxic fumes when heated
- Acids that could produce hydrogen gas as the result of a chemical reaction
- Any explosive or flammable residue

To identify the contents of containers check the label and then refer to the MSDS (Material Safety Data Sheet). The MSDS provides information about the chemical or material to help you determine if the material is hazardous. If the label is missing, or if you suspect the container has been used to hold materials other than what is on the label, do not proceed until the material is identified.

WARNING!   Do not heat, cut or weld on a container until its contents have been identified. Hazardous material could cause the container to violently explode.

If a container has held any hazardous materials, it must be cleaned before welding takes place. Clean containers by steam cleaning, flushing with water or washing with detergent until all traces of the material have been removed.

After cleaning the container, fill it with water or an inert gas such as argon or carbon dioxide (CO_2) for additional safety. Air, which contains oxygen, is displaced from inside the container by the water or inert gas. Without oxygen combustion cannot take place. When using water, position the container to minimize the air space. When using an inert gas, provide a vent hole.

Figure 2-8 shows using water in a container to minimize the air space.

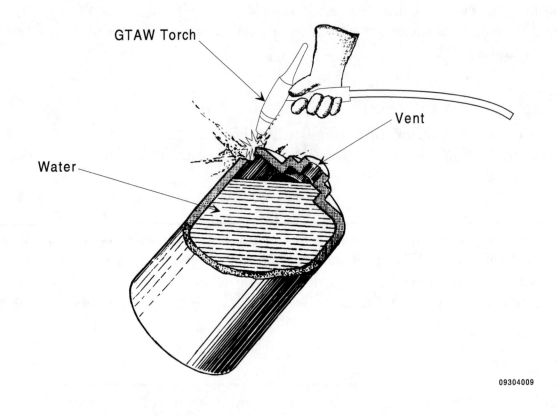

Figure 2-8. Using Water in a Container to Minimize the Air Space.

SELF-CHECK REVIEW 1

1. Name the two principal safety hazards associated with GTAW.
2. Name the types of fabric that should not be worn while welding and explain why.
3. What types of eye protection should welders wear during metal grinding and chipping?
4. Which shade filter plate is darker, shade 10 or shade 12?
5. Why is it important to wear ear protection while welding?
6. What should a welder do if it is impossible to provide adequate ventilation in the work area?
7. Why is positive ventilation a concern when performing GTAW in a confined space?
8. Why is the proper electrical grounding of welding equipment necessary?
9. What area checks must be made before welding can take place?
10. How can you protect other workers in the area from hazards while you are welding?
11. What is a hot work permit?
12. After cleaning a container, what else should be done to make it safe for heating, cutting or welding?

ANSWERS TO SELF-CHECK 1

1. Heat and arc flash. (2.0.0)
2. Polyester or other synthetic fabrics should not be worn while welding because sparks or intense heat will melt these materials, causing severe burns. (2.1.1)
3. Safety glasses and goggles. (2.2.1)
4. Shade 12. (2.2.2)
5. To prevent hearing loss due to excess noise and to keep sparks from entering the ear canal. (2.3.0)
6. Use an air-line respirator. (2.4.2)
7. GTAW shielding gas can displace the oxygen atmosphere and suffocation could result. (2.4.3)
8. To prevent dangerous electrical shock. (2.5.0)
9. Check the area for water, fire hazards and other workers. (2.6.0)
10. Enclose the welding area with screens (if possible) in order to avoid exposing others to harmful rays or flying debris. Inform everyone in the area that you are going to be welding so that they can exercise appropriate caution. (2.6.0)
11. An official authorization from a site manager to perform work which may pose a fire hazard. (2.7.0)
12. It should be filled with water or an inert gas such as argon or carbon dioxide (CO_2) to displace the oxygen. (2.8.0)

3.0.0   WELDING CURRENT

The welding arc required for GTAW is produced by electrical current from a welding power supply. Electrical current is the flow of electrons along a conductor. The current flows through two welding cables, one connected to the base metal (work) and the other connected to the torch which holds a virtually nonconsumable tungsten electrode. When welding, an arc is established in the air gap between the end of the tungsten electrode and the work. The air gap has high resistance. When the electrical current meets resistance, heat is generated. When the arc is struck across the air gap, the resistance generates intense heat of 6000 to 10,000 degrees melting the base metal and forming a weld.

Figure 3-1 shows the GTAW welding process.

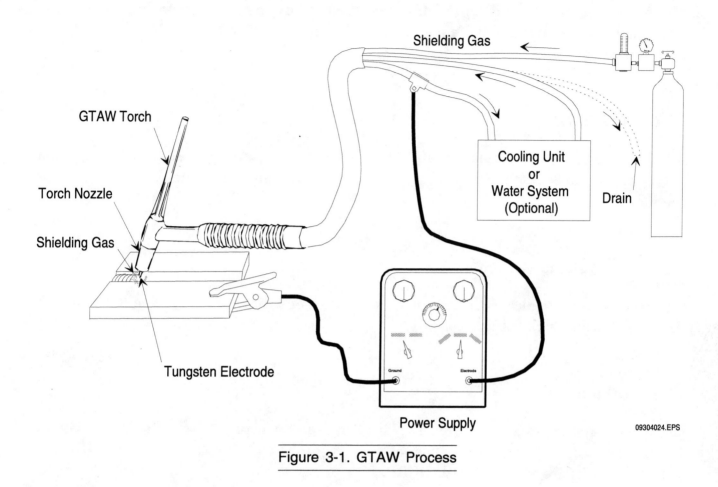

GTAW Torch

Torch Nozzle

Shielding Gas

Tungsten Electrode

Shielding Gas

Cooling Unit
or
Water System
(Optional)

Drain

Power Supply

09304024.EPS

Figure 3-1. GTAW Process

3.1.0   CHARACTERISTICS OF WELDING CURRENT

The current produced by a welding machine to perform welding has different characteristics than the current flowing through utility power lines. Welding current has low voltage and high amperage, while the power line current has high voltage and low amperage.

3.1.1   Voltage

Voltage is the measure of the electromotive force or pressure that causes current to flow in a circuit. There are two types of voltage associated with welding current: open circuit voltage and operating voltage. Open circuit voltage is the voltage present when the machine is on but no welding is taking place. For GTAW, there are usually ranges of open circuit voltages that can be selected up to about 80 volts. Operating voltage, or arc voltage, is the voltage after the arc is struck. With GTAW, this voltage is generally much lower than the open circuit voltage. The arc voltage drops rapidly with increasing current flow.

3.1.2 Amperage

Amperage is the electric current flow in a circuit. The unit of measurement for amperage is the Ampere (amp). The amount of amperes produced by the welding machine determines the intensity of the arc and the amount of heat generated. The higher the amperage, the hotter the arc.

3.2.0 TYPES OF WELDING CURRENT

There are two types of welding current: DC (direct current) and AC (alternating current). The type of current selected to be used depends on the equipment used and the type of welding to be done.

3.2.1 DC Welding Current

DC (direct current) is electrical current that flows in one direction only. The direction the current is flowing is called the polarity.

3.2.2 DC Polarity

Polarity only applies to DC current. The direction in which the current flows in a circuit determines polarity. There are two types of polarity: direct current straight polarity (DCSP), and direct current reverse polarity (DCRP).

Figure 3-2 shows direct DCSP and DCRP.

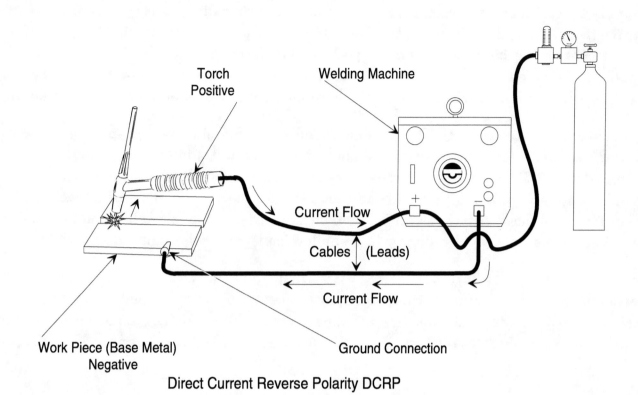

Direct Current Reverse Polarity DCRP

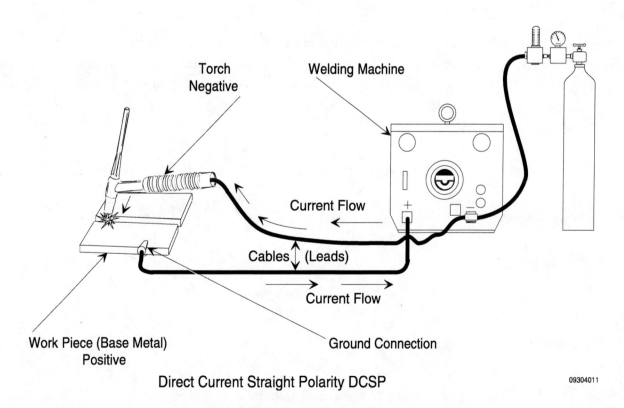

Direct Current Straight Polarity DCSP

09304011

Figure 3-2. DCSP and DCRP

When the welding circuit is set up as DCSP, the electrode (torch) is connected to the negative terminal of the welding machine and the work-piece (ground) is connected to the positive terminal. Direct current straight polarity (DCSP), also called direct current electrode negative (DCEN), produces the greatest amount of heat in the base metal. About 2/3 of the arc heat goes to the base metal and 1/3 goes to the electrode (tungsten). Because of this, a small size tungsten with a pointed end is usually used. The weld bead produced is narrow with deep penetration. DCSP is used to weld steels, stainless steels, nickel and titanium.

Direct current reverse polarity (DCRP), also called direct current electrode positive (DCEP), produces the least amount of heat in the base metal. About 1/3 of the arc heat goes to the base metal and 2/3 goes to the electrode (tungsten). Because of this, a large tungsten with a rounded end is required. The weld bead produced is wide with shallow penetration. This current type also has a strong cleaning action on the surface of oxidized base metals. DCRP is only used for special applications when shallow penetration is required.

Many welding machines have a polarity switch. Instead of physically having to disconnect the welding leads and reconnect them to change polarity, polarity is changed by turning a switch. When a welding machine has a polarity switch the welding cable terminals are generally labeled ELECTRODE and GROUND instead of NEGATIVE (-) and POSITIVE (+).

3.2.3   AC Welding Current

AC (alternating current) is electrical current which alternates (cycles) between positive and negative values. In the positive half of the cycle the current flows in one direction; during the negative half, the current reverses itself. The number of cycles completed in one second is called frequency. In the United States, AC current is almost always 60 cycles per second. *Figure 3-3* shows AC current.

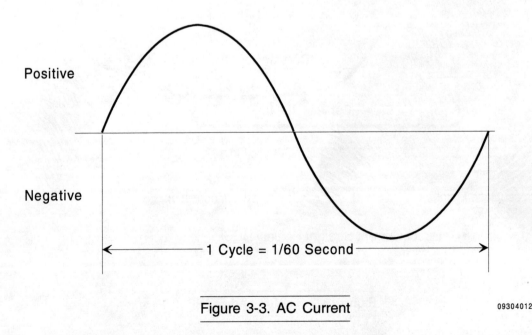

Figure 3-3. AC Current

WELDING THREE TRAINEE TASK MODULE 09304

Alternating current (AC) is DCSP for about half the time and DCRP for the remainder of the time. For this reason its characteristics fall in between those of DCSP and DCRP. About 1/2 the heat goes to the base metal and the other 1/2 to the electrode (tungsten). The weld bead size and penetration are midway between the beads produced by DCSP and DCRP. Some cleaning action is present with AC current. The AC tungsten has a small rounded tip. AC current is used with GTAW to weld aluminum, magnesium and their alloys.

3.2.4 High Frequency

High frequency is electrical current with a very high voltage (3,000 to 5,000) and very low amperage. It is called high frequency because it cycles at millions of times per second compared to standard current which has a frequency of 60 cycles per second. This high voltage, high frequency current creates a very stable arc able to jump a gap of about 1/2 inch. Because high frequency current has such low amperage it creates very little heat.

High frequency has two purposes, one is to stabilize AC current and the other is to allow a DC arc to be started without touching the tungsten electrode to the work.

When AC current is used with GTAW, each time the arc cycles it goes out and must be reestablish in the opposite direction. Even though this happens very fast (60 cycles per second) it can cause the arc to extinguish on the reverse polarity part of the AC cycle when the electron flow is weakest. If this happens, oxides are absorbed into the weld causing porosity which weakens the weld. To prevent the AC arc from extinguishing, a high frequency current is added to the AC welding current. The high frequency current with its high voltage and low amperage, establishes a path for the AC welding current to travel on preventing it from extinguishing as it cycles. When AC current is being used, the high frequency must be on continuously.

When using DC current for GTAW without high frequency, the tungsten must be touched to the work to establish the arc. This may not be desirable since it can cause contamination of the electrode and defects in the weld. To prevent this, high frequency current can be added to the DC welding current. When the tungsten electrode is brought close to the work, the high voltage, high frequency arc jumps the gap establishing a path for the DC welding current to travel without touching the tungsten to the work. Typically a timer will automatically shut off the high frequency about two to three seconds after the welding arc is established. Since DC current flows in only one direction, it is very stable. The high frequency is only needed for starting the arc.

Note   DC GTAW is often performed without high frequency for non-critical welds. The arc is touch started in the weld zone or on a piece of copper. Copper is less likely to contaminate the tungsten.

Welding power supplies specifically designed for GTAW will have a high frequency generator built into them. If conventional power supplies are used for GTAW, a separate auxiliary high frequency generator can be attached if required. They require 110 volt AC current to operate. Some auxiliary high frequency generators only have high frequency controls, others may have a combination of high frequency controls and gas and cooling water controls. Most will have a torch switch which plugs into the console. The switch attaches to the torch and can be used to turn the power, shielding gas and cooling water (if used) on or off. The auxiliary high frequency generator is connected to the power supply by the electrode cable. The GTAW torch is connected to the auxiliary high frequency generator. *Figure 3-4* shows an auxiliary high frequency generator.

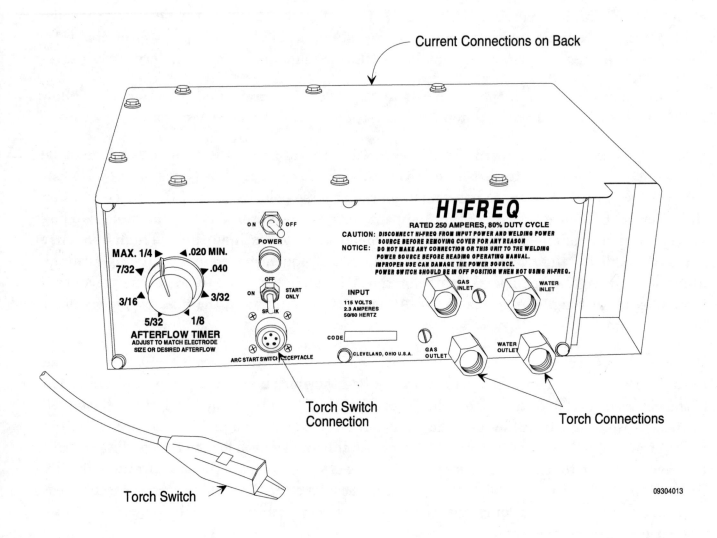

Figure 3-4. Auxiliary High Frequency Generator

4.0.0 WELDING POWER SUPPLY TYPES

GTAW is performed using constant current welding machines. The type of welding current (DCSP, DCRP or AC) depends upon the base metal type and thickness.

Welding current for GTAW may be supplied by many types of welding machines. Usually, any welding machine that produces constant current, such as those used for SMAW can be used. In addition there are special welding machines specially designed for GTAW. These special machines have built-in high frequency, and controls for shielding gas and cooling water. Types of welding machines that can be used for GTAW include:

- Transformer Welding Machines
- Transformer-Rectifier Welding Machines
- Motor Generator Welding Machines
- Engine-Driven Generator and Alternator Welding Machine

Direct current (DC) welding machines are usually designed to produce either constant current or constant voltage (constant potential) DC welding current. Constant current means that welding current is produced over a wide voltage range regardless of arc length. This is typical of a SMAW or GTAW welding machine. A constant voltage machine maintains a constant voltage as the output current varies. This is typical of a GMAW or FCAW welding machine. The output voltages and output currents of a welding machine can be plotted on a graph to form a curve. These curves show how the output voltage relates to the output current as either changes. *Figure 4-1* shows constant current and constant voltage output curves.

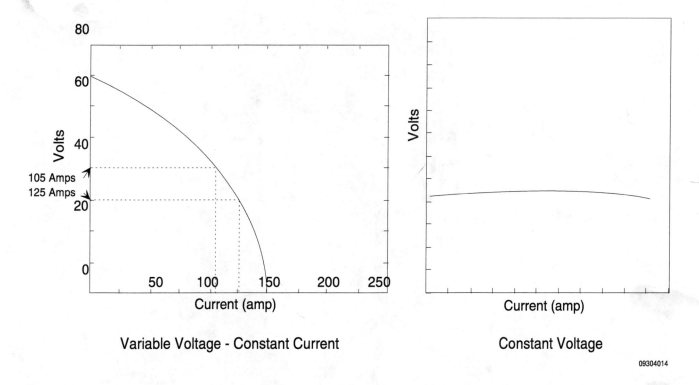

09304014

Figure 4-1. Constant Current and Constant Voltage Output Curves.

4.1.0 TRANSFORMER WELDING MACHINES

Transformer welding machines produce AC welding current only. They use a voltage step-down transformer which converts high voltage, low amperage current from commercial power lines to low voltage, high amperage welding current. The primary current required for a transformer welder can be 110- or 220-volt single-phase or 440-volt three-phase. Special light-duty transformer welding machines used for sheet metal work are designed to be plugged into a 110-volt outlet. However, most light-duty transformer welding machines require 220-volt primary current. Heavy duty industrial transformer welders require 440-volt three-phase primary current.

WARNING! Coming into contact with the primary current of a welding machine can cause death by electrocution. Ensure that welding machines are properly grounded to prevent injury.

Transformer welders are not as common as other types of welding machines on the job site but are used for special jobs. A transformer welder has an on/off switch, amperage control, and terminals for connecting the electrode cable and the ground cable. The terminals are marked ELECTRODE and GROUND. Transformer welding machines used for GTAW must have high frequency built-in, or use an auxiliary high frequency generator. *Figure 4-2* shows a typical transformer welding machine.

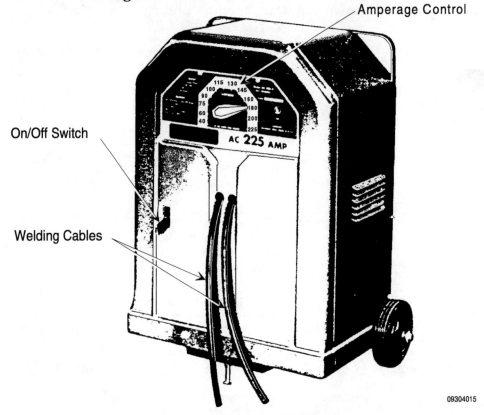

09304015

Figure 4-2. Transformer Welding Machine

4.2.0 TRANSFORMER-RECTIFIER WELDING MACHINES

The transformer-rectifier welding machine uses a transformer to transform the high voltage, low amperage primary current to low voltage, high amperage welding current and a rectifier to convert the alternating current (AC) to direct current (DC). Transformer-rectifier welding machines are designed to produce AC and DC welding current, or DC welding current only. Transformer-rectifiers which produce both AC and DC welding current are usually lighter duty than those which produce DC only. Transformer-rectifier welding machines that produce DC welding current only are sometimes called rectifiers. Depending on the size, transformer-rectifier welding machines may require 230-volt single-phase power, 230-volt three-phase power, or 460-volt three-phase power.

WARNING!   Coming into contact with the primary voltage of a welding machine can cause death by electrocution. Ensure that welding machines are properly grounded to prevent injury.

Transformer-rectifiers used for GTAW have an on/off switch and an amperage control. The welding cables (electrode cable and ground cable) are connected to terminals marked ELECTRODE and GROUND or POSITIVE (+) and NEGATIVE (-). They often have selector switches to select DCSP or DCRP. If there is no selector switch, the cables must be manually changed on the machine terminals to select the type of current desired. The most common welding power supply designed specifically for GTAW is the transformer-rectifier. Transformer-rectifiers designed specifically for GTAW will generally have:

- High frequency generator
- High frequency selector switch for DC START, CONTINUOUS (for AC current) or OFF (no high frequency)
- Controls to automatically start shielding gas flow when an arc is struck
- Controls to automatically start cooling water flow when an arc is struck
- Shielding gas post flow timer control to prevent weld contamination (keeps shielding gas flowing for a set time after arc is terminated)
- Hand or foot operated remote current control
- Remote/Local selector switch for remote control

GTAW is often used for critical welds on difficult to weld materials. For this reason many special advanced features and controls have been developed and are available on power sources designed for GTAW. These advanced features and controls include:

- High frequency intensity control for better arc starting
- High frequency stabilizer control to control tungsten spitting
- Balanced wave adjustment for penetration\cleaning control (sets percentage of time AC current remains on the straight or reverse side of the cycle)

- Pulse arc control (allows the welding current to be pulsed between a high current setting and a low current setting for puddle control)
- Hot start current control to enable a controlled surge of welding current to establish a puddle quickly
- Slope up control for DC hot start current (how quickly current raises to the hot start setting)
- Slope down control for DC weld termination to prevent crater cracks (slows the welding current drop from the welding setting to 0 (off) when the remote control is released at the termination of the weld)
- Optional spot arc timer control for GTAW spot welding

Note Refer to the machine's operating instruction manual for specific operating information on the controls for the welding machine being used.

Figure 4-3 shows a typical transformer-rectifier welding machine designed for GTAW.

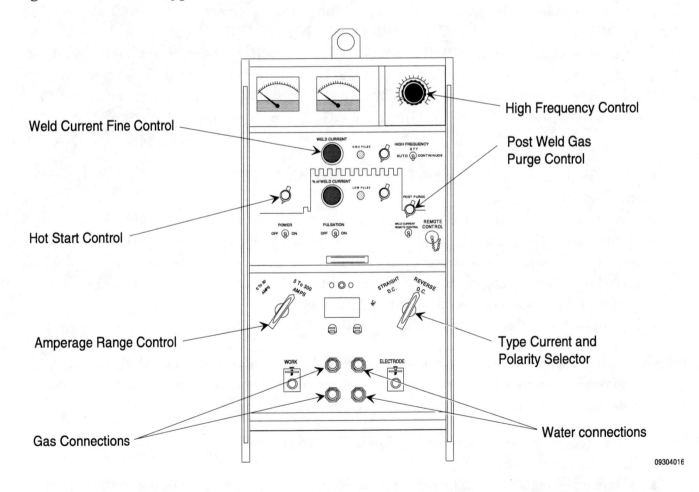

Figure 4-3. Transformer-Rectifier Welding Machine for GTAW

4.3.0 MOTOR-GENERATOR WELDING MACHINES

Motor-generator welding machines use an electric motor to turn a generator that produces DC welding current. The electric motor requires 440-volt, three-phase primary current. When motor generators are used for GTAW, the arc is established by touching the tungsten electrode to the work or by using an auxiliary high frequency generator. Motor generators used for GTAW have an on/off switch, voltage control and polarity switch. Some may have voltage and current gauges. They have welding cable terminals marked ELECTRODE and GROUND or POSITIVE (+) and NEGATIVE (-). *Figure 4-4* shows a typical motor-generator welding machine.

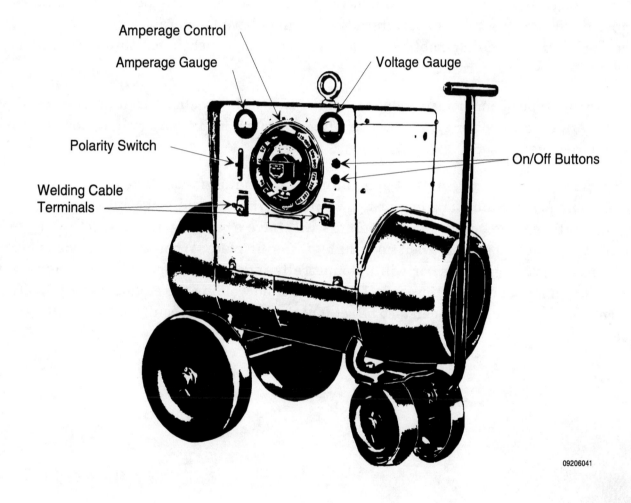

Figure 4-4. Motor-Generator Welding Machine.

Motor generators have fans which pull cooling air through the unit. In dusty conditions, dust builds up inside the unit. This buildup can cause overheating and excessive wear to the armature of the generator. Periodically clean motor generators by blowing them out with compressed air while the unit is running. Blowing out the units while they are running ensures the dust is discharged and not compacted inside the unit.

WARNING!   Motor generators use 460-volt three-phase primary power. Coming into contact with the primary current can cause death by electrocution. When blowing out motor generators, keep the end of the air hose well away from internal parts. Never remove guards.

4.4.0   ENGINE-DRIVEN GENERATOR AND ALTERNATOR WELDING MACHINES

Welding machines can also be powered by gasoline or diesel engines. The engine can be connected to a generator or to an alternator. Engine-driven generators produce DC welding current. Engine-driven alternators produce AC current which is fed through a rectifier to produce DC welding current.

The size and type of engine used depends on the size of the welding machine. Single-cylinder engines are used to power small rectifier alternators while six-cylinder engines are used to power larger generators.

To produce welding current, the generator or alternator must turn at a required rpm (revolutions per minute). The engines powering alternators and generators have governors to control the engine speed. Most governors will have a welding speed switch. The switch can be set to idle the engine when no welding is taking place. When the electrode is touched to the base metal, the governor will automatically increase the speed of the engine to the required rpm for welding. After about 15 seconds of no welding the engine will automatically return to an idle. The switch can also be set for the engine to run continuously at the welding speed.

Figure 4-5 shows an example of an engine-driven generator welding machine.

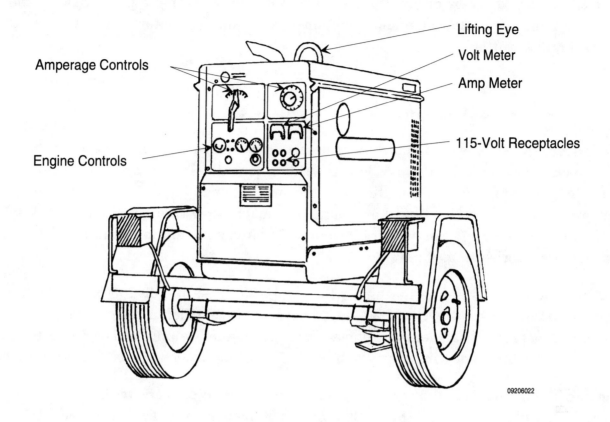

Figure 4-5. Engine Driven Welding Machine.

Engine-driven generators and alternators often have an auxiliary power unit to produce 115 VAC for lighting, power tools, and other electrical equipment. When 115 VAC is required, the engine-driven generator or alternator must run continuously at the welding speed.

Engine-driven generators and alternators have engine controls and welding current controls. The engine controls vary with the type and size, but normally include:

- Starter
- Voltage Gauge
- Temperature Gauge
- Fuel Gauge
- Hour Meter

The welding current controls include:

- Amperage Control
- Current Range Switch
- Amperage and Voltage Gauge
- Polarity Switch

The advantage of engine-driven generators and alternators is that they are portable and can be used in the field where electricity is not available to power other types of welding machines. The disadvantage is that engine-driven generators and alternators are costly to purchase, operate and maintain.

4.5.0   POWER SUPPLY RATINGS

The rating (size) of a welding machine is determined by the amperage output of the machine at a given duty cycle. The duty cycle of a welding machine is based on a ten-minute period. It is the percentage of ten minutes that the machine can continuously produce its rated amperage without overheating. For example, a machine with a rated output of 300 amps at 60 percent duty cycle can deliver 300 amps of welding current for six minutes out of every ten without overheating.

The duty cycle of a welding machine will be 20-, 30-, 40-, 60-, or 100-percent. A welding machine having a duty cycle of 20- to 40-percent is considered a light- to medium-duty machine. Most industrial, heavy-duty machines for manual welding will be 60- or 100-percent duty cycle. Machines designed for automatic welding operations are 100 percent duty cycle.

The maximum amperage that a welding machine will produce is always higher than its rated capacity. A welding machine rated 300 amps at 60 percent duty cycle will generally put out a maximum of 375 to 400 amps. But, since the duty cycle is a function of its rated capacity, the duty cycle will decrease as the amperage is raised over 300 amps. Welding at 375 amps with a welding machine rated 300 amps at 60 percent duty cycle will lower the duty cycle to about 30 percent. If welding continues for more than three out of ten minutes, the machine will overheat.

Note   Most welding machines have a heat-activated circuit breaker which will shut off the machine automatically when it overheats. The machine cannot be turned back on until it has cooled.

If the amperage is set below the rated amperage, the duty cycle increases. Setting the amperage at 200 amps for a welding machine rated 300 amps at 60 percent duty cycle will increase the duty cycle to 100 percent.

Figure 4-6 shows the relationship between amperage and duty cycle.

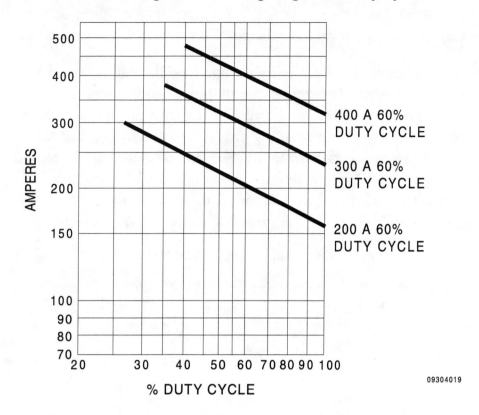

Figure 4-6. Amperage and Duty Cycle.

5.0.0   WELDING CABLE

Cables used to carry welding current are designed for maximum strength and flexibility. The conductors inside the cable are made of fine strands of copper wire. The copper strands are covered with layers of rubber reinforced with nylon or dacron cord. *Figure 5-1* shows the construction of a welding cable.

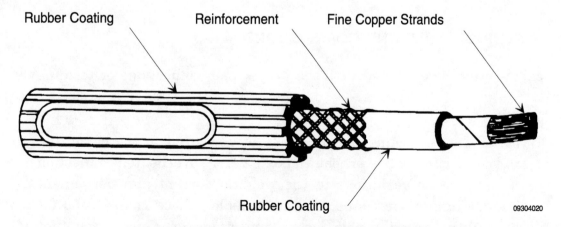

Figure 5-1. Construction of a Welding Cable.

The size of a welding cable is based on the number of copper strands it contains. Large diameter cable has more copper strands and can carry more welding current. Typically, the smallest cable size is number 8 and the largest is number 3/0 (three aught).

When selecting a welding cable size, the amperage load as well as the distance the current will travel must be considered. The longer the distance the current has to travel, the larger the cable must be to reduce voltage drop and heating caused by the electrical resistance in the welding cable. When selecting welding cable, use the rated capacity of the welding machine for the cable amperage requirement. For the distance, measure both the electrode and ground leads and add the two lengths together. To identify the size welding cable required, refer to a table which recommends welding cable sizes. These tables are furnished by most welding cable manufacturers. *Figure 5-2* shows a typical welding cable size table.

RECOMMENDED CABLE SIZES FOR MANUAL WELDING

Machine Size In Amperes	Duty Cycle (%)	Copper Cable Sizes for Combined Lengths of Electrodes Plus Ground Cable				
		Up to 50 Feet	50 - 100 Feet	150 - 150 Feet	150 - 200 Feet	200 - 250 Feet
100	20	#8	#4	#3	#2	#1
180	20	#5	#4	#3	#2	#1
180	30	#4	#4	#3	#2	#1
200	50	#3	#3	#2	#1	#1/0
200	60	#2	#2	#2	#1	#1/0
225	20	#4	#3	#2	#1	#1/0
250	30	#3	#3	#2	#1	#1/0
300	60	#1/0	#1/0	#1/0	#2/0	#3/0
400	60	#2/0	#2/0	#2/0	#3/0	#4/0
500	60	#2/0	#2/0	#3/0	#3/0	#4/0
600	60	#3/0	#3/0	#3/0	#4/0	* * *
650	60	#3/0	#3/0	#4/0	* *	* * *

* * Use Double Strand of #2/0
* * * Use Double Strand of #3/0

09304021

Figure 5-2. Welding Cable Size Table.

5.1.0   WELDING CABLE END CONNECTIONS

Welding cables must be equipped with the proper end connections or terminals to be used efficiently.

CAUTION   If the end connection is not tightly secured to the cable, the connection will overheat and oxidize. An overheated connection will cause variations in the welding current and permanent damage to the connector and/or cable. Check connections for tightness and repair loose connections or connections that overheat.

Lugs are used at the end of the welding cable to connect the cable to the welding machine current terminals. The lugs come in various sizes to match the welding cable size and are mechanically crimped onto the welding cable.

Quick disconnects are also mechanically connected to the cable ends. They are insulated and serve as cable extensions for splicing two lengths of cable together. Quick disconnects are connected or disconnected with a half twist. When using quick disconnects, care must be taken to ensure that they are tightly connected to prevent overheating or arcing in the connector. *Figure 5-3* shows typical lugs and quick disconnects used as welding cable connectors.

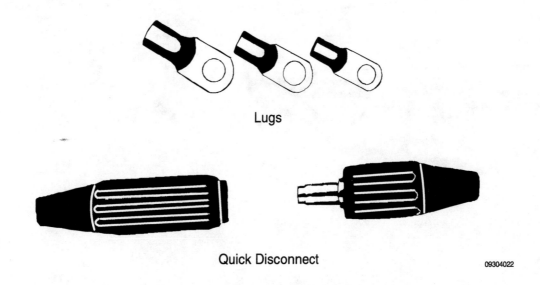

Lugs

Quick Disconnect

09304022

Figure 5-3. Lugs and Quick Disconnects.

The ground clamp provides the connection between the end of the ground cable and the work-piece. Ground clamps are mechanically connected to the welding cable and come in a variety of shapes and sizes. The size of a ground clamp is the rated amperage that it can carry without overheating. When selecting a ground clamp be sure it is rated at least the same as the rated capacity of the power source it will be used on.

Figure 5-4 shows examples of various styles of ground clamps.

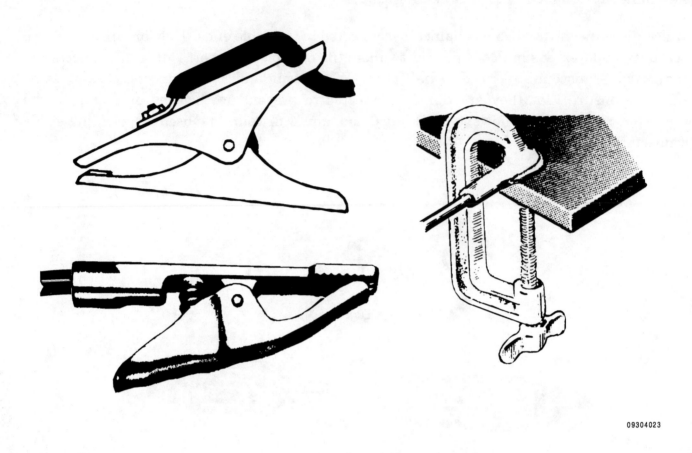

09304023

Figure 5-4. Ground Clamps.

6.0.0 GTAW EQUIPMENT

In addition to a constant current power supply (and high frequency generator for AC welding) GTAW requires:

- GTAW torch
- Torch nozzle
- Nonconsumable tungsten electrode
- Shielding gas

In addition, a remote amperage control is often used with GTAW.

Figure 6-1 shows a basic GTAW welding system.

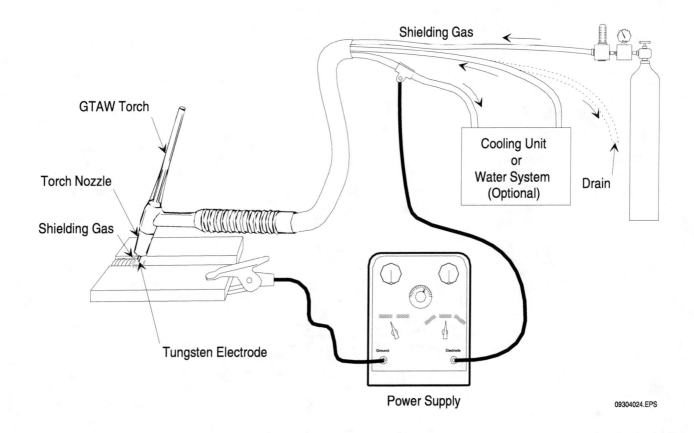

Figure 6-1. Basic GTAW welding system.

6.1.0   GTAW TORCHES

The GTAW torch provides an electrical path for the welding current between the welding current lead and the electrode (tungsten). There are many different manufacturers of torches but all are basically the same (although parts are not interchangeable). All torches have a collet holder (built in or replaceable) and electrode collet. The electrode collet comes in a variety of sizes to match the size electrode being used. The electrode collet holds the electrode in place. The electrode collet is secured by the electrode cap. The electrode cap which is threaded onto the torch is loosened to change the electrode or to adjust its stickout. The reserve length of electrode that extends from the back of the collet is insulated and protected by the electrode cap. Caps are made in several lengths to cover tungstens of different lengths. The shorter caps are used with short tungstens in confined spaces where torch clearance is inadequate. The torch also contains passages for the shielding gas and cooling fluid or water, if used. The shielding gas exits the torch through an insulated nozzle or cup that surrounds the tungsten.

Figure 6-2 shows a GTAW torch.

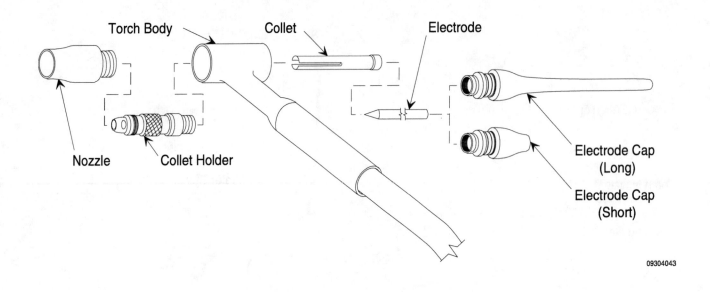

Figure 6-2. GTAW Torch.

To remove heat from GTAW torches, the torches are cooled with either the shielding gas (referred to as air cooled) or water. Two styles of air cooled torches are common. One style uses a single plastic hose. The single hose has a power cable inside it. The shielding gas flows around the power cable cooling it as well as the torch before it is discharged to shield the electrode and weld. A special power cable lug attaches to the power terminal on the power supply. The single hose connects to one side of the lug. This connection also makes the electrical connection. The shielding gas supply hose connects to the other side of the lug. The second type of air cooled torch has a separate power cable and shielding gas hose. Either the single or double hose torch can be purchased with a gas control valve in the torch to manually control the shielding gas flow. This is necessary when the power supply being used does not have a solenoid valve to automatically start and stop the shielding gas flow.

Figure 6-3 shows air cooled GTAW torches.

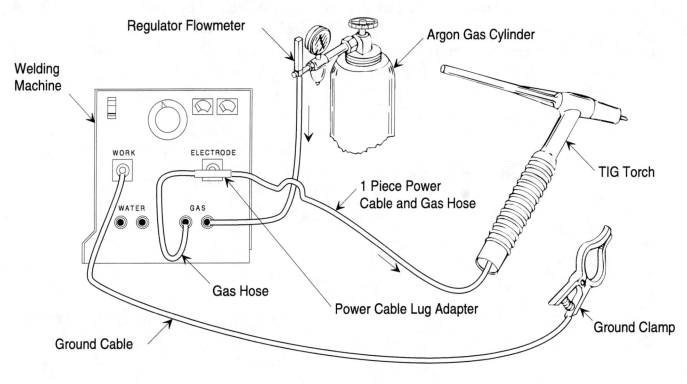

Regulator Flowmeter

Argon Gas Cylinder

Welding Machine

WORK   ELECTRODE

WATER   GAS

1 Piece Power Cable and Gas Hose

TIG Torch

Gas Hose

Power Cable Lug Adapter

Ground Clamp

Ground Cable

SINGLE HOSE GAS COOLED TORCH SET-UP

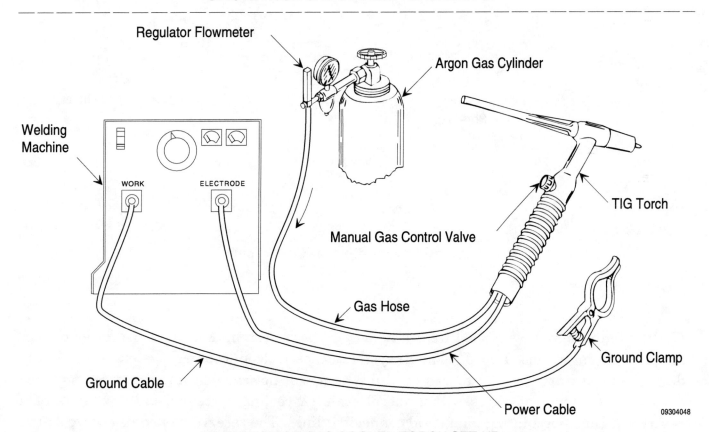

Regulator Flowmeter

Argon Gas Cylinder

Welding Machine

WORK   ELECTRODE

TIG Torch

Manual Gas Control Valve

Gas Hose

Ground Clamp

Ground Cable

Power Cable

09304048

DOUBLE HOSE GAS COOLED TORCH SET-UP

Figure 6-3. Air Cooled GTAW Torches.

Water cooled torches have three hoses. One hose supplies shielding gas, the second hose is the cooling water supply to the torch and the third hose contains the power cable and is also the cooling water return or discharge hose. The cooling water may be supplied from the domestic water system, or from a closed loop water cooling system. Regardless of the system, it should include a fusible link in the torch cooling line. A fusible link is a safety device that will open up and stop the welding current flow if the coolant flow to the torch is interrupted. *Figure 6-4* shows a water cooled GTAW torch.

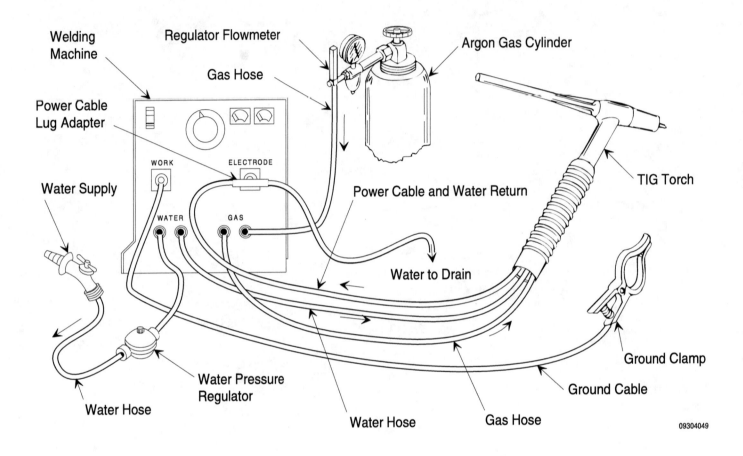

Figure 6-4. Water Cooled GTAW Torch.

Cooling liquid may be supplied from a domestic water system and discarded after use, or supplied from a closed loop cooling system. Closed loop cooling systems recirculate demineralized water or special cooling fluids that will not corrode internal torch surfaces or plug the cooling passages with mineral scale. A cooling system usually consists of a reservoir (tank), circulating pump and connecting lines. The reservoir may also serve as the unit's base and be equipped with wheels for easy transport. Heavy duty cooling systems have large reservoirs and may also be equipped with a heat transfer coil (radiator) and cooling fan to cool the fluid in the cooling system.

Figure 6-5 shows a closed loop cooling system.

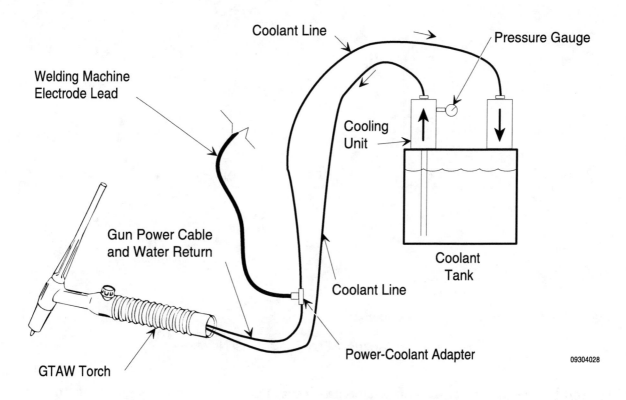

Figure 6-5. Closed Loop Cooling System.

CAUTION   Using a water cooled torch without cooling water flowing can severely damage the torch and power cable.

6.2.0   TORCH NOZZLES

The GTAW torch nozzle shapes and directs the shielding gas flow as it exits the torch. Nozzles (sometimes called cups) are usually threaded onto the torch to form a gas-tight joint. They are made of ceramic material, chrome plated steel, pyrex glass or plastic. Ceramic nozzles can be used up to about 300 amperes. Above that, water cooled metal-coated ceramic or water cooled ceramic nozzles must be used.

Nozzles are available in different lengths and diameters. Nozzle length is determined by the job requirements such as limited clearance or deep grooves. Nozzle size is the exit orifice inside diameter and is dependent upon the torch type and size and the electrode diameter. Nozzle size may be specified in fractions of an inch (1/4", 3/8", 7/16", 1/2", 3/4") or in millimeters (6mm, 10mm, 11mm, 13mm, 19mm). Sometimes the diameter is given as a size number, such as 4, 6, 7, 8, or 12. These numbers are the nozzle opening diameters in multiples of 1/16 of an inch. For example, a number 6 nozzle is 6 times 1/16 (or 6/16) or 3/8 inch in diameter.

Figure 6-6 shows various style torch nozzles.

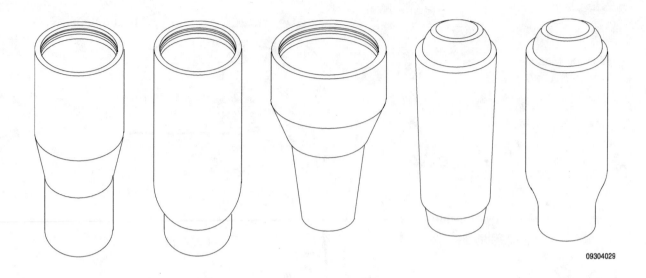

09304029

Figure 6-6. Various Style Torch Gas Nozzles.

A GTAW torch may also be equipped with a gas lens. A gas lens is an assembly of fine screens that straightens the gas flow from the nozzle to eliminate turbulence and cause the gas to flow smoothly past the tungsten electrode. This prevents the chance that the turbulence could pull atmosphere into the weld zone causing contamination and weld defects. *Figure 6-7* shows a typical gas lens.

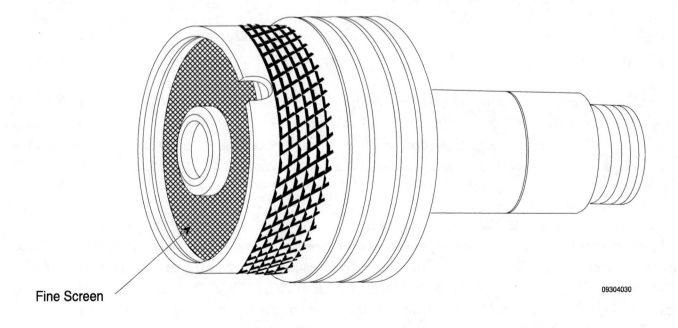

Fine Screen

09304030

Figure 6-7. Typical Gas Lens.

6.3.0   TUNGSTEN ELECTRODES

Tungsten electrodes used for GTAW are manufactured in different formulations and sizes to meet the requirements of American Welding Society specification A5-12. Electrode diameters range from 0.01 inch up to 1/4 inch, for current ratings of 5 to 1000 amperes. They are made in lengths of 3, 6, 7, 12, 18 and 24 inches. Manual welding is done with 7-inch or shorter electrodes.

Electrodes are manufactured with two different finishes, chemically cleaned or ground. The ground finish is the more expensive.

Three basic types of tungsten electrodes are manufactured. They are:

- Pure tungsten
- Zirconiated tungsten
- Thoriated tungsten

Pure tungsten electrodes are used for AC welding. They provide good arc stability and good resistance against contamination. If used for DC welding they are easily contaminated. Although expensive, pure tungsten is still the least expensive of the common types of GTAW electrodes.

Zirconiated tungstens are also used for AC welding. They have a small percentage of zirconium added in the tungsten. They are used for AC welding where tungsten inclusions cannot be tolerated. The zirconium also gives them easy arc starting characteristics. Zironiated tungstens are the most expensive type of tungsten electrode.

Thoriated tungsten electrodes are used for DC welding. They have a small percentage of thorium added to:

- Make the arc easier to start
- Increase the current range
- Help prevent tip melt
- Reduce tendency to stick or freeze to work
- Increase resistance to contamination when properly used

Thoriated tungsten electrodes are manufactured in three concentrations of thorium:

- EWTh-1 (approximately 1 percent thoria)
- EWTh-2 (approximately 2 percent thoria)
- EWTh-3 (approximately 1/2 percent thoria)

Tungsten electrodes are identified by a color band at one end. *Table 6-8* shows GTAW electrode color codes and AWS classifications:

Table 6-8

GTAW Electrode Color Codes And AWS Classifications.

ELECTRODE	Color Band	Electrode AWS Classification
Pure Tungsten	Green	EWP (Minimum 99.5% Tungsten)
1% Thoriated	Yellow	EWTh-1 (0.8% - 1.2% Thorium)
2% Thoriated	Red	
		EWTh-2 (1.7% - 2.2% Thorium)
3% Thoriated	Blue	EWTh-3 (0.35% - 0.55% Thorium)
Zirconiated	Brown	EWZr (0.15% - 0.40% Zirconium)

6.4.0   SHIELDING GAS

The GTAW process always uses inert shielding gasses to displace atmosphere from the weld zone to prevent oxidation and contamination of the tungsten electrode, weld puddle and filler metal. The two principal shielding gases used for GTAW are argon (Ar) and helium (He). These gases are used alone, as mixtures or sometimes small percentages of other gases are added for special conditions.

Each of the shielding gases and gas mixtures has distinctive individual performance characteristics. Each affects the arc differently and produces different weld characteristics. Mixtures of gases often have the best features of the individual gases. The following sections explain shielding gas characteristics and principal uses.

6.4.1   Argon

Argon is the most common shielding gas used with GTAW. It provides a smooth quiet arc requiring a lower arc voltage than other shielding gases for a given arc length. This means that it gives the welder the greatest tolerance for arc gap variation. Argon also works well with AC and provides better base metal cleaning (electrical cathode base metal oxide removal) than helium. Argon is ten times heavier than helium so it forms a better gas shield than helium which tends to float (raise) at the same flow rates. Argon is used both alone and in combination with other shielding gases.

When compared with helium, argon has the following advantages:

- A smoother quieter arc
- Lower arc voltage for a given arc length
- Easier arc starting
- Better cathode cleaning on aluminum and magnesium with AC
- Lower shielding flow rate
- Better shielding in cross drafts

Small amounts of hydrogen are sometimes mixed with argon for deeper penetration when welding stainless steel. Hydrogen cannot be used with aluminum or carbon steels because it produces porosity and underbead cracking. The most common argon-hydrogen mixture is 85 percent argon and 15 percent hydrogen.

Nitrogen is sometimes added to argon to stabilize austenitic stainless steel, and to increase penetration when welding copper.

6.4.2 Helium

Helium is used when deeper penetration and higher travel speed are required. Arc stability is not as good with helium as with argon. But, helium is capable of delivering more heat on the base metal than is possible with argon although it does require higher amperages. This makes helium better for welding thick sections of high heat conductivity metals such as aluminum, and copper.

When compared with argon, helium has the following advantages:

- Deeper penetration into the weld joint
- Increased welding speed
- Welding of high heat conductivity metals

Helium is often mixed with argon. A common mixture is helium-argon mixed at 75 percent helium and 25 percent argon. This mixture gives good cathode cleaning action and deep weld penetration, the good characteristics of both gases.

6.4.3 Cylinder Safety

Shielding gases may be supplied in liquid bulk tanks, liquid cylinders, or in high pressure cylinders of various sizes. The most common is high pressure cylinders which are portable and can be easily moved where needed.

Gas cylinders must always be handled with great care because of the dangerously high pressure (2000 to 4000 psi) gases they contain. If the valve is broken off a high pressure gas cylinder, the gas will escape with explosive force, which can cause severe injury with the blast and blown debris. The cylinder itself may become a deadly rocketing missile.

When transporting and handling cylinders, always observe the following rules:

- The safety cap should always be installed over the valve except when the cylinder has been secured for use.
- When in use, the cylinder must always be secured to prevent it from falling. It should be chained or clamped to the welding machine or to a post, beam or pipe.
- Always use a cylinder cart to transport cylinders.
- Never hoist cylinders with a sling. Cylinders can slip out of the sling. Always use a hoisting basket or similar device.
- Open the cylinder valve slowly. Once open, the valve should be completely opened to prevent leakage from around the valve stem.

WARNING!   Do not remove the protective cap until the cylinder is secured. If the cylinder falls over and the valve breaks off, the cylinder will shoot like a rocket, and will cause severe injury or death to anyone in its way.

6.4.4 Gas Regulator/flowmeters

A gas regulator/flowmeter is required to supply the shielding gas to the torch at the proper pressure and flow rate. A typical regulator/flowmeter consists of a preset pressure regulator with a cylinder valve spud, a flowmetering (needle) valve and a flow rate gauge. The pressure regulator is usually equipped with a tank pressure gauge to indicate the cylinder gas pressure. The metering valve is used to adjust the gas flow to the torch nozzle. The flow gauge shows the gas flow rate in cubic feet per minute or liters per second.

Figure 6-9 shows a shielding gas regulator/flowmeter.

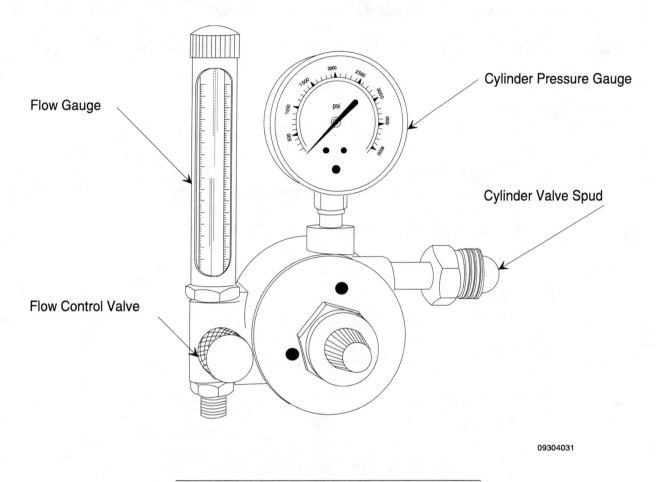

Figure 6-9. Shielding Gas Regulator/flowmeter.

Gas flow to the torch is started and stopped either by a manually operated valve or by an electric solenoid valve. The manual valve may be on the torch or it may be in the gas line to the torch. The solenoid is controlled by welding current flow or a manual switch.

6.5.0   REMOTE CURRENT CONTROL

With GTAW, control of the welding current is often performed with remote controls such as foot or hand operated switches and potentiometers. Remote controls are generally designed for use with a specific welding machine so they can not be used universally on all machines. Remote controls, with only a switch, control (turn on and off) one or more of the following:

- High frequency
- Shielding gas
- Cooling water
- Welding current (on and off only)

Remote controls with only a potentiometer can be used to continuously vary the welding current (up or down) as needed from 0 to the maximum current set on the power supply.

The most useful remote control has both switches and a potentiometer built into it. These control:

- High frequency
- Shielding gas
- Cooling water
- Welding current on and off
- Welding current up or down as needed (0 to maximum set on the power supply)

Figure 6-10 shows foot and hand operated remote controls.

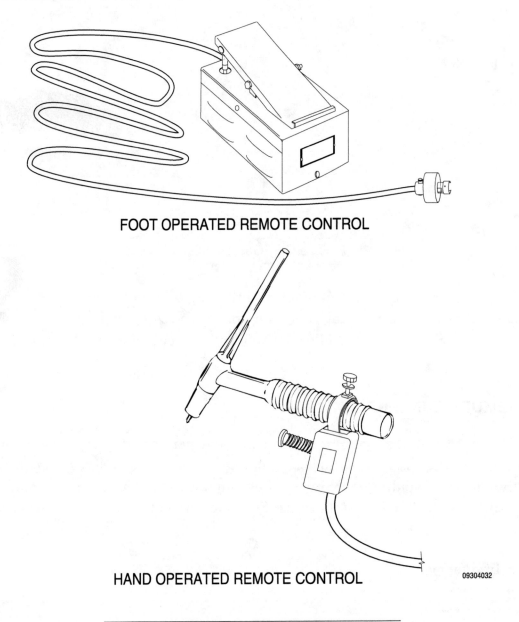

FOOT OPERATED REMOTE CONTROL

HAND OPERATED REMOTE CONTROL

09304032

Figure 6-10. Foot and Hand Remote Controls

7.0.0 GTAW FILLER METALS

Filler metal for manual GTAW is generally supplied in 36-inch lengths from 1/16- to 1/4-inch diameters. There are some automatic and manually operated wire feeders used with manual GTAW, but they are usually found only in high production facilities. Filler rods for GTAW are drawn from high-grade pure alloys compounded for specific applications. Generally the rods are not coated (bare without flux) except for a corrosion resistant copper electro-plating on some carbon steel rods. However, some special purpose rods may have a flux coating.

Several industry organizations and the U.S. government publish specification standards for filler rod. The most common is the American Welding Society (AWS). The purpose of the AWS specifications is to set standards that all manufacturers must follow when manufacturing welding consumables. This ensures consistency for the user regardless of who manufactured the product. The specifications set standards for the:

- Classification system, identification and marking
- Chemical composition of the deposited weld metal
- Mechanical properties of the deposited weld metal

The AWS specifications that pertain to gas tungsten arc welding filler rods are:

- AWS A5.7 *Specification for Copper and Copper Alloy Bare Welding Rods and Electrodes*
- AWS A5.9 *Specification for Corrosion Resisting Chromium and Chromium-Nickel Steel Bare and Composite Metal Cored and Stranded Welding Electrodes and Welding Rods*
- AWS A5.10 *Specification for Bare Aluminum and Aluminum Alloy Welding Electrodes and Rods*
- AWS A5.14 *Specification for Nickel and Nickel Alloy Base Welding Electrodes and Rods*
- AWS A5.16 *Specification for Titanium and Titanium Alloy Electrodes and Rods*
- AWS A5.18 *Specification for Carbon Steel filler Metals for Gas Shielded Arc Welding*
- AWS A5.19 *Specification for Magnesium Alloy Welding Rods and Bare Electrodes*
- AWS A5.28 *Specification for Low Alloy Steel Filler Metals for Gas Shielded Arc Welding*

Note: The specification number is generally followed by a dash and a two digit number such as A5.18-79 or A5.28-79. The dash and following number indicate the year that the specification was last revised.

Filler rods and wire are graded for three major areas of use:

- General Use - Wire or rod meets specifications. No record of chemical composition, strength and so forth is supplied to the user with the wire purchase.
- Rigid Control Fabrication - A "Certificate Of Conformance" is supplied with the wire or rod at purchase. The stock is identified by heat numbers, or code numbers located on the roll package.

- Critical Use - A "Certified Chemical Analysis" report is supplied, and records are kept by the fabricator of welds and processes for later reference on aircraft, nuclear reactors and pressure vessels.

Factors that affect the selection of a GTAW filler rod include:

- Base metal chemical composition
- Base metal mechanical properties
- Weld joint design
- Service or specification requirements
- Shielding gas used

Commonly used filler rod types include the following:

- Carbon steel filler rods
- Stainless steel filler rods
- Aluminum and aluminum alloy filler rods
- Copper and copper alloy filler rods

7.1.0 CARBON STEEL AND LOW ALLOY FILLER RODS

Mild steel filler rods are identified by specification AWS A5.18. Low alloy steel filler rods are identified by specification AWS A5.28. The rod classification (number) is located on a label on the packaging.

Typical AWS carbon steel filler rod classifications are defined in *Figure 7-1*.

MILD STEEL FILLER ROD

ER 70S-2

Electrode   Rod   Minimum Tensile Strength of Weld (in psi) Times 1,000   Solid Wire   Chemical Composition and Shielding Gas

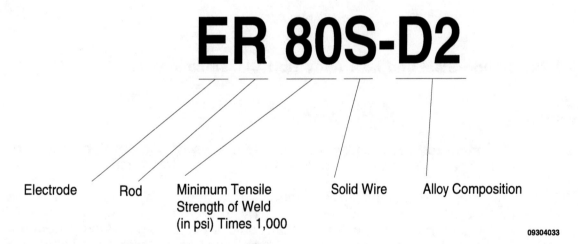

LOW ALLOY FILLER ROD

ER 80S-D2

Electrode   Rod   Minimum Tensile Strength of Weld (in psi) Times 1,000   Solid Wire   Alloy Composition

09304033

Figure 7-1. AWS Classification for Carbon Steel Filler Rods.

Note   The letter E preceding the R means that the wire or rod is classified as both an electrode and a filler rod. Electrode wire is used in GMAW. Filler rods are used in oxyfuel welding and in GTAW (TIG).

All carbon steel filler rods contain alloys such as: silicon, manganese, aluminum and carbon. Other alloys such as nickel, chromium and molybdenum are also often added. The purpose of the alloys are:

- Silicon (Si) - Concentrations of 0.40-1.00% are employed to deoxidize the puddle and to strengthen the weld. Silicon above 1% may make the welds crack sensitive.
- Manganese (Mn) - Concentrations of 1-2% are also employed as a deoxidizer and to strengthen the weld. Manganese also decreases hot crack sensitivity.
- Aluminum (Al), titanium (Ti) and zirconium (Zr) - One or more may be added in very small amounts for deoxidizing. Strength may also be increased.
- Carbon (C) - Concentrations of 0.05-0.12%, are employed to add strength without adversely affecting ductility, porosity or toughness.
- Nickel (Ni), chromium (Cr) and molybdenum (Mo) - May be added in small amounts to improve corrosion resistance, strength and toughness.

Table 7-2 lists AWS carbon steel and low alloy rod classifications and their uses.

Table 7-2.

AWS Carbon Steel and Low Alloy Rod Classifications and Their Uses

Rod Classification	Metals Welded
ER70S-2	Mild steel rod recommended for most grades of carbon steel. Contains extra deoxidizers which improves rust and mill scale tolerance.
ER80S-D2	Low alloy rod for making high quality, high strength X-ray quality welds on pipe, construction equipment and alloy applications where porosity, high sulfur or carbon in the base metal could be a problem. Also recommended for parts that are stress relieved because of its addition of molybdenum.

7.2.0 STAINLESS STEEL FILLER RODS

Stainless steel filler rods are identified by specification AWS A5.9. A typical AWS stainless steel rod classification is defined in *Figure 7-3*.

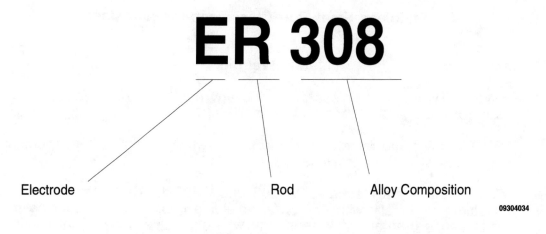

Figure 7-3. Typical Stainless Steel Rod Classification.

Stainless steel filler rod should be selected to closely match the alloy composition of the base metal. Stainless steel filler rods also require specific shielding gases or gas mixtures.

Table 7-4 lists AWS stainless steel rod classifications and uses.

Table 7-4

AWS Stainless Steel Rod Classifications and Uses

Rod Class	Metals Welded
ER308	Welds Type 308 stainless steel and any chromium-nickel alloy containing approximately 19% chromium and 9% nickel.
ER308L	Welds the same metals as ER308 but contains less carbon, reducing the chances of carbon precipitation.
ER308L Si	Welds the same metals as ER308 and ER308L. The higher silicon improves wetting, especially with Ar-1% O_2 shielding gas. The high silicone content can cause greater crack sensitivity with extensive base metal dilution.
ER308H	Welds the same metals as ER308 but the higher range of carbon ensures the creep strength at the high end of the 308 range. Used for welding Type 304H stainless steel.
ER309	Welds Type 309 and 309S stainless steels, Type 304 stainless steel for severe corrosion service and joining mild steel to 304 stainless steel.
ER309 ELC	Has extra low carbon content for resistance to carbide precipitation. Used for Type 309L stainless steel and overlay work.
ER310	Welds Type 310 stainless steel.
ER312	Welds stainless steels to mild steels. Welds high strength steels that are difficult to weld with ferritic electrodes. Used to weld carbon steel mounts to stainless steel vessels.
ER316	Welds Type 316 and 319 stainless steels. Used in high temperature applications for creep resistance.
ER316L	Welds Type 316L. More resistant to intergranular corrosion than ER316.
ER316L Si	Contains a higher amount of silicone, which markedly improves wetting characteristics for use in applications demanding a good cosmetic appearance.
ER316H	Welds Type 316H stainless steel. This electrode can be used in place of ER316. The restricted range of carbon ensures that the creep strength of the weld metal is at the high end of the 316 range.
ER320LR	For welding Carpenter 20Cb-3 type stainless steel.
ER347	For welding Type 321 and 347 stainless steel where the service temperature will be below 600 degrees F and maximum resistance to corrosion is required.

7.3.0   ALUMINUM AND ALUMINUM ALLOY FILLER RODS

Aluminum rods are covered by AWS Specification A5.10. Aluminum filler rods usually contain magnesium, manganese, zinc, silicon and copper, for increased strength. Corrosion resistance and weldability are also considerations. Aluminum filler rods are designed for specific types of aluminum and should be selected to closely match the base metal's chemistry. The most widely used aluminum rods are ER4043 (contains silicon) and ER5356 (contains magnesium). A typical AWS aluminum filler rod classification is defined in *Figure 7-5.*

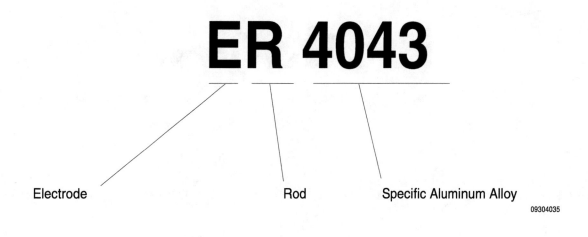

Electrode   Rod   Specific Aluminum Alloy

09304035

Figure 7-5. Typical Aluminum Filler Rod Classification.

Table 7-6 lists recommended filler rods for common aluminum alloys.

Table 7-6

Recommended Filler Rods for Common Aluminum Alloys

Aluminum Base Alloy	For Max Strength	For Max Ductility	For Max Salt Corrosion Resistance	For Least Cracking
1100	ER4043	ER1100	ER1100	ER4043
3003	ER4043	ER1100	ER1100	ER4043
5052	ER5556	ER5654	ER5554	ER5356
5083	ER5556	ER5356	ER5183	ER5356
5086	ER5556	ER5356	ER5356	ER5356
5454	ER5356	ER5554	ER5554	ER5356
5456	ER5556	ER5356	ER5556	ER5356

7.4.0 COPPER AND COPPER ALLOY FILLER RODS

Copper rods are covered by AWS Specification A5.6. Most copper filler rods contain other elements to increase strength, deoxidize the weld metal and match the base metal composition.

A typical AWS copper filler rod classification is defined in *Figure 7-7*.

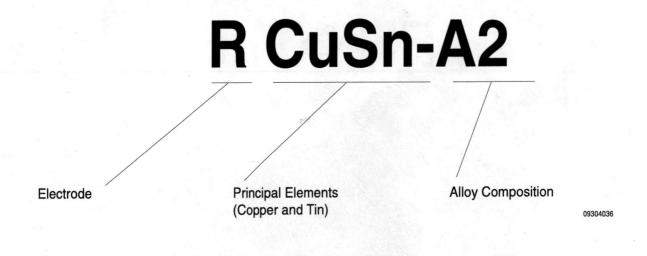

Figure 7-7. Typical Copper Filler Rod Classification.

Table 7-8 lists some AWS copper rod identifiers and their principal uses.

Table 7-8

AWS Copper Rod Classifications and Principal Uses

Rod Classification	Metals Welded
RCuSnA	Red Brasses
RCuAl-A2	Yellow Brasses
RCuSi	
RCuAl-A2	Manganese Bronze
RCuSi	
RCuSiA	Silicon Bronze
RCuSiA	Phosphor Bronze
RCuSnC	
RCuAl-A2	Aluminum Bronze

1. How does the power line current and voltage supplied to a welding machine differ from the current and voltage supplied to the welding cables?
2. What is operating voltage?
3. What determines the intensity of an arc and the amount of heat available for welding?
4. Is a constant current or a constant voltage (potential) welding machine best for GTAW?
5. What type of current does a transformer welder produce?
6. What type of welding machine can produce both AC and DC welding current?
7. What type of welding machine uses an electric motor to turn a generator to produce DC welding current?
8. What is duty cycle?
9. Which welding cable size can carry the most current; #6 or 2/0?
10. What will happen to a welding cable end connection that is not tight?
11. What purpose does the torch collet assembly serve?
12. What does the nozzle size specify?
13. What does a gas lens do?
14. What does the specification EWTh-2 indicate?
15. What are the two principal shielding gases used for GTAW?
16. What is the most common shielding gas used with GTAW?
17. Which shielding gas is best for thick sections of high heat conductivity metals such as aluminum, and copper?
18. Why should a cylinder always be secured before its protective valve cap is removed?
19. What is the name of the organization (most common) that publishes standards for GTAW filler rods?
20. What type of GTAW filler rod is ER308?

ANSWERS TO SELF-CHECK 2

1. Power line voltage is much higher and the current is much lower than welding voltage and current. (3.1.0)
2. Operating voltage is the welding voltage after the arc is established. (3.1.1)
3. The welding amperes produced by the welding machine. (3.1.2)
4. A constant current machine is best for GTAW. (4.0.0)
5. AC welding current. (4.1.0)
6. A transformer-rectifier welding machine. (4.2.0)
7. A motor generator welding machine. (4.3.0)
8. It is the percentage of ten minutes that the welding machine can continuously produce its rated amperage without overheating. (4.5.0)
9. 2/0. (5.0.0)
10. It will overheat and oxidize. (5.1.0)
11. It holds the electrode and provides a current path between the welding current lead to the electrode. (6.1.0)
12. The diameter of the exit orifice. (6.2.0)
13. Straightens and smooths the shielding gas flow from the nozzle to eliminate turbulence. (6.2.0)
14. A tungsten with 2% thorium. (6.3.0)
15. Argon and helium. (6.4.0)
16. Argon. (6.4.1)
17. Helium. (6.4.2)
18. If the cylinder falls over and the valve breaks off, the cylinder will shoot like a rocket, and will cause severe injury or death to anyone in its way. (6.4.3)
19. American Welding Society (AWS). (7.0.0)
20. Stainless steel. (7.2.0)

8.0.0   WELDING EQUIPMENT SET-UP

In order to weld safely and efficiently the welding equipment must be properly set up. The following sections will explain the steps for setting up GTAW equipment.

8.1.0   SELECTING A GTAW POWER SUPPLY

Select the power supply and current type for the base metal to be welded. To select a power supply (welding machine) the following factors must be considered.

- GTAW requires a constant current power source. SMAW power supplies and welding machines are commonly used for GTAW.

- The welding current type required, AC or DC. (Generally, carbon steels, stainless steels and alloy steels are welded with DC, and aluminum and magnesium are welded with AC and high frequency.)
- The maximum amperage required.
- The primary power requirements. (Are there electrical receptacles to plug a welding machine into or does an engine driven generator/alternator need to be used?)

Table 8-1 shows welding current requirements for different base metals.

Table 8-1.

Base Metal Welding Current Recommendations

Base Metal	ACHF Current	DCSP Current	DCRP Current
Aluminum	1	NR	2
Aluminum Casting	1	NR	NR
Aluminum Bronze	1	NR	2
Beryllium Copper	1	NR	2
Brass Alloys	2	1	NR
Copper Base Alloys	2	1	NR
Cast Iron	2	1	NR
Deoxidized Copper <0.09"	NR	1	NR
Dissimilar Metals	2	1	NR
Hard-Facing	1	2	NR
Hastelloy Alloys	2	1	NR
High Carbon Steel 0.015 - 0.030"	2	1	NR
High Carbon Steel 0.030 - and thicker	2	1	NR
Low Carbon Steel 0.015 - 0.030"	2	1	NR
Low Carbon Steel 0.030 - 0.125"	NR	1	NR
Magnesium Casting	1	NR	2
Magnesium <1/8"	1	NR	2
Magnesium >3/16"	1	NR	NR
Nickel & Nickel Alloys	2	1	NR
Silicon Bronze	NR	1	NR
Silver	2	1	NR
Silver Cladding	1	NR	NR
Stainless Steel <0.05"	1	2	NR
Stainless Steel >0.05"	2	1	NR
Titanium	2	1	NR

1 = recommended or first choice
2 = acceptable alternate or second choice
NR = not recommended
< = Less than
> = Greater than

8.2.0 LOCATING THE GTAW EQUIPMENT

Because of the limited length of GTAW torch cables, the shielding gas supply must be located reasonably close to the welding site. The power supply can be some distance away, but unless remote controls are installed, it will not be convenient to operate.

Select a site where the GTAW equipment will not be in the way but will be protected from welding or cutting sparks. There should be good air circulation to keep the welding machine cool. The environment should be free from explosive or corrosive fumes and as free as possible from dust and dirt. Welding machines have internal cooling fans which will pull these materials into the welding machine if they are present. The site should also be free of standing water or water leaks. If an engine-driven generator or alternator is used, locate it so it can be easily refueled and serviced.

There should also be easy access to the site so the equipment can be started, stopped or adjusted as needed. If the machine is to be plugged into an outlet, be sure the outlet has been properly installed by a licensed electrician to ensure it is grounded. Also be sure to identify the location of the electrical disconnect for the outlet before plugging the welding machine into it.

8.3.0 MOVING WELDING POWER SUPPLIES

Large engine-driven generators are mounted on a trailer frame and can be easily moved by a pickup truck or tractor using a trailer hitch. Other types of welding machines will have a skid base or be mounted on steel or rubber wheels. When moving welding machines that are mounted on wheels by hand, use care. Some machines will be top-heavy and may fall over in a tight turn or if the floor or ground is uneven or soft.

WARNING!	If the gas cylinder is attached to the power supply, check to be sure it is secure or remove the gas cylinder before attempting to move the power supply. Severe injury could result if the cylinder fell over.

WARNING!	If a welding machine starts to fall over do not attempt to hold it. Welding machines are very heavy and severe crushing injuries can occur if a welding machine falls on you.

WELDING THREE TRAINEE TASK MODULE 09304

Most welding machines have a lifting eye. The lifting eye is used to move machines mounted on skids or to lift any machine. Before lifting a welding machine check the equipment specifications for the weight. Be sure the lifting device and tackle will handle the weight of the machine. Always use a shackle or a sling. Never attempt to lift a machine by placing the lifting hook directly in the machine's lifting eye. Also, before lifting or moving a welding machine be sure the welding cables are secure. *Figure 8-2* shows lifting a welding machine.

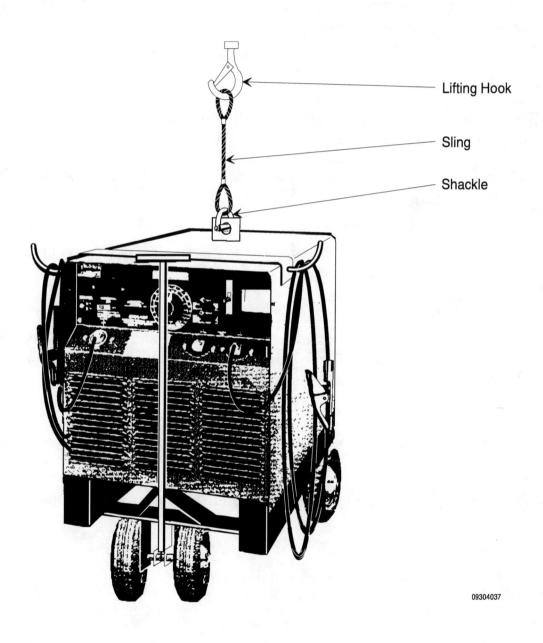

Lifting Hook

Sling

Shackle

09304037

Figure 8-2. Lifting A Welding Machine.

8.4.0 CONNECTING THE SHIELDING GAS

The hose from the shielding gas regulator/flowmeter connects to the welding machine gas solenoid if the welding machine is designed for GTAW. If a standard welding machine is used, the shielding gas hose connects to the torch cable. To connect the shielding gas follow these steps:

Step 1 Identify the shielding gas required by referring to the WPS (Welding Procedure Specification) or site quality standard.

Step 2 Locate a cylinder of the correct gas or mixture and secure it nearby.

CAUTION Be sure to secure the cylinder so it cannot fall over.

Step 3 Remove the cylinder's protective cap and momentarily crack open the cylinder valve to blow out any dirt.

Step 4 Install the regulator/flowmeter on the cylinder.

Step 5 Connect the gas hose to the flowmeter and to the gas solenoid on the welding machine or to the end of the torch cable.

Note If the gas hose is connected to the torch cable and the torch does not contain a gas shut off valve, a valve should be installed in the line between the torch cable and the regulator/flowmeter.

Step 6 Check to be sure the flowmeter adjusting valve is released (screwed out).

Step 7 Slowly crack open the cylinder valve and then open it completely.

Figure 8-3 shows connecting the shielding gas.

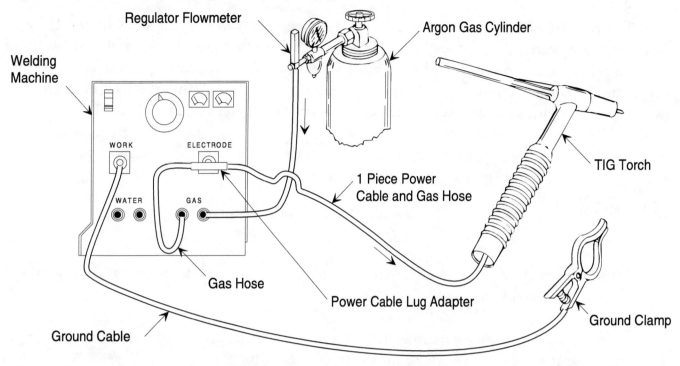

Regulator Flowmeter

Welding Machine

Argon Gas Cylinder

WORK ELECTRODE

WATER GAS

1 Piece Power Cable and Gas Hose

TIG Torch

Gas Hose

Power Cable Lug Adapter

Ground Cable

Ground Clamp

SINGLE HOSE GAS COOLED TORCH SET-UP

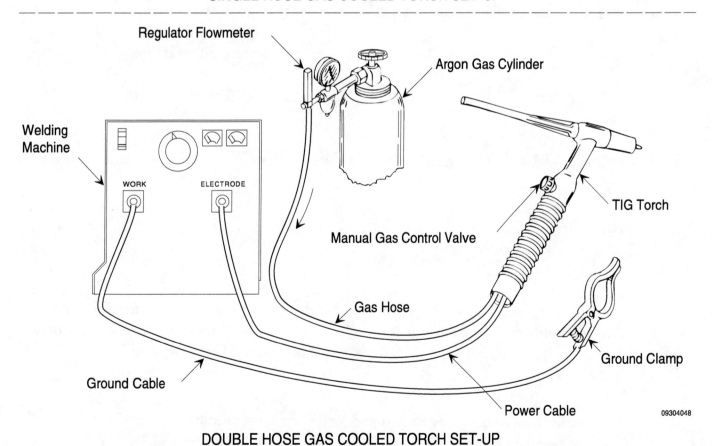

Regulator Flowmeter

Argon Gas Cylinder

Welding Machine

WORK ELECTRODE

Manual Gas Control Valve

TIG Torch

Gas Hose

Ground Cable

Ground Clamp

Power Cable

09304048

DOUBLE HOSE GAS COOLED TORCH SET-UP

Figure 8-3. Connecting the Shielding Gas.

8.5.0   SET SHIELDING GAS FLOW RATE

Gas flow rate from the torch nozzle is important because it affects the quality and the cost of a weld. Too low a flow rate will not shield the weld zone adequately and will result in a poor quality weld. An excessively high gas flow rate wastes expensive gas and can generate turbulence in the gas column. The turbulence can pull atmosphere into the weld zone causing oxidation and weld contamination. The nozzle must be large enough to gently flood the weld pool with inert gas. As a general rule, larger nozzles require higher flow rates. Also helium requires a higher flow rate than argon because it rises rapidly due to its very low density. Welding specifications specify shielding gas flow rates.

Other factors that may affect the shielding gas flow rate include:

- Drafts - The flow rate must be increased in a drafty location to maintain the gas shield around the weld zone.

- Specific gas used - For example, helium usually requires a higher flow rate than argon because it rises much faster due to its low density.

- Welding current - High welding currents require higher flow rates.

- Nozzle size (exit opening) - Larger nozzles require higher flow rates.

- Weld joint type - Welds on flat surfaces require higher flow rates than welds in deep grooves or fillets.

- Welding speed - Fast advance speeds require higher flow rates than slower advance speeds.

- Weld position - Vertical and horizontal position welds require higher flow rates than flat or overhead welds.

To set the shielding gas flow rate be sure the shielding gas valve is open (manual valve at the torch or solenoid on the welding machine) and adjust the flow rate with the flowmeter valve. Adjust the flow for the type and thickness base metal being welded.

Note   Some flowmeters are equipped with several scales of different calibrations around the same sight tube for monitoring the flows of different types (densities) of gases. Be sure to rotate the scales or read the correct side for the gas type being used.

The following sections provide general recommendations for flow rates for various types of base metals.

8.5.1 Gas Flow Rates For Carbon Steel

When using GTAW on low and medium carbon steels and low alloy steels, argon flow rates typically vary from 15 cfh (cubic feet per hour) for light welding at 60 amperes to 20 cfh for heavier welding at 200 or more amperes.

8.5.2 Gas Flow Rates For Stainless Steel

When using GTAW on stainless steel, argon flow rates typically vary from 11 cfh for light welding at 80 amperes to 15 cfh for heavy welding at 200 or more amperes.

8.5.3 Gas Flow Rates For Aluminum

When using GTAW on aluminum, argon flow rates typically vary from 15 cfh for light welding at 60 amperes to 30 cfh for heavy welding at 200 or more amperes. Welding specifications specify shielding gas flow rates.

8.5.4 Gas Flow Rates For Copper

When using GTAW on deoxidized copper up to 1/8 inch thick, a typical argon flow rate is 15 cfh from 110 to 250 amperes. With copper over 1/8 thick, helium is used at 30 cfh from 190 to 400 amperes.

8.6.0 SELECT THE ELECTRODE

The GTAW electrode (commonly called the tungsten) must be selected and the end properly prepared before it can be installed in the torch. Correct preparation of the electrode is absolutely essential. It the tip is improperly shaped, it will not produce the required arc shape and characteristics.

When selecting a GTAW electrode, choose the type recommended for the welding current required by the base metal. The type of electrode will be specified in the WPS or Site Quality Standards. General recommendations are:

- **Pure Tungsten (EWP),** used for AC welding aluminum and magnesium using AC current and high frequency.
- **Zirconiated Tungsten (EWZr),** used for welding aluminum and magnesium using AC current and high frequency for welds where tungsten inclusions are not tolerated.
- **Thoriated Tungsten (EWTh-1, EWTh-2, or EWTh-3),** used for mild steels, alloy steels, and stainless steels using DCEN (DCSP) current.

Select the size electrode rated for the amperage to be used. *Table 8-4* lists electrode current ranges by electrode size using argon shielding gas.

Table 8-4

Electrode Current Ratings

Electrode Dia. (inches)	DCSP EWP EWTh-1 EWTh-2 EWTh-3	DCRP EWP EWTh-1 EWTh-2 EWTh-3	ACHF EWP	ACHF EWTH-1 EWTh-2 EWZr	ACHF EWTh-3
0.020	5-20	-	5-15	5-20	-
0.040	15-80	-	10-60	15-80	10-80
1/16	70-150	10-20	50-100	70-150	50-150
3/32	150-250	15-30	100-160	140-230	100-230
1/8	250-400	25-40	150-210	225-325	150-325
5/32	400-500	40-55	200-275	300-400	200-400
3/16	500-750	55-80	250-350	400-500	250-500
1/4	750-1000	80-125	325-450	500-630	325-630

8.7.0   PREPARING THE ELECTRODE

The different types of GTAW welding current require different shaped electrode ends (tips). For manual DCSP welding, a taper ground point (pencil end) is the best shape. For manual DCRP welding, a rounded end is the best shape. For manual AC welding, a rounded end (hemispherical or balled) is the best shape.

Figure 8-5 shows GTAW electrode end shapes for different current types.

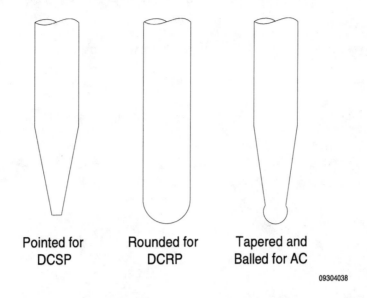

Pointed for
DCSP

Rounded for
DCRP

Tapered and
Balled for AC

09304038

Figure 8-5, GTAW Electrode End Shapes For Different Current Types.

Note   Do not prepare the end of the electrode that contains the color code band or you will not be able to identify the electrode's type later.

Electrode tips must be formed on all new electrodes and on used electrodes with damaged or contaminated tips. On contaminated electrodes, the contaminated section must be cleanly broken off before the new tip is formed. This can be done by striking the electrode with a hammer while it is held against a sharp metal edge, or by breaking it between two pair of pliers.

8.7.1   Pointing The Electrode Tip

For DCSP welding, electrodes may be pointed by dipping the red hot end into a special chemical powder or tapered by grinding or sanding.

To use a chemical power to sharpen a tungsten follow these steps:

Step 1   Place the tungsten in the torch so it extends about one inch from the gas nozzle.

Step 2   Heat the tungsten by shorting it on the work or by striking an arc. Be sure shielding gas is flowing to protect the tungsten.

Step 3   When the end of the electrode is cherry red up the electrode a distance of about four times the electrode diameter, place the end of the electrode in the chemical powder.

Step 4 Hold the end of the electrode in the powder for the time recommended on the container for the diameter electrode being pointed. The chemical powder will dissolve the tungsten forming a sharp point at the proper angle.

Figure 8-6 shows sharpening a tungsten with chemical powder.

09304039

Figure 8-6, Sharpening A Tungsten With Chemical Powder

More often tungsten electrodes are sharpened by grinding or sanding. Since tungsten electrodes are very hard, they are best ground with silicon carbide or alumina oxide grinding wheels or sanding belts. Use 80 grit for fast shaping and 120 grit for finishing.

To point the end of a tungsten electrode with a grinder follow these steps:

Step 1 With the color band end held farthest from the grinder, hold the electrode against the grinder so that the electrode is in the plane of the wheel and makes an angle of about 30 degrees with the wheel or belt surface.

Figure 8-7 shows the proper grinding angle and electrode orientation.

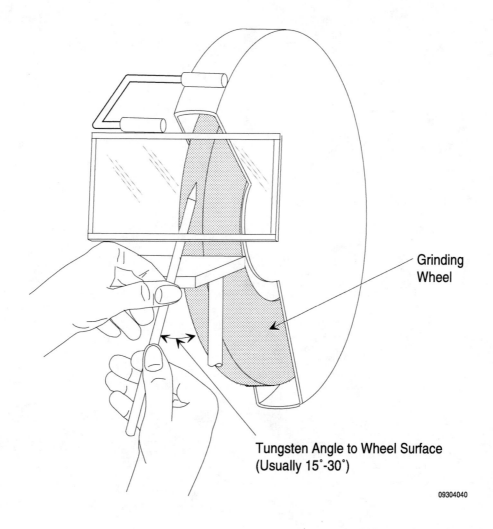

Figure 8-7, Proper Grinding Angle And Electrode Orientation.

Step 2   Slowly rotate the electrode as it is ground to keep the point on the centerline of the axis. The grind marks must be orientated toward the point, along the long dimension of the electrode and not around the point.

Note   For DCSP (DCEN) electrodes, the taper length should be between two and three diameters in length. On larger electrodes, the point should be slightly blunted.

Step 3   Check the ground end to make sure the taper length is correct and that the point is in the center of the electrode and not toward one side (asymmetrical). Correct the shape if necessary.

Figure 8-8 shows properly and improperly prepared pointed electrode ends.

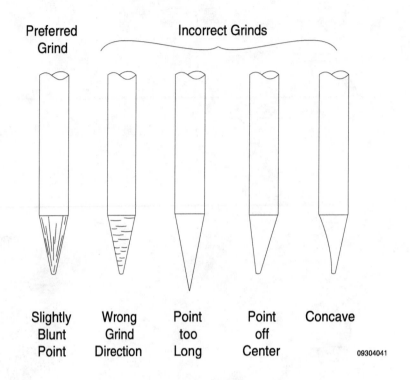

Figure 8-8, Properly and Improperly Prepared Pointed Electrode Ends.

8.7.2   Balling The Electrode Tip

Only pure tungsten (EWP) or zirconiated tungsten (EWZr) can be balled. The thorium in thoriated tungstens inhibits the formation of a ball. If an attempt is made to ball a thoriated tungsten, instead of a ball forming, the thoriated tungsten will form a number of small protrusions at the tip. These protrusions cause the arc to be unstable.

If the electrode is to be used for lower amperage DCRP (DCEP) welding, the electrode end should be tapered about two diameters in length and be quite blunt. A ball will form on the blunt end when welding.

For AC welding, the end is balled (rounded) without tapering. Balling is done by arcing the electrode against clean copper or other clean metal.

CAUTION   Do not arc against carbon as it will contaminate the electrode.

AC or DCEP (DCRP) can be used to ball an electrode. When balling an electrode, start off with a low current and then gradually increase it until the end of the electrode starts to melt forming a hemispherical end or slight ball. The ball should be no more than one and one half the electrode diameter. Using DCEP (DCRP) will cause a ball to form at a much lower amperage setting than using AC current.

CAUTION   Using excess amperage will cause the ball to melt and drop off the electrode. Be especially careful when using DCRP. DCRP will generate much more heat on the electrode than with the same AC current setting.

Figure 8-9 shows properly balled electrode ends.

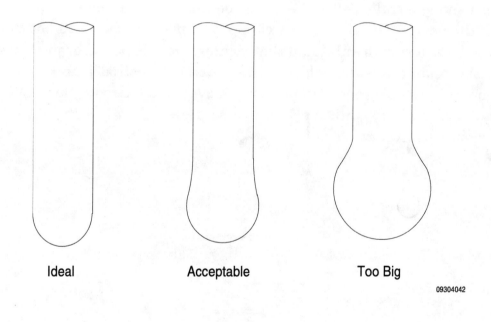

Ideal   Acceptable   Too Big

09304042

Figure 8-9, Properly Balled Electrode Ends.

8.8.0   SELECTING AND INSTALLING THE NOZZLE

Gas nozzles are designed to fit specific model torches. Usually one manufacture's nozzles will not fit another manufacturer's torches. Typically, a range of nozzle sizes and styles are available for a given torch to accommodate different electrode sizes and welding applications. Nozzles are made for close clearances, for long reaches, for reaching deep into narrow places and for covering wide beads. Nozzles with chips, cracks or metal build up on the end should be discarded because they can affect the gas flow pattern and produce weld defects.

The nozzle must be sized for the electrode diameter to be used and for any unique welding requirement, such as reaching into a V-groove. The nozzle diameter must be large enough to cover the entire weld area with shielding gas. A wide easy flow of gas is generally preferable to a narrow jet which is apt to create turbulence and draw atmosphere into the weld zone. Small diameter nozzles tend to overheat and to break easily. However, a small diameter nozzle is useful for steadying the torch and arc for the root pass in a V-groove weld.

Since nozzles are sized differently by various manufactures, refer to the manufactures nozzle recommendations for the torch being used to select the proper size and shape nozzle. In general, the nozzles inside diameter should be about four to six times the diameter of the electrode being used.

8.9.0 INSTALLING THE ELECTRODE

The electrode is usually installed after the nozzle because the electrode stickout is measured from the end of the nozzle. The electrode is clamped in place in the torch body by the collet assembly. The collet is a cylindrical clamp that tightens as it is pulled or pushed (depending on the torch design) into a tapered holder as the threaded electrode cap is screwed on and tightened. To adjust the electrode stickout or remove the electrode, the cap assembly is loosened. Collet designs vary with torch manufacturers. *Figure 8-10* shows the collet assembly of a typical GTAW torch.

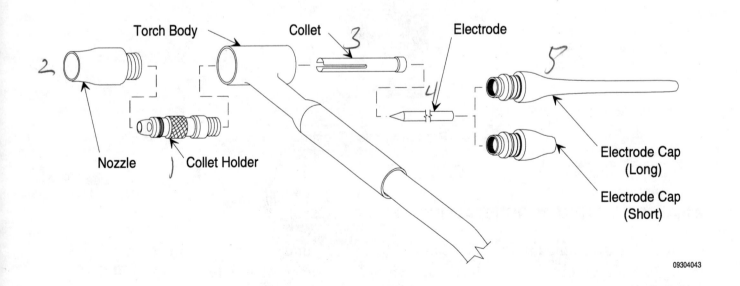

09304043

Figure 8-10, Collet Assembly of a Typical GTAW Torch.

WELDING THREE TRAINEE TASK MODULE 09304

To install the electrode follow these steps:

Step 1 Remove the electrode cap from the rear of the torch.

Step 2 Check the collet size, if it is not the correct size for the electrode to be installed, remove and replace it with the correct size.

Step 3 Replace the electrode cap but do not tighten. Screw the cap only about one turn to leave the collet clamp loose.

Step 4 Insert the electrode into the collet from the nozzle end of the torch and adjust it so that it extends from the nozzle about 1-1/2 times the electrode diameter. Insert the color coded end first to prevent destroying the electrode markings.

Step 5 Tighten the electrode cap to lock the electrode in place.

8.10.0 LOCATING THE GROUND CLAMP

The ground clamp must be properly located to prevent damage to surrounding equipment. If the electrical welding current travels through a bearing, seal, valve or contacting surface it could cause severe damage from heat and arcing. This would require that these items be replaced. Carefully check the area to be welded and position the ground clamp so the welding current will not pass through any contacting surface. If in doubt, ask your supervisor for assistance before proceeding.

CAUTION   Welding current can severely damage bearings, seals, valves or contacting surfaces. Position the ground clamp to prevent welding current from passing through them.

CAUTION   Welding current passing through electrical or electronic equipment will cause severe damage. Before welding on any type of mobile equipment, the ground lead at the battery must be disconnected to protect the electrical system.

WARNING!   Do not weld near batteries. A welding spark could cause a battery to explode, showering the area with battery acid.

CAUTION   The slightest spark of welding current can destroy electronic or electrical equipment. Have an electrician check the equipment and, if necessary, isolate the system before welding.

Ground clamps must never be connected to pipes carrying flammable or corrosive materials. The welding current could cause overheating or sparks, resulting in an explosion or fire.

The ground clamp must make a good electrical contact when it is connected. Dirt and paint will inhibit the connection and cause arcing resulting in overheating of the ground clamp. Dirt and paint also affect the welding current and can cause defects in the weld. Clean the surface before connecting the ground clamp. If the ground clamp is damaged and does not close securely onto the surface, replace it.

8.11.0   ENERGIZE THE POWER SUPPLY

Electrically powered welding machines are energized by plugging them into an electrical outlet. The electrical requirements, (primary current) will be on the equipment specification tag displayed prominently on the machine. Most machines will require single-phase 230-volt current or 460-volt three-phase current. Machines requiring single-phase 230-volt power will have a three-prong plug. Machines requiring three-phase 460-volt power will have a four-prong plug. *Figure 8-11* shows welding machine electrical plugs.

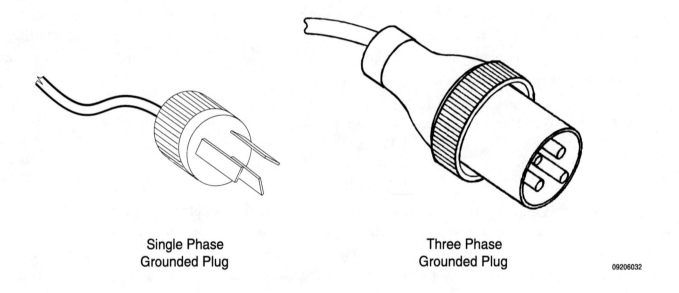

Single Phase
Grounded Plug

Three Phase
Grounded Plug

09206032

Figure 8-11, Welding Machine Electrical Plugs.

If a welding machine does not have a power plug, an electrician must connect it. The electrician will add a plug or hard-wire the machine directly into a electrical box.

8.12.0 STARTING ENGINE-DRIVEN GENERATORS OR ALTERNATORS

Before welding can take place with an engine-driven generator or alternator the engine must be checked and then started. As with a car engine, the engine powering the generator or alternator must also have routine maintenance performed.

8.12.1 Pre-Start Checks

Many sites will have pre-start check lists which must be completed and signed prior to starting or operating an engine-driven generator or alternator. Check with your supervisor. If your site has such a check list, complete and sign it. If your site does not have a pre-start check list, perform the following checks before starting the engine.

- Check the oil using the engine oil dipstick. If the oil is low, add the appropriate grade oil for the time of year.
- Check the coolant level in the radiator if the engine is liquid-cooled. If the coolant level is low, add coolant.

CAUTION　　Do not add plain water to radiators that contain antifreeze. Antifreeze not only protects radiators from freezing in cold weather. It also has rust inhibitors and additives to aid cooling. If the antifreeze is diluted, it will not function properly. If the weather turns cold the system may freeze, causing damage to the radiator, engine block and water pump.

- Check the fuel. The unit may have a fuel gauge or a dipstick. If the fuel is low, add the correct fuel (diesel or gasoline) to the fuel tank. The type of fuel required should be marked on the fuel tank. If it is not marked, contact your supervisor to verify the fuel required and have the tank marked.

CAUTION　　Adding gasoline to a diesel engine or diesel to a gasoline engine will cause severe engine problems. It can also cause a fire hazard. Always be sure to add the correct fuel to the fuel tank.

- Check the battery water level unless the battery is sealed. Add de-mineralized water if the battery water level is low.
- Check the electrode holder to be sure it is not grounded. If the electrode holder is grounded it will arc and overheat the welding system when the welding machine is started. This is a fire hazard and will cause damage to the equipment.
- Open the fuel shut-off valve if the equipment has one. If there is a fuel shut-off valve it will be located in the fuel line between the fuel tank and the carburetor.
- Record the hours from the hour meter if the equipment has one. An hour meter records the total number of hours the engine runs. This information is used to determine when the engine needs to be serviced. The hours will be displayed on a gauge similar to a odometer.
- Clean the unit. Use a compressed air hose to blow off the engine and generator or alternator. Use a rag to remove heavier deposits that cannot be removed with the compressed air.

WARNING!   Always wear eye protection when using compressed air to blow dirt and debris from surfaces. NEVER point the nozzle at yourself or anyone else.

Note   Cleaning may not be required on a daily basis. Clean the unit as required.

8.12.2   Starting the Engine

Most engines will have an on/off ignition switch and a starter. They may be combined into a key switch similar to the ignition on a car. To start the engine, turn on the ignition switch and press the starter. Release the starter when the engine starts. The engine speed will be controlled by the governor. If the governor switch is set for idle the engine will slow to an idle after a few seconds. If the governor is set to welding speed the engine will continue to run at the welding speed.

Small engine-driven alternators may have an on/off switch and a pull cord. These are started by turning on the ignition switch and pulling the cord similar to starting a lawn mower.

Engine-driven generators and alternators should be started about five to ten minutes before they are needed for welding. This will allow the engine to warm up before a welding load is placed on it.

8.12.3 Stopping the Engine

If no welding is required for thirty or more minutes, stop the engine by turning off the ignition switch. If you are finished with the welding machine for the day, also close the fuel valve if there is one.

8.12.4 Preventive Maintenance

Engine-driven generators and alternators require regular preventive maintenance to keep the equipment operating properly. Most sites will have a preventive maintenance schedule based on the hours that the engine operates. In severe conditions such as very dusty or cold weather, maintenance may have to be performed more frequently.

CAUTION   To prevent equipment damage, perform preventive maintenance as recommended by the site procedures or manufacturer's
maintenance schedule in the equipment manual.

The responsibility for performing preventive maintenance will vary by site. Check with your supervisor to determine who is responsible for performing preventive maintenance.

When performing preventive maintenance follow the manufacturer's guidelines in the equipment manual. Typical items to be performed as a part of preventive maintenance include:

- Changing the oil
- Changing the gas filter
- Changing the air filter
- Checking/changing the antifreeze
- Greasing the undercarriage
- Re-packing the wheel bearings

9.0.0 TOOLS FOR WELD CLEANING

The tools used for weld cleaning include hand-held tools such as chipping hammers, wire brushes and pliers, and power tools such as pneumatic weld flux chippers and needle scalers.

9.1.0 HAND TOOLS

Wire brushes are used to clean welds and to remove paint or surface corrosion. Wire brushing will remove light to medium corrosion, but will not remove tight corrosion. Tight corrosion must be removed by filing, grinding, or sand blasting.

Chipping hammers are used to remove cutting and welding slag. The head of a chipping hammer has a point at one end and a chisel at the other. When chipping hammers become dull they can be sharpened on a grinder. When sharpening chipping hammers use care not to overheat the head. The head is hardened by tempering and overheating will remove the temper, causing the head to become soft. If the temper is removed from a chipping hammer head it will mushroom and wear out much faster than a tempered head. Prevent overheating by plunging the chipping hammer's head into a pail of water every few seconds while grinding.

Welders should also have pliers to handle hot metal. Hot metal should not be handled with leather welding gloves. The heat will cause the leather to shrivel and become hard. *Figure 9-1* shows wire brushes, chipping hammers and pliers.

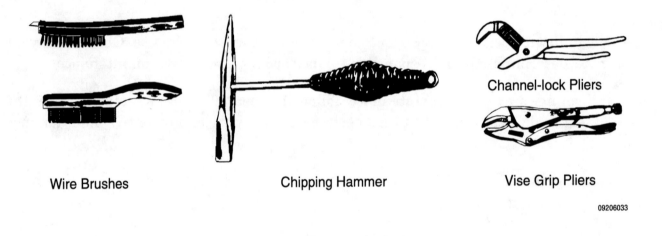

Wire Brushes   **Chipping Hammer**   **Channel-lock Pliers**

Vise Grip Pliers

09206033

Figure 9-1. Wire Brushes, Chipping Hammer and Pliers.

9.2.0 WELD FLUX CHIPPERS AND NEEDLE SCALERS

Weld flux chippers and needle scalers are pneumatically powered. They are used by welders to clean surfaces and to remove slag from cuts and welds. Weld flux chippers and needle scalers are also excellent for removing paint or hardened dirt but are not very effective for removing surface corrosion. Weld flux chippers have a single chisel, and needle scalers have about 18 to 20 blunt steel needles approximately ten inches long. Most weld flux chippers can be converted to needle scalers with a needle scaler attachment.

WELDING THREE TRAINEE TASK MODULE 09304

Figure 9-2 shows a weld flux chipper and a needle scaler.

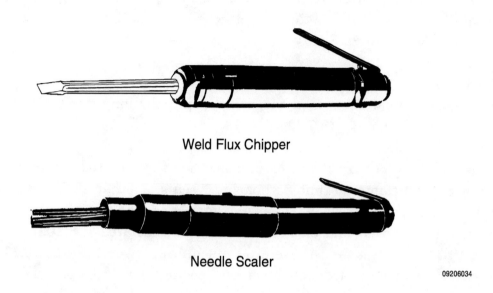

Weld Flux Chipper

Needle Scaler

09206034

Figure 9-2. Weld Flux Chipper and Needle Scaler.

SELF-CHECK REVIEW 3

1. What must be done before attempting to move a GTAW welding machine?
2. Why is it important not to use an excessively high shielding gas flow rate?
3. What type of tungsten should be used to weld stainless steel?
4. How should the end of the tungsten be prepared for AC welding?
5. When grinding a point on an electrode tip, which direction should the grind marks run?
6. As a general rule, what should the inside diameter of the gas nozzle be?
7. Before installing an electrode what should the collet be checked for?
8. When placing a ground clamp, how do you avoid damaging bearings, seals and other heat or current damageable components with the welding current?
9. What should you locate before using an electrically powered welding machine?
10. What should you always check before starting an engine-driven welding machine?
11. Why should you never fuel a diesel engine with gasoline or fuel a gasoline engine with diesel fuel?

ANSWERS TO SELF-CHECK 3

1. The shielding gas cylinder must be removed or safely secured. (8.3.0)
2. An excessively high gas flow rate wastes expensive gas and can generate turbulence in the gas column. The turbulence can pull atmosphere into the weld zone causing oxidation and weld contamination. (8.5.0)
3. Thoriated. (8.6.0)
4. Balled. (8.7.0)
5. The grind marks must be toward the point, along the long dimension of the electrode and not around the point. (8.7.1)
6. About four to six times the electrode diameter. (8.8.0)
7. Check the collet to be sure it the correct size for the electrode to be used. (8.9.0)
8. Locate the ground clamp so that no bearings, seals or other easily damaged items are in the electrical path between the weld zone and the clamp. Batteries should be disconnected or removed and electronics should be isolated. (8.10.0)
9. The electrical disconnect for the welding machine's power circuit. (8.11.0)
10. Always check the engine lube oil level and coolant levels before starting an engine. Operating an engine with insufficient oil or coolant can destroy the engine. (8.12.1)
11. Operating a diesel engine on gasoline or a gasoline engine on diesel fuel will cause severe engine problems. (8.12.1)

SUMMARY

Gas tungsten arc welding can be dangerous if the proper safety precautions are not followed. Before proceeding be sure you understand and can follow the safety precautions presented in this module. Before GTAW can take place, the appropriate equipment must be selected and set up. By following the recommendations in this module, you will be able to select and safely set up the appropriate welding equipment.

References

For advanced study of topics covered in this Task Module, the following works are suggested:

Welding Skills, Giachino and Weeks, American Technical Publishers Inc., Homewood, IL, 1985, 1-800-323-3471.

Welding Principles and Applications, Jeffus and Johnson, Delmar Publishers, Inc., 2 Computer Drive West, Box 15-015, Albany, N.Y. 12212, Phone 1-800-347-7707.

Modern Welding, Althouse, Turnquist, Bowditch, and Bowditch, Goodheart-Willcox Publishers, South Holland, IL, 1988. Phone 1-800-323-0440.

Basic Tig & Mig Welding (GTAW & GMAW), Griffin, Roden and Briggs, Delmar Publishers, Inc., 2 Computer Drive West, Box 15-015, Albany, N.Y. 12212, Phone 1-800-347-7707.

GTAW Handbook, William H. Minnick, Goodheart-Willcox Publishers, South Holland, IL, 1988. Phone 1-800-323-0440.

PERFORMANCE / LABORATORY EXERCISES

None

The NCCER makes every effort to keep these manuals up-to-date and free of technical errors. We appreciate your help in this process. If you have an idea for improving this manual, or if you find an error, a typographical mistake, or an inaccuracy in the *Wheels of Learning*, please write us, using this form or a photocopy. Be sure to include the exact module number, page number, a description of the problem, and the correction, if possible. We'll do our best to correct it in later editions. Thank you for your assistance.

Write: *Wheels of Learning*
National Center for Construction Education and Research
P.O. Box 141104
Gainesville, FL 32614-1104
Fax: 352-334-0932

WHEELS OF LEARNING USER UPDATE

Please let us know if you have found an inaccuracy, error, or other problem in a *Wheels of Learning* manual. Use this form or write us a letter. Please be sure to tell us the exact module name and module number, the page number, and the problem. Thanks for your help.

Craft _____   Module Name _____

Module Number _____   Page Number(s) _____

Description of Problem _____

_____

_____

(Optional) Correction of Problem _____

_____

_____

(Optional) Your Name and Address _____

_____

_____

GTAW -- Plate

Module 09305

NATIONAL CENTER FOR CONSTRUCTION EDUCATION AND RESEARCH

GAS TUNGSTEN ARC WELDING (GTAW) - PLATE

OBJECTIVES

Upon completion of this module, the trainee will be able to:

1. Pad in all positions with stringer beads using GTAW and carbon steel filler rod.
2. Make multi-pass V-butt open-groove welds on mild steel plate in the 1G (flat) position using GTAW and carbon steel filler rod.
3. Make multi-pass V-butt open-groove welds on mild steel plate in the 2G (horizontal) position using GTAW and carbon steel filler rod.
4. Make multi-pass V-butt open-groove welds on mild steel plate in the 3G (vertical) position using GTAW and carbon steel filler rod.
5. Make multi-pass V-butt open-groove welds on mild steel plate in the 4G (overhead) position using GTAW and carbon steel filler rod.

Prerequisites

Successful completion of the following module(s) is required before beginning study of this module:

- *Weld Quality # 09107*
- *Gas Tungsten Arc Welding - Equipment and Filler Metals # 09304*

Required Student Materials

Each trainee will need:

1. Personal protective equipment
2. Leather welding gloves
3. Welding shield
4. GTAW equipment
   - Power supply
   - Torch
   - Torch nozzles
   - Tungsten electrodes (EWTh-2, 3/32- and 1/8-inch)

5. Welding table with positioning arm
6. Carbon steel filler rod, 3/32- and 1/8-inch diameter
7. Welding grade argon shielding gas
8. Cutting goggles
9. Chipping hammer
10. Wire brush
11. Pliers
12. Tape measure
13. Soapstone
14. Mild steel plate, 1/4 inch to 3/4 inch thick

Each trainee will need access to:

1. Oxyfuel cutting equipment
2. Framing square
3. Grinders
4. Bench grinder (for Tungsten)

Course Map Information

This course map shows all of the *Wheels of Learning* task modules in the third level of the Welding curricula. The suggested training order begins at the bottom and proceeds up. Skill levels increase as a trainee advances on the course map. The training order may be adjusted by the local Training Program Sponsor.

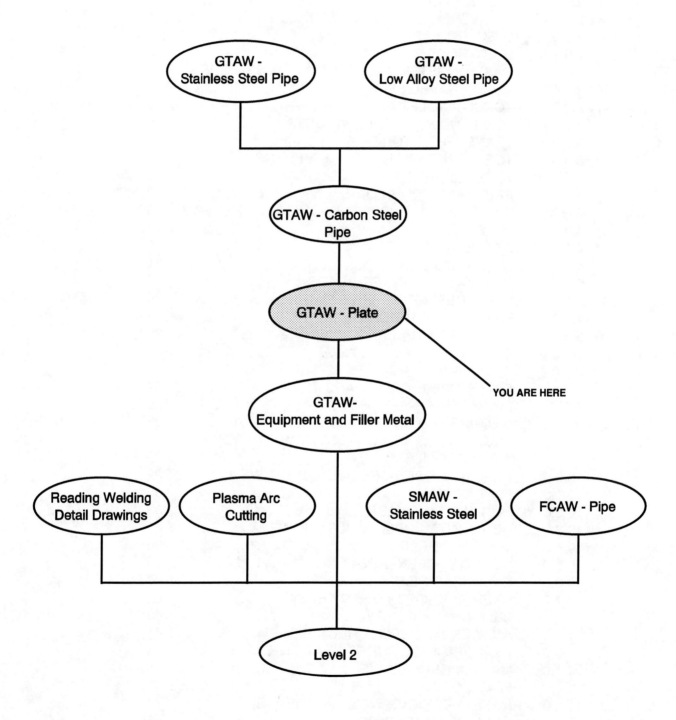

TABLE OF CONTENTS

Trade Terms Introduced In This Module

Arc Voltage: The voltage across the welding arc.

cfh: Abbreviation for cubic feet per hour.

Direct Current Electrode Negative (DCEN): Also called Direct Current Straight Polarity (DCSP). The DC welding lead arrangement where the base metal (workpiece) lead is connected to the positive (+) welding machine post and the torch lead is connected to the negative (-) welding machine post.

Direct Current Electrode Positive (DCEP): Also called Direct Current Reverse Polarity (DCRP). The DC welding lead arrangement where the base metal (workpiece) lead is connected to the negative (-) welding machine post and the torch lead is connected to the positive (+) welding machine post.

Electrode Stickout: The length of electrode that extends beyond the torch nozzle (gas cup).

Hot Pass: The first filler pass after the root pass.

Low-Carbon Steel: Steel with a carbon content of 0.05 to 0.25 percent.

Mild Steel: A common name that includes the low- and medium-carbon steels with carbon contents at or below 0.5 percent.

Out-Of-Position Weld: Any of the welding positions except the flat welding position where both the axis and the face of the weld are in the horizontal plane.

Penetration: The distance that the weld fusion line extends below the surface of the base metal.

Weld Coupon: The metal to be welded as a test or practice.

Welding Variables: Items that affect the quality of a weld such as weld voltage, weld current, weld travel rate, shielding gas type, torch angle and electrode stickout.

1.0.0 INTRODUCTION

Gas Tungsten Arc Welding (GTAW), is an arc welding process that uses an arc between a virtually nonconsumable tungsten electrode and the base metal to melt the base metal. The electrode, the arc and the molten base metal are shielded from atmospheric contamination by a flow of inert gas from the torch nozzle. Filler metal is an uncoated rod that is usually hand-held and manually feed into the leading edge of the weld puddle. GTAW produces high quality welds without slag or oxidation. Since there is no flux, there can be no corrosion due to flux entrapment and no post weld cleaning is necessary. *Figure 1-1* shows the GTAW process.

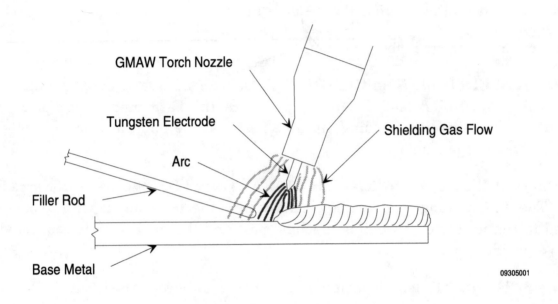

GMAW Torch Nozzle

Tungsten Electrode

Shielding Gas Flow

Arc

Filler Rod

Base Metal

09305001

Figure 1-1. GTAW Process.

This module explains how to set up GTAW equipment and perform open-groove V-butt welds on mild steel plate with carbon steel filler rod in the 1G, 2G, 3G and 4G welding positions.

Figure 1-2 shows the 1G, 2G, 3G and 4G welding positions.

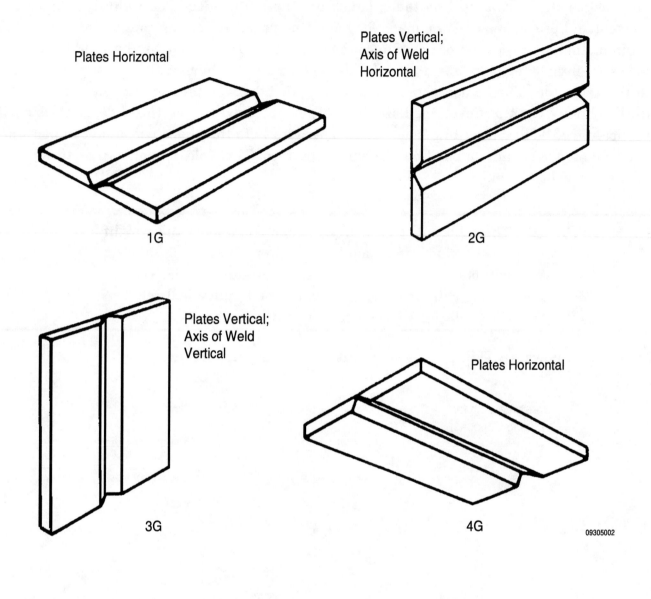

Plates Horizontal

1G

Plates Vertical;
Axis of Weld
Horizontal

2G

Plates Vertical;
Axis of Weld
Vertical

3G

Plates Horizontal

4G

09305002

Figure 1-2. 1G, 2G, 3G and 4G Welding Positions.

2.0.0   GTAW WELDING EQUIPMENT SET-UP

Before welding can take place the work area has to be made ready, the welding equipment set up and the metal to be welded prepared. The following sections will explain how to prepare the area and set up the equipment to perform GTAW welding of mild or low-carbon steel plate.

2.1.0   PREPARE WELDING AREA

To practice welding, a welding table, bench or stand is needed. The welding surface must be steel and provisions must be made for mounting practice welding coupons out of position. A simple mounting for out of position coupons is a pipe stand that can be welded vertically to the steel table top. To make a pipe stand, weld a three- or four-foot length of pipe vertically to the table top. Then cut a short section pipe to slide over the vertical pipe. Drill a hole in the slide. Weld a nut over the hole so a bolt can be used to lock the slide in place on the vertical pipe. Weld a piece of pipe or angle horizontally to the slide. The slide can be rotated and adjusted vertically to position the horizontal arm. The weld coupon can be tack-welded to the horizontal arm.

WARNING!   The table must be heavy enough to support the weight of the welding coupon extended on the horizontal arm without falling over. It must also support the coupon during chipping or grinding. Serious injury will result if the table falls onto someone in the area.

Figure 2-1 shows a welding table with out-of-position support arm.

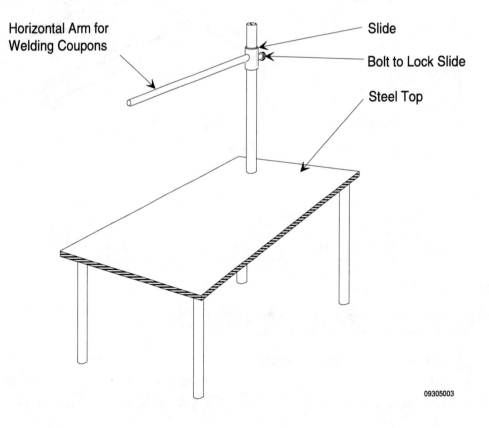

Horizontal Arm for Welding Coupons

Slide

Bolt to Lock Slide

Steel Top

09305003

Figure 2-1. Welding Table With Out-Of-Position Support Arm.

To set up the area for welding follow these steps.

Step 1   Check to be sure the area is properly ventilated. Make use of doors, windows and fans.

Step 2   Check the area for fire hazards. Remove any flammable materials before proceeding.

Step 3   Check the location of the nearest fire extinguisher. Do not proceed unless the extinguisher is charged and you know how to use it.

Step 4   Position a welding table within convenient distance to the GTAW welding machine.

Step 5   Set up flash shields around the welding area.

2.2.0   PREPARE WELDING COUPONS

Cut the welding coupons from 1/4 to 3/4 inch thick mild or low-carbon steel. Use a wire brush or grinder to remove any heavy mill scale or corrosion. Prepare welding coupons to practice the welds indicated as follows:

- Stringer Beads: The coupons can be any size or shape that can be easily handled.
- Open V-Butt Joint: The metal should be cut into three- by eight-inch rectangles with one long edge beveled at 30 degrees. Grind a 1/16-inch land on the beveled edges. Two of these rectangles will be needed to make one joint.

Note   The land can be from a feather (no land) up to 1/8-inch. Adjust the land size as needed during practice welding to obtain the best results.

Open root welds generally use a bevel angle of 30 degrees (forms a 60-degree included angle) or 37-1/2 degrees (forms a 75-degree included angle).

Note   The examples used in this student manual show a 30-degree bevel. Depending on your site requirements, a 37-1/2 degree bevel can be substituted. Check with your instructor to determine if you should use the 37-1/2 degree bevel instead of the 30-degree bevel.

Tack-weld the coupon beveled edges together with a 3/32 inch root space. Use a piece of 3/32 inch rod to space the coupons while tacking.

Note   The root opening can be 1/8-inch plus or minus 1/16-inch wide (1/16- to 3/16-inch). Adjust the root opening as needed during practice welding to obtain the best results.

Figure 2-2 shows a open V-butt joint welding coupon.

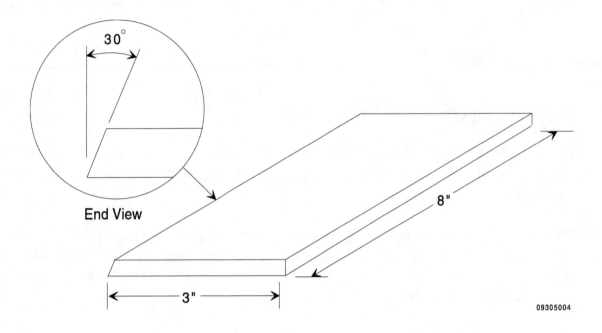

Figure 2-2. Open V-Butt Joint Welding Coupon.

Note   Steel for practice welding is expensive and difficult to obtain. Every effort should be made to conserve and not waste the material that is available. Reuse weld coupons until all surfaces have been welded on. Weld on both sides of the joint and then cut the weld coupon apart and reuse the pieces. Use material that can not be cut into weld coupons to practice running beads.

2.3.0 FILLER ROD

Filler rods are selected to be compatible with the base metal to be welded. Often, the WPS or job site standards will specify the filler rod type and size to be used.

CAUTION   When the WPS or Site Quality Standards specify a filler rod it must be used. Severe penalties will result from not following a WPS or Site Quality Standard if available.

For the welding exercises in this module, 3/32- and/or 1/8-inch carbon steel filler rods classification ER70S-2 or equivalent should be used to weld the mild steel coupons. Remove only a small number of filler rods at a time. Keep the remainder in its package to keep it clean. Before using the filler rod, check it for contamination such as corrosion, dirt, oil or grease, or burnt ends. All of these forms of contamination will cause weld defects. Clean the filler rod with a clean oil free rag and snip burnt ends. If the filler rod cannot be cleaned, do not use it.

2.4.0 GTAW WELDING EQUIPMENT

Locate the following GTAW equipment:

- Constant current power supply (specifically designed for GTAW if possible)
- GTAW torch and cable assembly
- Collet assembly (3/32- and 1/8-inch)
- Tungsten electrodes, EWTh-2 (3/32- and/or 1/8-inch)
- Gas nozzles for torch
- Shielding gas cylinder and regulator-flow meter

Figure 2-3 shows a complete GTAW system.

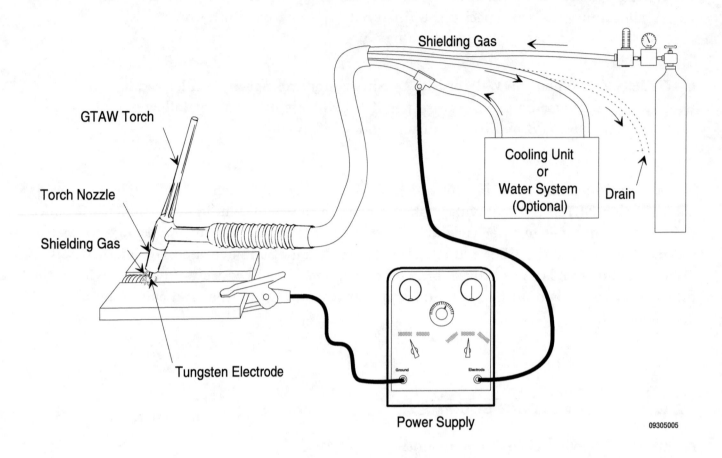

Figure 2-3. Complete GTAW System.

Perform the following steps to set up the GTAW equipment.

Step 1 Verify that the welding machine is a constant current type DC or AC/DC power supply.

Step 2 Check to be sure that the welding machine is properly grounded through the primary current receptacle.

Step 3 Verify the location of the primary current disconnect.

Step 4 Plug the power source into the power receptacle.

Step 5 Locate a GTAW torch, and if the power supply is not a dedicated GTAW machine, make sure the GTAW torch is adapted to connect to the power supply (special power cable lug).

Step 6 Place the 1/8-inch collet in the torch.

Note Depending on site conditions, a 3/32-inch electrode can be used. Check with your instructor to determine if you should use a 3/32-inch electrode instead of the 1/8-inch.

Step 7 Check the nozzle to be sure it is the correct size for the electrode to be used and that it is in good condition (no cracks or chips).

Note Since nozzle numbering systems vary by manufacture, refer to the torch manufactures data to identify the correct nozzle number (size) for the electrode being used.

Step 8 Connect the torch to the power supply.

- Power cable
- Shielding gas
- Cooling water (if used)

Step 9 If used, connect the remote foot control cable to the power supply and set the remote/local switch to remote.

Step 10 Set up the welding machine for direct current straight polarity (DCSP).

Step 11 Set the high frequency (if equipped) to "START".

Step 12 Locate a cylinder of welding grade argon gas and secure it in an upright position near the power supply.

Step 13 Remove the protective cap and clear the valve of visible debris. Momentarily crack the cylinder valve to blow out any remaining dirt.

Step 14 Install the gas regulator/flowmeter and connect the shielding gas hose to the "Gas In" fitting on the welding machine or to the torch shielding gas supply line (power lug) if using a non-GTAW power supply.

Step 15 Turn on the welding machine.

Step 16 Fully open the shielding gas cylinder valve and then open the flow meter and activate the purge control on the welding machine to purge the torch line and torch for a few seconds. After purging, adjust the gas flowmeter to the specified flow rate (15-20 cfh).

Step 17 Set the welding amperage as follows:

- Root pass   70-95 amperes

- Filler passes   90-110 amperes

  (Use higher end of ranges for 1G and lower end of ranges for 3G)

Note If a remote current control is being used, the amperage set on the power supply will be the maximum amperage available when the remote control is fully depressed. Set the amperage slightly higher so additional current will be available if required for the start of the weld, misalignment or at tacks.

Step 18 Set any other applicable variables. Refer to the manufacturer's specifications and instructions for setting:

- Hot start current
- Gas preflow (prepurge) duration
- Gas postflow (post purge) duration
- High frequency intensity
- Pulse current
- Current slope up
- Current slope down

2.5.0 ELECTRODE PREPARATION

The EWTh-2 tungsten electrodes used with DCSP should be taper ground with a dulled point. The taper length should be between two and three times the electrode diameter. The electrode should be ground so that the grind marks are along the direction of the electrode axis. Do not grind around the point. Also, do not grind the color-coded end of the tungsten or the identification code will be lost.

Note Either EWTh-1 or EWTh-3 electrodes can be used for carbon steel welding if EWTh-2 electrodes are not available.

Figure 2-4 shows a GTAW tungsten electrode properly ground for DCSP.

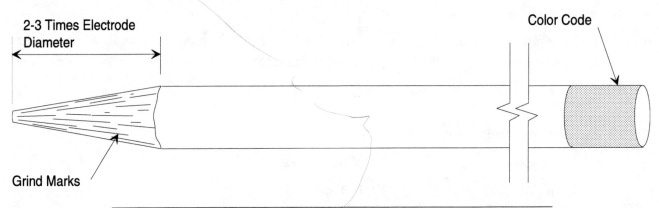

Figure 2-4. GTAW Tungsten Electrode Properly Ground For DCSP.

09305008

2.5.1 Electrode Stickout

Electrode stickout is measured from the end of the gas nozzle to the end of the electrode. It is normally two to three times the electrode diameter. For example, normal stickout for a 1/8 inch diameter electrode would be 1/4 to 3/8 inch (2 x 1/8 inch = 1/4 inch, 3 x 1/8 inch = 3/8 inch). *Figure 2-5* shows GTAW tungsten electrode stickouts.

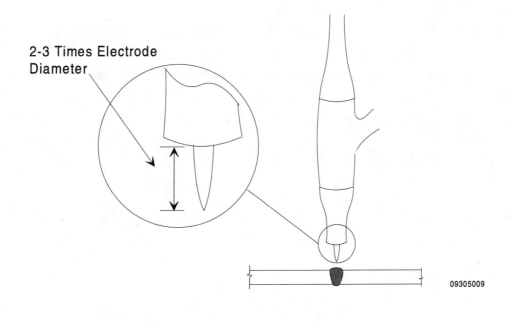

2-3 Times Electrode Diameter

09305009

Figure 2-5. GTAW Tungsten Electrode Stickout.

3.0.0   GAS TUNGSTEN ARC WELDING TECHNIQUES

GTAW weld bead characteristics and quality are affected by several factors resulting from the way the individual performing the welding handles the torch. These factors include:

- Torch travel speed
- Torch angle
- Torch handling techniques

3.1.0   WELD TRAVEL SPEED

Weld travel speed affects the size of the weld puddle and the penetration of the weld. A slow travel speed allows more heat to concentrate and forms a larger, more deeply penetrating puddle. Faster travel speeds prevent heat buildup and forms smaller and shallower puddles.

3.2.0 TORCH ANGLES

There are two basic torch angles that must be controlled when performing GTAW. These are the longitudinal torch angle and the transverse torch angle.

3.2.1 Longitudinal Torch Angles

The longitudinal torch angle is the angle between the torch and the weld axis. GTAW normally uses push longitudinal torch angles. With a push angle, the torch tip points somewhat ahead of the torch body in the direction of weld advancement. Except for vertical welds, GTAW uses longitudinal push angles between 15 and 20 degrees when welding carbon steel. Carbon steel is welded vertical up with a 30 degree push angle. Too great a push angle (torch slanted too much) will tend to draw atmosphere from under the back edge of the nozzle, where it will mix with the shielding gas stream and contaminate the weld.

Figure 3-1 shows correct longitudinal GTAW torch angles for carbon steel.

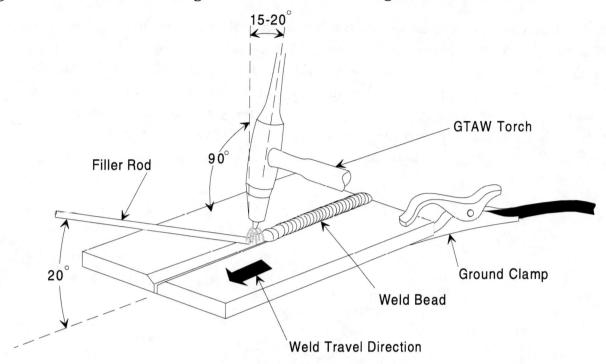

For Position 1G, 2G, and 4G

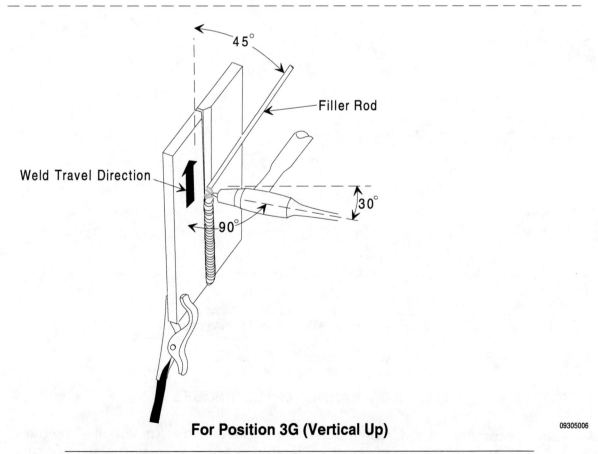

For Position 3G (Vertical Up)

09305006

Figure 3-1. Correct Longitudinal GTAW Torch Angles For Carbon Steel.

3.2.2 Transverse Torch Angles

Transverse torch angle is the angle the torch makes with a line perpendicular to the plate surface across the weld. This angle is adjusted for each bead in order to tie the bead into the side of a previous bead or a beveled edge. For the root bead or first filler bead (hot pass) of a flat groove weld, the transverse angle would is 0 degrees (90 degrees to either plate surface). At times, the torch may have to be transversely angled to heat one side of a V-groove more than the other, especially if the plate thicknesses are not equal. *Figure 3-2* shows various GTAW torch transverse angles.

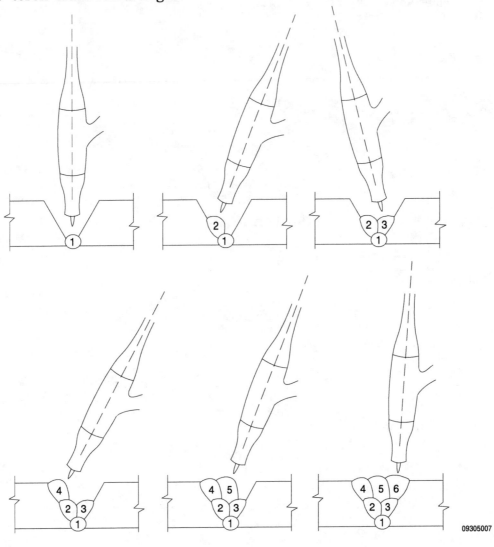

09305007

Figure 3-2. Various GTAW Torch Transverse Angles.

3.3.0 TORCH AND FILLER ROD HANDLING TECHNIQUES

The two basic techniques used when performing GTAW are known as "freehand" and "walking-the-cup". Try both techniques and use the one which gives the best results. Both techniques are explained in the following sections.

3.3.1 Freehand Technique

In the freehand technique the torch electrode tip is held just above the weld puddle or base metal. The torch is supported by the welder's hand which is usually steadied by resting some part of it on or against the base metal to maintain the proper arc length (electrode tip to puddle distance). If required, the welder can move the torch tip in a small circular motion within the molten puddle to maintain the puddle size and advance the puddle. Filler metal is added as it is needed.

The GTAW filler rod is held in the hand not occupied by the torch. For flat, horizontal and overhead position welding, the rod is held at an angle of about 20 degrees above the base metal surface and in line with the weld. Vertical welding is done vertical-up and the rod is held above the torch at an angle of 45 degrees with the base metal surface. In all cases, the tip of the rod is always kept within the shielding gas envelope to protect the rod from atmospheric contamination and keep it preheated. Filler rod is added into the leading edge of the weld puddle using extreme care not to touch the tungsten electrode with the end of the filler rod. If the tungsten electrode touches the filler rod or weld puddle it will become contaminated with filler metal. The electrode must then be removed and cleaned by grinding (or chemically) before proceeding. A technique to prevent contamination of the electrode by the filler rod is to move the electrode to the back edge of the weld puddle as the filler rod is advanced into the leading edge of the weld puddle. As the filler rod end approaches the molten puddle, the end of the filler metal should melt off and flow into the molten puddle. Do not insert the end of the filler rod into the molten puddle and then attempt to melt it off. This can cause hard spots and weld defects. *Figure 3-3* shows the freehand technique.

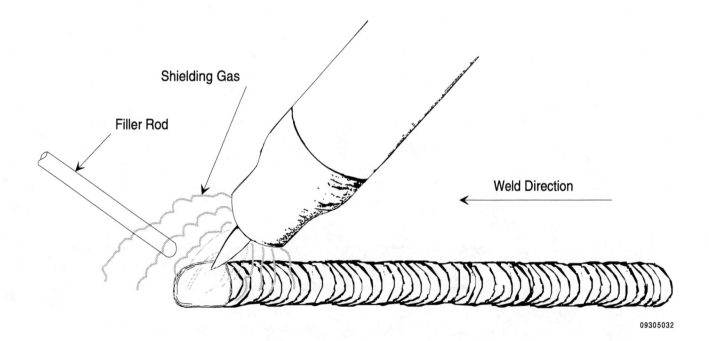

Shielding Gas

Filler Rod

Weld Direction

09305032

Figure 3-3. Freehand Technique.

3.3.2 Walking-The-Cup Technique

In the walking-the-cup technique, the welder rests the edge of the torch nozzle (cup) against the base metal or groove edges to steady the torch and maintain a constant arc length (electrode tip to weld pool distance). The torch is rocked from side to side on the edge of the cup as it is advanced to maintain the puddle size and heat both sides of the groove.

Filler metal is added in the same manner as with the freehand technique using care not to contaminate the electrode.

4.0.0 BEAD TYPES

Two basic bead types may be made with GTAW. They are:

- Stringer beads
- Weave beads

4.1.0 STRINGER BEADS

Stringer beads are made with little or no side-to-side movement of the torch. The bead is generally no more than three times the diameter of the electrode being used. *Figure 4-1* shows a stringer bead.

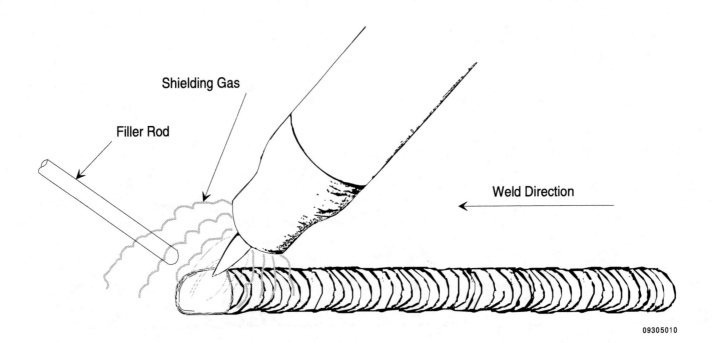

Figure 4-1. Stringer Bead.

4.2.0   WEAVE BEADS

Weave beads are made by working the weld puddle back and forth across the axis of the weld. This produces a weld bead much wider than the puddle width. The width of a weave bead is determined by the amount of cross motion. Weave beads put more heat into the base metal. For this reason, they are not recommended for welding metals sensitive to heat input such as most types of stainless steels.

CAUTION   Check the WPS or Site Quality Standards to determine if stringer or weave beads should be used. Failure to comply with the WPS or Site Quality Standards will result in severe disciplinary action.

Figure 4-2 shows a weave bead.

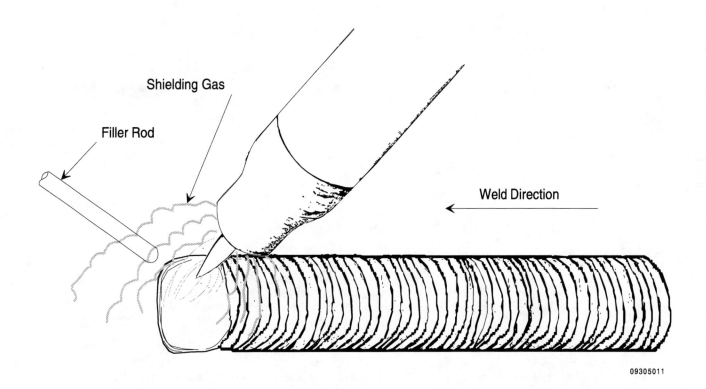

09305011

Figure 4-2. Weave Bead.

Weave beads will not be practiced or used in this training module.

Practice running stringer beads in the flat position. Practice both the freehand and walking-the-cup techniques. Flat position stringer beads should be run with a 15 to 20 degree longitudinal torch push angle. Practice stringer beads without filler rod until you are comfortable with the torch and get consecutive uniform rippled beads.

To run stringer beads with filler rod, the filler rod should be held in front of the weld bead with the end of the filler rod within the torch gas envelope. The rod should be held at a 20 degree angle with the plate surface. *Figure 5-1* shows the flat position torch and filler rod angles.

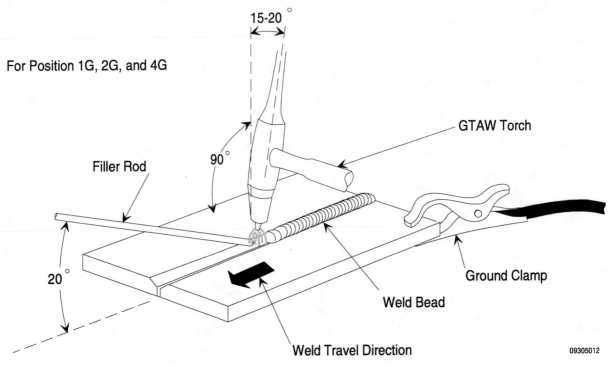

Figure 5-1. Flat Position Torch And Filler Rod Angles.

Follow these steps to run flat position stringer beads.

Step 1 Position the torch at a 15 to 20 degree push angle with the electrode tip directly over the point where the weld is to begin.

Step 2 Energize the electrode and start the shielding gas flowing by pressing the remote control.

Note If the equipment being used does not have remote control the electrode will be energized when the power source is turned on. Start the shielding gas flow by opening the manual shielding gas valve on the torch or in the line.

Step 3 Initiate the arc by bringing the electrode close to the base metal (with high frequency) or by lightly touching the electrode to the base metal (no high frequency).

Step 4 Work the arc slightly until the weld puddle begins to form.

Step 5 While using either the freehand or the walking-the-cup technique, slowly advance the torch and add filler rod as needed.

Step 6 Continue to weld until a bead about 2 to 3 inches long is formed, and then stop the arc without withdrawing the torch.

Note If the equipment being used does not have remote control, quickly lift the torch to break the arc and then immediately return it to shield the weld and the end of the filler rod.

Step 7 Continue to hold the torch in place so that the postflow gas protects the hot weld and filler rod until they are sufficiently cool to prevent oxidation damage.

Step 8 Remove the torch and close the shielding gas valve if the equipment does not have remote control.

Step 9 Inspect the bead for:

- Straightness of the bead
- Uniform rippled appearance of the bead face
- Smooth flat transition with complete fusion at the toes of weld
- No porosity
- No undercut

Step 9 Continue practicing stringer beads until acceptable beads are produced consistently.

If the coupon gets too hot between passes, cool it in water. Cooling with water is only done on practice coupons. Never cool test coupons or any other weld with water. Cooling with water can cause weld cracks and affect the mechanical properties of the base metal.

WARNING! Use pliers to handle the hot practice coupons. Wear gloves when placing the practice coupon in water. Steam will rise off the coupon and can burn or scald unprotected hands.

6.0.0 WELD RESTARTS

A restart is the junction where a new weld connects to continue the bead of a previous weld. Restarts are important because an improperly made restart will create a weld defect. A restart must be made so that it blends smoothly with the previous weld and does not stand out. If possible, avoid restarts by running a bead the full length of the weld joint to be made. Follow these steps to make a restart.

Step 1 Clean the area of the restart of any glass silicon beads that may have formed from the previous weld. Use a chipping hammer and/or wire brush.

Step 2 Hold the torch at the proper angle and arc distance and then restart the arc directly over the center of the crater. (The welding codes do not allow arc strikes outside the area to be welded.)

Step 3 Move the electrode tip in a small circular motion over the crater and add filler rod when the molten puddle is the same size as the crater.

Step 4 As soon as the puddle fills the crater, advance the puddle slightly and continue to add filler metal as needed.

Step 5 Inspect the restart. A properly made restart will blend into the bead, making it hard to detect.

If the restart has undercut, not enough time was spent in the crater to completely melt it or not enough filler rod was added. If the restart is higher than the rest of the bead, too much filler rod was added.

Continue to practice restarts until they are correct. Use the same techniques for making restarts whenever performing GTAW.

7.0.0 WELD TERMINATIONS

A weld termination is made at the end of a weld. When making a termination, the welding codes require that the crater must be filled to the full cross-section of the weld. This can be difficult since most terminations are at the edge of a plate where welding heat tends to build up. This makes filling the crater more difficult. Filling the crater is made much easier if remote control equipment with a potentiometer is used. The potentiometer allows the welding current to be reduced as the end of the weld is approached.

Follow these steps to make a termination using remote control equipment and a potentiometer.

Step 1 As the end of the weld is approached slowly back off (reduce) the welding current. The amount to reduce the welding current is determined by the weld puddle width. If the puddle starts to become wider there is too much heat, if the puddle starts to become narrower there is not enough heat.

Step 2 Continue to back off the welding current while adding filler rod until the crater is filled.

Step 3 Stop the arc and hold the torch in place until the gas postflow (post purge) cools the weld metal and filler rod.

Follow these steps to make a termination without a potentiometer.

Step 1 As the end of the weld is approached, add filler rod at a faster pace. The filler rod will absorb excess heat.

Step 2 Continue to add filler rod until the crater is filled and then stop the arc.

Step 3 Hold the torch in place until the gas postflow (post purge) cools the weld metal and filler rod.

CAUTION Do not remove the torch or filler rod from the torch shielding gas flow until the puddle has solidified and cooled to a temperature where it will not be affected by the atmosphere. The gas post flow that continues after the welding stops protects the molten metal. If the torch and shielding gas are removed before the weld has solidified, crater porosity or cracks can occur. Also the filler rod will be contaminated and the contaminated end will have to be discarded.

Inspect the termination for:

- Crater filled to the full cross-section of the weld
- No crater cracks
- No crater porosity

8.0.0 OVERLAPPING BEADS

Overlapping beads are made by depositing overlapping weld beads parallel to one another. Each successive bead overlaps the toe of the previous bead to form an approximately flat surface. This is also called padding. Overlapping beads are used to build up surfaces (padding) and to make multi-pass welds.

Properly overlapped beads, when viewed from their ends, will form a nearly flat surface. *Figure 8-1* shows proper and improper overlapping beads.

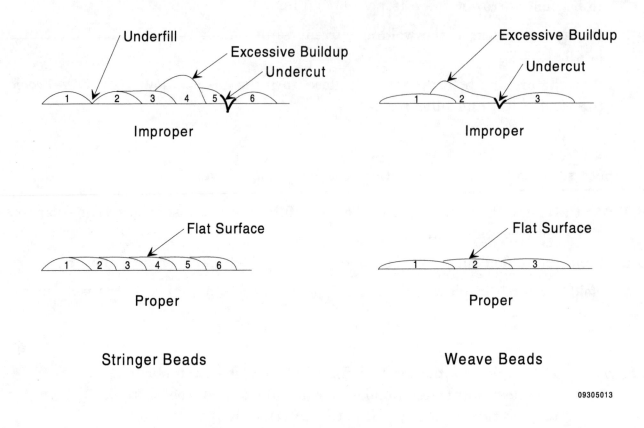

Figure 8-1. Proper and Improper Overlapping Beads.

Follow these steps to make GTAW overlapping stringer beads using 3/32 or 1/8 inch carbon steel filler rod.

Step 1 Mark out a 4-inch square on a piece of steel.

Step 2 Weld a stringer bead along one edge.

Step 3 Clean the weld of glass silicon beads.

Step 4 Run a second bead along the previous bead. Be sure to overlap the previous bead to obtain a good tie-in.

Step 5 Continue running overlapping stringer beads until the face of the metal square is covered.

Step 6 Continue building layers of stringer beads (pads), one on top of the other, until the technique is perfected.

9.0.0 GROOVE WELDS

The open-groove V-butt joint is a common groove weld normally made on plate and pipe. Practicing the open V-butt weld on plate will prepare the welder to make the more difficult pipe welds.

The performance qualification requirements at the end of this module for visual and destructive testing are based on the ASME Boiler and Pressure Vessel code welder certification test.

9.1.0 OPEN V-BUTT ROOT PASS TECHNIQUES

The most difficult part of welding an open-groove V-butt joint is the root pass. The root must have complete penetration but not an excessive amount of penetration or burn-through. The penetration is controlled by using the correct amperage and torch travel speed for the root opening and land being used. Root reinforcement (bead extension from back of the weld) should be flush to 1/8-inch, depending upon site specifications.

There are two techniques for running the root pass. These are:

- On-the-wire
- From-either-side

Regardless of the technique used, always start the root pass from a tack weld.

9.1.1 On-The-Wire Root Pass Technique

With the on-the-wire root pass technique, the filler rod is held in the bottom of the root opening (groove), parallel with the plates. It is fused in place by moving the torch along the groove using either the freehand or walking-the-cup torch techniques. The torch is advanced slowly enough to thoroughly fuse the filler rod with the plate edges, but not so slowly that it can burn through. This technique can be used with a filler rod that is the same size as the root opening is wide or slightly larger. If the filler rod is slightly larger it will be easier since the filler rod will be supported on the top side of the root land. With this technique the filler rod should not be moved into or out of the weld puddle. This will ensure a uniform weld bead thickness.

Figure 9-1 shows the on-the-wire root pass technique.

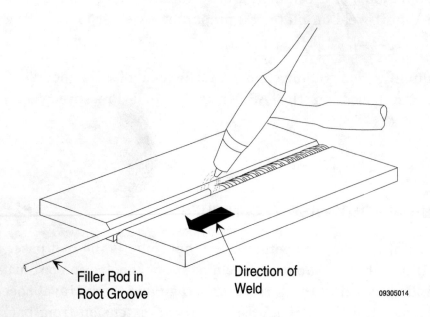

Filler Rod in
Root Groove

Direction of
Weld

09305014

Figure 9-1. On-The-Wire Root Pass Technique.

9.1.2   From-Either-Side Root Pass Technique

With the from-either-side root pass technique, the filler rod is held at an angle to the plate surfaces and fed into the groove root from either the weld side (front) or the opposite-weld side (back), depending upon the welder's preference. When making a root pass, start on a tack weld. When a weld puddle is established, the end of the filler rod is fed into the groove and the leading edge of the weld puddle. When a small amount of the filler melts off and flows into the weld puddle, the filler rod is moved forward away from the weld puddle. The torch is advanced using either the walking-the-cup or the freehand technique. The filler rod is moved back into the leading edge of the molten puddle and forward out of the molten puddle to add filler rod as needed. But, the filler rod is always kept in the shielding gas stream to keep the filler rod preheated and protected from atmospheric contamination.

CAUTION   When adding filler metal, always add filler metal to a molten
                 puddle created by the base metal. This prevents cold lap and
                 lack of fusion caused when filler metal is melted onto solid base
                 metal in an attempt to fuse the two together.

WELDING THREE TRAINEE TASK MODULE 09305

Figure 9-2 shows the from-either-side root pass technique.

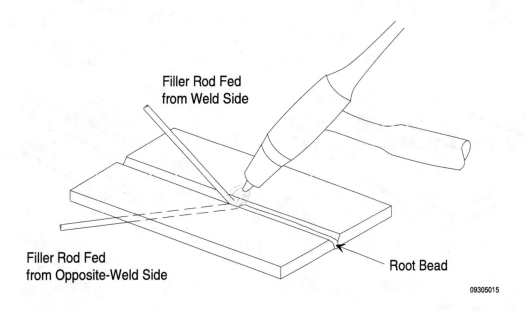

Figure 9-2. From-Either-Side Root Pass Technique.

After the root pass is completed, it should be cleaned and inspected. Clean the root pass of the small glass silicon beads that tend to form along the joint. Use a chipping hammer and/or wire brush. Inspect the root pass for:

- Uniform smooth rippled face and root
- Excessive root reinforcement
- Excessive build-up
- Undercut
- Tungsten inclusions
- Porosity

9.2.0 GROOVE WELD POSITIONS

Open groove welds can be made in all positions. The weld position is determined by the axis of the weld. Groove weld positions are flat (1G), horizontal (2G), vertical (3G) and overhead (4G).

Figure 9-3 shows the groove weld positions.

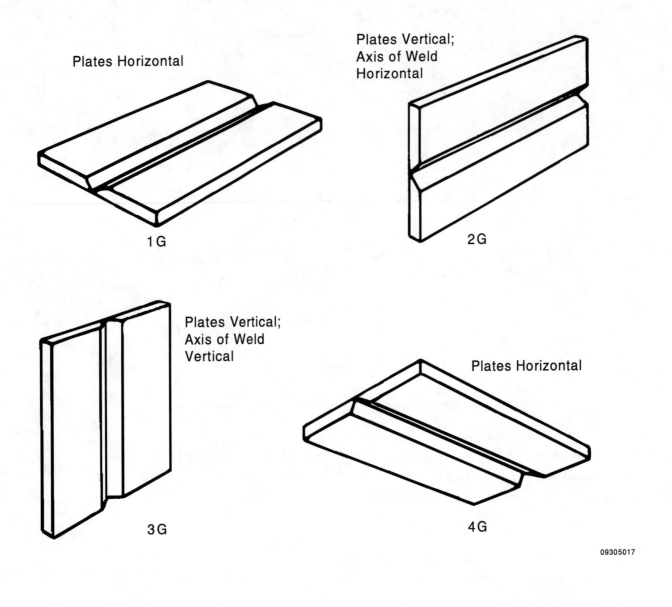

Plates Horizontal

1G

Plates Vertical;
Axis of Weld
Horizontal

2G

Plates Vertical;
Axis of Weld
Vertical

3G

Plates Horizontal

4G

09305017

Figure 9-3. Groove Weld Positions.

9.3.0   ACCEPTABLE AND UNACCEPTABLE GROOVE WELD PROFILES

Groove welds should be made with slight reinforcement (not exceeding 1/8 inch) and a gradual transition to the base metal at each toe. Groove welds must not have excess convexity, insufficient throat, excessive undercut or overlap. If a groove weld has any of these defects it must be repaired.

CAUTION Refer to your site's WPS (Welding Procedure Specifications) for specific requirements on groove welds. The information in this manual is provided as a general guideline only. The site WPS or Quality Specifications must be followed for all welds. Check with your supervisor if you are unsure of the specifications for your application.

Figure 9-4 shows acceptable and unacceptable groove weld profiles.

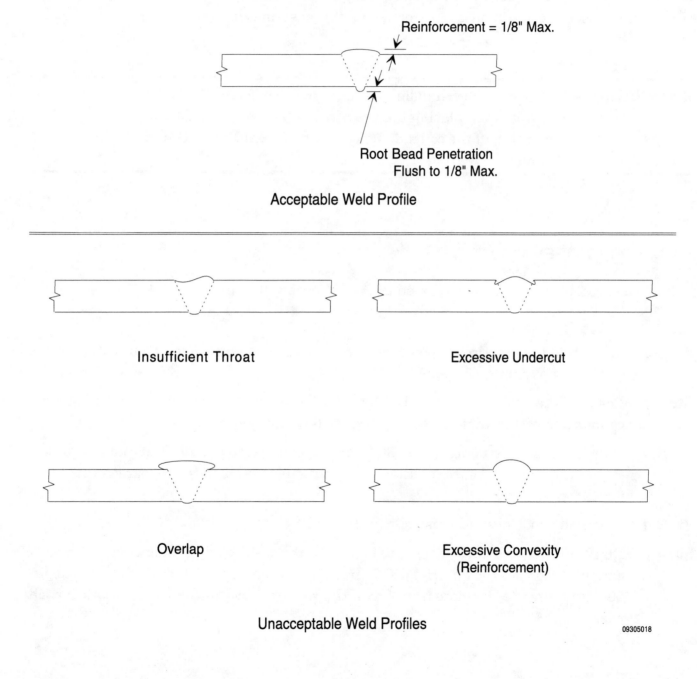

Reinforcement = 1/8" Max.

Root Bead Penetration
Flush to 1/8" Max.

Acceptable Weld Profile

Insufficient Throat

Excessive Undercut

Overlap

Excessive Convexity
(Reinforcement)

Unacceptable Weld Profiles

09305018

Figure 9-4. Acceptable and Unacceptable Groove Weld Profiles.

10.0.0 PRACTICE OPEN-GROOVE V-BUTT WELDS

Practice making GTAW open-groove V-butt welds in the 1G, 2G, 3G and 4G positions using either 3/32- or 1/8-inch carbon steel filler rod for the root pass, filler and cover passes with stringer beads.

Pay particular attention at the termination of the weld to fill the crater. If the coupon gets too hot between passes, cool it in water. Cooling with water is only done on practice coupons. Never cool test coupons or any other weld with water. Cooling with water can cause weld cracks and affect the mechanical properties of the base metal.

WARNING!   Use pliers to handle the hot practice coupons. Wear gloves when placing the practice coupon in water. Steam will rise off the coupon and can burn or scald unprotected hands.

10.1.0 FLAT OPEN V-BUTT WELD

Follow these steps to make GTAW open-groove V-butt welds in the flat (1G) position.

Step 1   Tack-weld the practice coupon together following the example given in section 2.2.0, Preparing Welding Coupons.

Step 2   Position the weld coupon in the flat position above the welding table. Space the coupon above the table to allow clearance under the root.

Step 3   Run the root pass using either 3/32- or 1/8-inch carbon steel filler rod and either the on-the-wire or the from-either-side technique. Use a 15 to 20 degree push angle for the root pass.

Step 4   Clean the root pass of glass silicon beads.

Step 5   Run the filler and cover passes to complete the weld using either 3/32 or 1/8 inch carbon steel filler rod and 15 to 20 degree push torch angle. Hold the filler rod about 20 degrees above the plate face. Clean the weld of glass silicon beads between each pass.

Figure 10-1 show the flat position groove weld bead sequence and transverse torch angles.

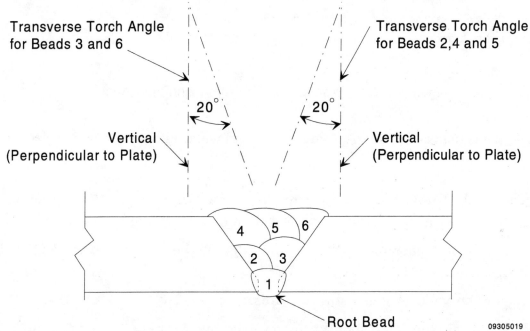

Note: The Actual Number of Weld Beads Will
Vary Depending on the Plate Thickness

Transverse Torch Angle
for Beads 3 and 6

Transverse Torch Angle
for Beads 2,4 and 5

20°   20°

Vertical
(Perpendicular to Plate)

Vertical
(Perpendicular to Plate)

Root Bead

09305019

Figure 10-1. Flat Position Groove Weld Bead Sequence.

10.2.0   HORIZONTAL BEADS

Before welding a joint in the horizontal (2G) position, practice running horizontal stringer beads by building a pad in the horizontal position. Use 3/32 or 1/8 inch carbon steel filler rod to build the pad. Use a torch push angle of 15 to 20 degrees and hold the filler rod in front of the weld at a 20 degree angle with the plate face.

Figure 10-2 shows horizontal position torch and rod angles.

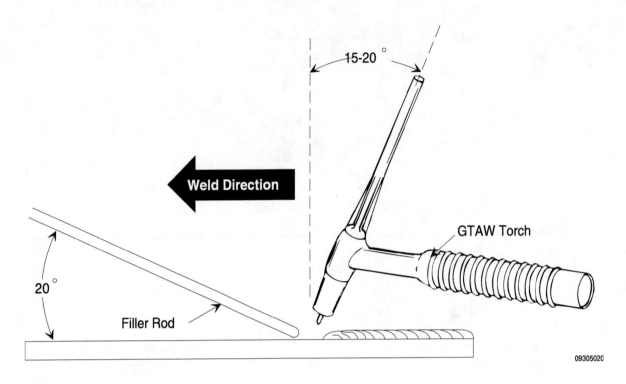

Figure 10-2. Horizontal Position Torch And Rod Angles.

Follow these steps to practice overlapping horizontal welding beads in the horizontal position.

Step 1 Tack-weld the welding coupon to the positioning arm in the horizontal position. Position the arm for comfortable welding.

Step 2 Run the first pass along the bottom edge of the coupon either using 3/32 or 1/8 inch carbon steel filler rod.

Step 3 Clean the weld of glass silicon beads.

Step 4 Run the second bead along the top edge overlapping the first bead. Adjust the transverse torch angle to obtain a good tie-in.

Step 5 Continue running beads to complete the pad. Be sure to clean the weld of glass silicon beads between each pass.

Figure 10-3 shows building a pad in the horizontal position.

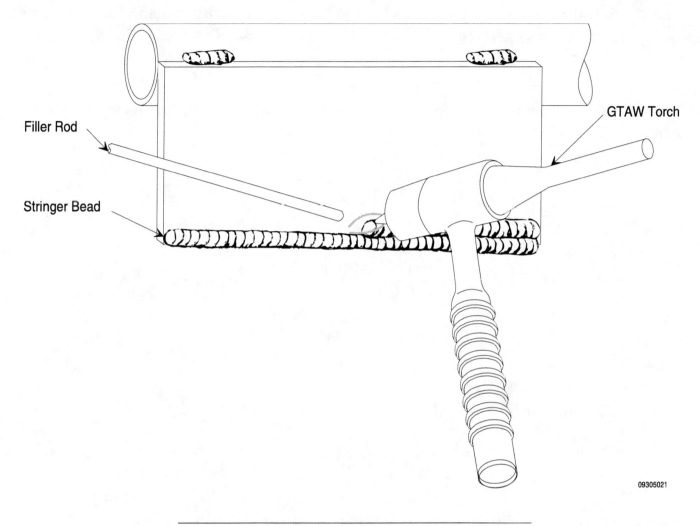

Figure 10-3. Building a Pad in the Horizontal Position.

10.3.0   HORIZONTAL OPEN V-BUTT JOINT

Follow these steps to practice welding open-groove V-butt welds in the horizontal (2G) position.

Step 1   Tack-weld the practice coupon together following the example given in section 2.2.0, Preparing Welding Coupons.

Step 2   Tack weld the coupon to the positioning arm of the welding table in the horizontal (2G) position. Position the arm and coupon for comfortable welding.

Step 3   Run the root pass using either the on-the-wire or the from-either-side welding technique and 3/32- or 1/8-inch carbon steel filler rod. Use a 15 to 20 degree torch push angle and a 20 degree filler rod angle with the plate surface.

Step 4   Clean the weld of glass silicon beads.

Step 5   Run the filler and cover passes to complete the weld using either 3/32 or 1/8 inch carbon steel filler rod and a 15 to 20 degree torch push angle. Clean the weld between each pass.

Figure 10-4 shows the horizontal position groove weld bead sequence and transverse torch angles.

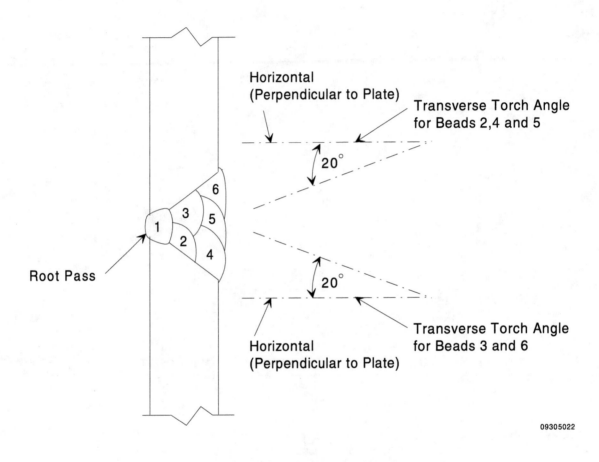

Figure 10-4. Horizontal Groove Weld Bead Sequence.

10.4.0   VERTICAL BEADS

Before welding a joint in the vertical position practice running vertical stringer beads by building a pad in the vertical position. Use 3/32 or 1/8 inch carbon steel filler rod to build the pad. Weld vertical up with a torch push angle of 30 degrees and a filler rod angle of about 40 degrees with the plate surface.

Figure 10-5 shows vertical position torch and rod angles.

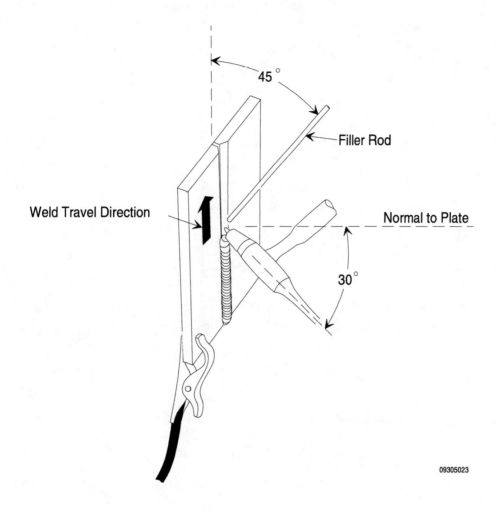

45°

Filler Rod

Weld Travel Direction

Normal to Plate

30°

09305023

Figure 10-5. Vertical Position Torch And Rod Angles.

Follow these steps to practice overlapping vertical beads in the vertical position.

Step 1   Tack-weld the welding coupon in the vertical position.

Step 2   Run the first bead vertically up along a vertical edge of the coupon using either 3/32 or 1/8 inch carbon steel filler rod.

Step 3   Clean the weld of glass silicon beads.

Step 4   Run the second bead overlapping the edge of the first bead. Adjust the transverse torch angle to obtain a good tie-in.

Step 5   Continue running beads to complete the pad. Clean the weld between each pass.

Figure 10-6 shows building a pad in the vertical position.

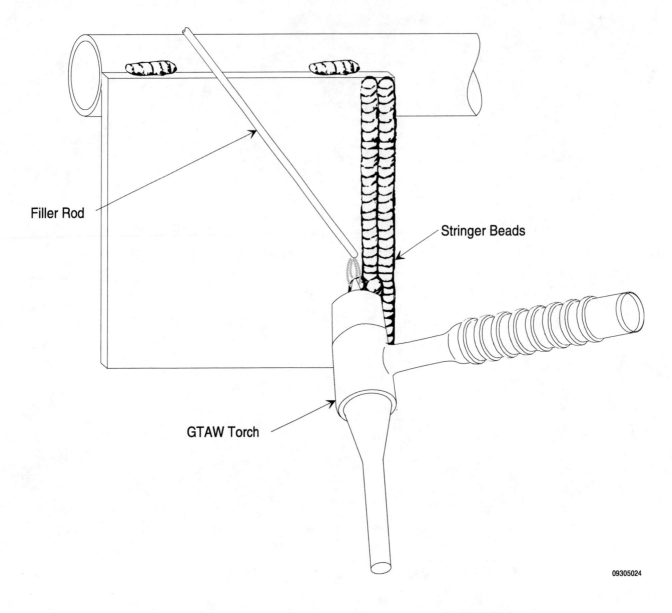

Figure 10-6. Building a Pad in the Vertical Position.

10.5.0 VERTICAL OPEN V-BUTT JOINT

Follow these steps to practice welding open-groove V-butt welds in the vertical (3G) position.

Step 1   Tack-weld the practice coupon together following the example given in section 2.2.0, Preparing Welding Coupons.

Step 2   Tack weld the coupon to the positioning arm of the welding table in the vertical (3G) position. Adjust the arm and coupon to a comfortable welding position.

Step 3 Run the root pass vertical up using either the _Tech_ on-the-wire or the from-either-side welding technique and either 3/32- or 1/8-inch carbon steel filler rod. Use a torch push angle of 30 degrees and a filler rod angle of 40 degrees with the plate surface.

Step 4 Clean the weld of glass silicon beads.

Step 5 Run the filler and cover passes vertical up to complete the weld. Use either 3/32 or 1/8 inch carbon steel filler rod with a 30 degree push torch angle. Clean the weld between each pass.

Figure 10-7 shows the vertical position groove weld bead sequence and transverse torch angles.

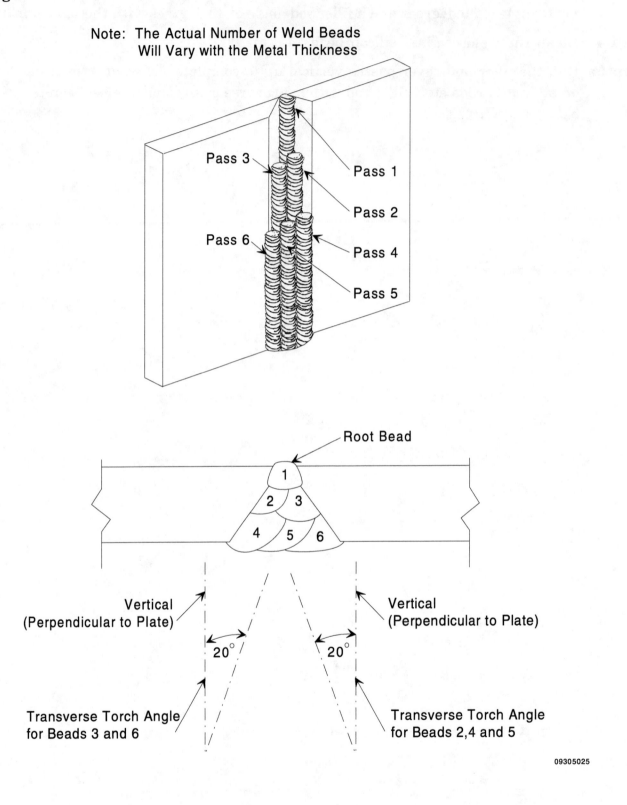

Figure 10-7. Vertical Position Groove Weld Bead Sequence.

10.6.0 OVERHEAD BEADS

Before welding a joint in the overhead position, practice running overhead stringer beads by building a pad in the overhead position. Use 3/32 or 1/8 inch carbon steel filler rod to build the pad. Weld overhead stringer beads with a torch push angle of 15 to 20 degrees. Hold the filler rod ahead of the weld at about 20 degrees with the plate surface. Follow these steps to practice overlapping vertical beads in the vertical position.

Step 1 Tack-weld the welding coupon to the positioning arm in the overhead position. Adjust the arm and coupon to a comfortable welding position.

Step 2 Run the first bead along one edge of the coupon using either 3/32 or 1/8 inch carbon steel filler rod.

Step 3 Clean the weld of glass silicon beads.

Step 4 Run the second bead overlapping the edge of the first bead. Adjust the transverse torch angle to obtain a good tie-in.

Step 5 Continue running beads to complete the pad. Be sure to clean the weld between each pass.

Figure 10-8 shows building a pad in the overhead position.

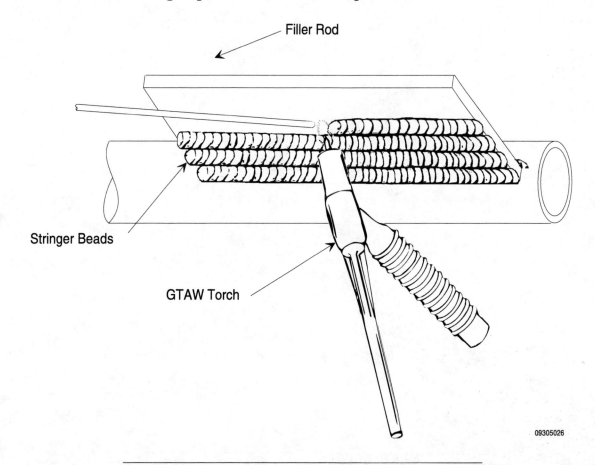

Filler Rod

Stringer Beads

GTAW Torch

09305026

Figure 10-8. Building A Pad In The Overhead Position.

10.7.0 OVERHEAD OPEN V-BUTT JOINT

Follow these steps to practice welding open-groove V-butt welds in the overhead (4G) position.

Step 1   Tack-weld the practice coupon together following the example given in section 2.2.0, Preparing Welding Coupons.

Step 2   Tack weld the coupon to the positioning arm of the welding table in the overhead (4G) position. Position the arm and coupon for comfortable welding.

Step 3   Run the root pass using either the on-the-wire or the from-either-side welding technique and 3/32- or 1/8-inch carbon steel filler rod. Use a 15 to 20 degree torch push angle and a 20 degree filler rod angle with the plate surface.

Step 3   Clean the weld of glass silicon beads.

Step 5   Run the filler and cover passes to complete the weld using either 3/32 or 1/8 inch carbon steel filler rod and a 15 to 20 degree torch push angle. Clean the weld between each pass.

Figure 10-9 shows the overhead position groove weld bead sequence and transverse torch angles.

Note: The Actual Number of Weld Beads Will
       Vary with the Metal Thickness

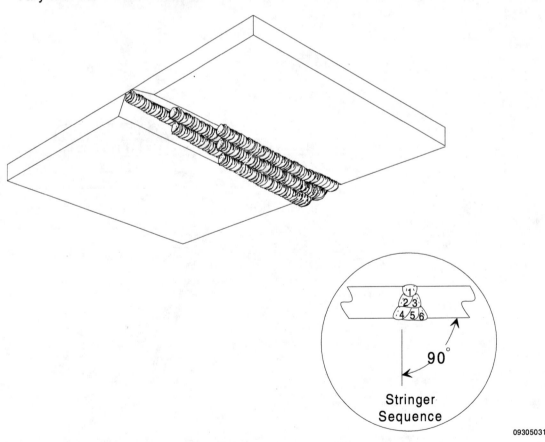

09305031

Figure 10-9. Overhead Position Groove Weld Bead Sequence.

SELF CHECK REVIEW

1. What fire safety item should you positively identify and locate before starting to weld?

2. What should the land size be for a open V-butt joint?

3. Can filler rod be substituted if the one specified in the WPS is not available?

4. Before plugging in the power supply, what must be done?

5. What polarity is used to weld mild steel with GTAW?

6. What is the normal electrode stickout range for a GTAW tungsten electrode?

7. Except for vertical welds, what torch angles are normally used for performing GTAW on mild steel?

8. What are the two basic techniques for running a bead with a GTAW torch?

9. While running a GTAW bead, where should the filler rod tip be kept when filler metal is not being added to the weld puddle?

10. What do welding codes specify about weld terminations?

11. What is padding?

12. What is the most difficult part to weld of an open-groove V-butt joint?

13. Where is the filler rod placed when using the on-the-wire GTAW root pass technique?

14. Where is the filler rod placed when using the from-either-side GTAW root pass technique?

15. What is the maximum weld reinforcement height for a groove weld?

ANSWERS TO SELF CHECK REVIEW

1. The nearest fire extinguisher. Do not proceed unless the extinguisher is charged and you know how to use it. (2.1.0)

2. From a feather up to 1/8-inch. (2.2.0)

3. No. (2.3.0)

4. Check that the welding machine is properly grounded through the primary current receptacle and locate the primary current disconnect. (2.4.0)

5. DCSP. (2.4.0)

6. The electrode stickout range is from two to three times the electrode diameter. (2.4.1)

7. Push angles of 15 to 20 degrees are generally used for welding carbon steel in all positions except the vertical. (3.2.1)

8. The freehand technique and the walking-the-cup technique. (5.0.0)

9. The rod tip is always kept within the torch shielding gas envelope to prevent oxidation and contamination of the rod. (5.0.0)

10. Welding codes require that weld termination craters must be filled to the full cross-section of the weld. (7.0.0)

11. Padding is a layer of overlapping weld beads. (8.0.0)

12. The most difficult part is the root pass. (9.1.0)

13. The filler rod is laid in the root opening, parallel with the plates. (9.1.1)

14. The rod is held at an angle to the plate surfaces and may be fed into the groove root from either the weld side (front) or the opposite-weld side (back). (9.1.2)

15. The maximum reinforcement for a groove weld is 1/8 inch. (9.3.0)

11.0.0   PERFORMANCE ACCREDITATION TASKS

The following tasks are designed to evaluate your ability to run groove welds on carbon steel plate in all positions with GTAW equipment. Perform each task when you are instructed to do so by your instructor. As you complete each task, show it to your instructor for evaluation. Do not proceed to the next task until directed to do so by your instructor.

11.1.0   MAKE A GROOVE WELD IN THE (1G) FLAT POSITION

Using GTAW and carbon steel filler rod, make a groove weld on carbon steel plate in the flat (1G) position as shown in *Figure 11-1*.

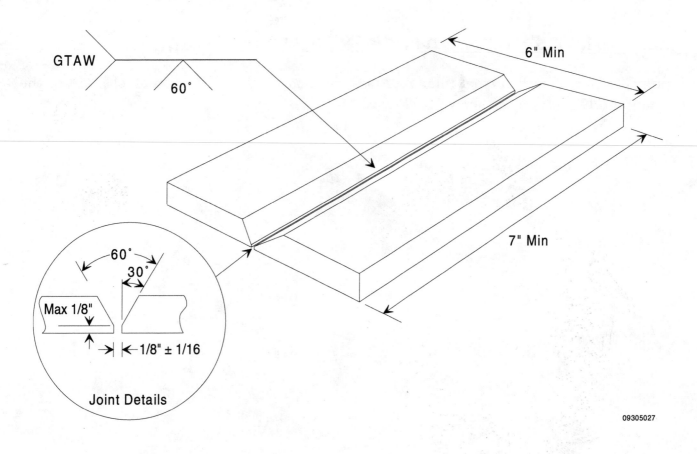

Figure 11-1. Groove Weld in the 1G Flat Position.

Criteria For Acceptance:

- Uniform rippled appearance on the bead face
- Craters and restarts filled to the full cross section of the weld
- Uniform weld size plus or minus 1/16 inch

- Acceptable weld profile in accordance with the *ASME Boiler and Pressure Vessel Code, Section IX.*
- Smooth transition with complete fusion at the toes of the weld
- Complete uniform root penetration at least flush with the back side of the plate to a maximum buildup of 1/8 inch.
- No porosity
- No oxidation
- No undercut
- No tungsten inclusions
- No pinholes (fisheyes)
- Acceptable guided bend test results

11.2.0 MAKE A GROOVE WELD IN THE (2G) HORIZONTAL POSITION

Using GTAW and carbon steel filler rod, make a groove weld on carbon steel plate in the horizontal (2G) position as shown in *Figure 11-2.*

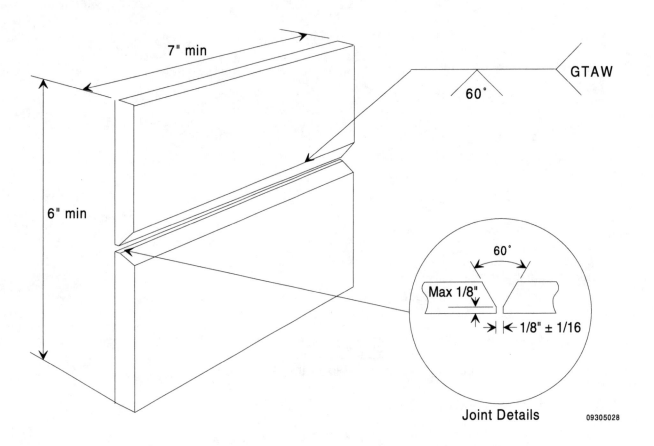

Figure 11-2. Groove Weld in the 2G Horizontal Position.

WELDING THREE TRAINEE TASK MODULE 09305

Criteria For Acceptance:

- Uniform rippled appearance on the bead face
- Craters and restarts filled to the full cross section of the weld
- Uniform weld size plus or minus 1/16 inch
- Acceptable weld profile in accordance with the *ASME Boiler and Pressure Vessel Code, Section IX.* ~~condition you to do it over over over~~
- Smooth transition with complete fusion at the toes of the weld
- Complete uniform root penetration at least flush with the back side of the plate to a maximum buildup of 1/8 inch.
- No porosity
- No oxidation
- No undercut
- No tungsten inclusions
- No pinholes (fisheyes)
- Acceptable guided bend test results

11.3.0   MAKE A GROOVE WELD IN THE (3G) VERTICAL POSITION

Using GTAW and carbon steel filler rod, make a groove weld on carbon steel plate in the vertical (3G) position as shown in *Figure 11-3*.

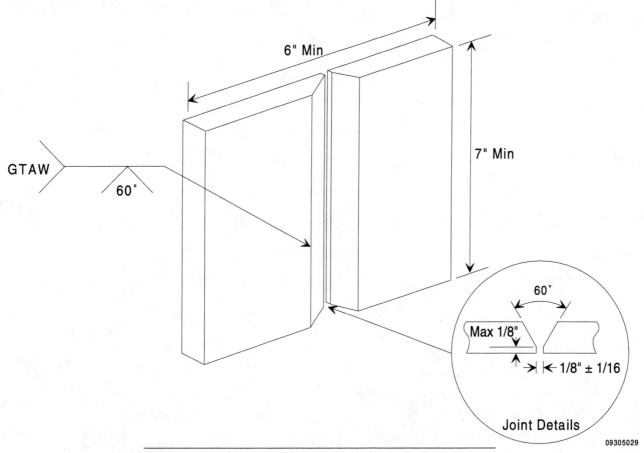

Figure 11-3. Groove Weld in the 3G Vertical Position.

Criteria For Acceptance:

- Uniform rippled appearance on the bead face
- Craters and restarts filled to the full cross section of the weld
- Uniform weld size plus or minus 1/16 inch
- Acceptable weld profile in accordance with the *ASME Boiler and Pressure Vessel Code, Section IX.*
- Smooth transition with complete fusion at the toes of the weld
- Complete uniform root penetration at least flush with the back side of the plate to a maximum buildup of 1/8 inch.
- No porosity
- No oxidation
- No undercut
- No tungsten inclusions
- No pinholes (fisheyes)
- Acceptable guided bend test results

11.4.0 MAKE A GROOVE WELD IN THE (4G) OVERHEAD POSITION

Using GTAW and carbon steel filler rod, make a groove weld on carbon steel plate in the overhead (4G) position as shown in *Figure 11-4.*

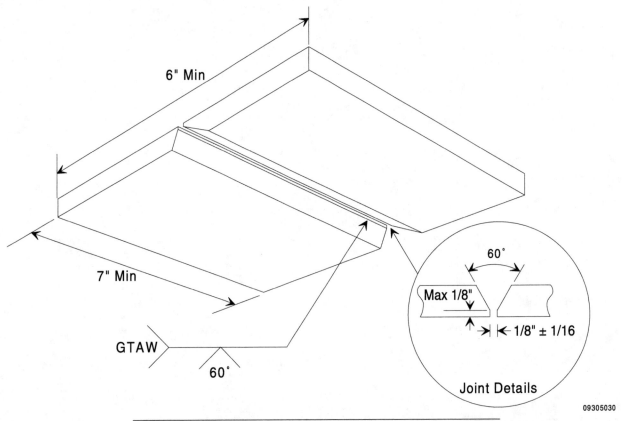

Figure 11-4. Groove Weld in the 4G Overhead Position.

Criteria For Acceptance:

- Uniform rippled appearance on the bead face
- Craters and restarts filled to the full cross section of the weld
- Uniform weld size plus or minus 1/16 inch
- Acceptable weld profile in accordance with the *ASME Boiler and Pressure Vessel Code, Section IX.*
- Smooth transition with complete fusion at the toes of the weld
- Complete uniform root penetration at least flush with the back side of the plate to a maximum buildup of 1/8 inch.
- No porosity
- No oxidation
- No undercut
- No tungsten inclusions
- No pinholes (fisheyes)
- Acceptable guided bend test results

SUMMARY

Setting up GTAW equipment, preparing the welding work area, running stringer beads and making acceptable groove welds in all positions are essential skills a welder must have to perform basic welding jobs or to progress on to more difficult welding procedures. Practice these welds until you can consistently produce acceptable welds as defined in the criteria for acceptance.

REFERENCES

For advanced study of topics covered in this Task Module, the following works are suggested:

GTAW Handbook (1988), William H. Minnick, Goodheart-Willcox Publishers, South Holland, IL. Phone 1-800-323-0440.

OSHA Requirements On Electrical Grounding.

Welding Skills, Giachino and Weeks, American Technical Publishers Inc., Homewood, IL, 60430. Phone 1-800-323-3471.

Welding Principles and Applications (1985), Jeffus and Johnson, Delmar Publishers, Inc., 2 Computer Drive West, Box 15-015, Albany, N.Y. 12212. Phone 1-800-347-7707.

Modern Welding (1988), Althouse, Turnquist, Bowditch, and Bowditch, Goodheart-Willcox Publishers, South Holland, IL. Phone 1-800-323-0440.

PERFORMANCE / LABORATORY EXERCISES

1. Practice GTAW flat overlapping stringer beads using carbon steel filler rod.

2. Weld plate, open V-butt joint, using GTAW and carbon steel filler rod in the 1G position.

3. Practice GTAW horizontal overlapping stringer beads using carbon steel filler rod.

4. Weld plate, open V-butt joint, using GTAW and carbon steel filler rod in the 2G position.

5. Practice GTAW vertical overlapping stringer beads using carbon steel filler rod.

6. Weld plate, open V-butt joint, using GTAW and carbon steel filler rod in the 3G position.

7. Practice GTAW overhead overlapping stringer beads using carbon steel filler rod.

8. Weld plate, open V-butt joint, using GTAW and carbon steel filler rod in the 4G position.

The NCCER makes every effort to keep these manuals up-to-date and free of technical errors. We appreciate your help in this process. If you have an idea for improving this manual, or if you find an error, a typographical mistake, or an inaccuracy in the *Wheels of Learning*, please write us, using this form or a photocopy. Be sure to include the exact module number, page number, a description of the problem, and the correction, if possible. We'll do our best to correct it in later editions. Thank you for your assistance.

Write: *Wheels of Learning*
National Center for Construction Education and Research
P.O. Box 141104
Gainesville, FL 32614-1104
Fax: 352-334-0932

WHEELS OF LEARNING USER UPDATE

Please let us know if you have found an inaccuracy, error, or other problem in a *Wheels of Learning* manual. Use this form or write us a letter. Please be sure to tell us the exact module name and module number, the page number, and the problem. Thanks for your help.

Craft _____ Module Name _____

Module Number _____ Page Number(s) _____

Description of Problem _____

_____

_____

(Optional) Correction of Problem _____

_____

_____

(Optional) Your Name and Address _____

_____

_____

WELDING THREE TRAINEE TASK MODULE 09305

GTAW -- Carbon Steel Pipe

Module 09306

NATIONAL
CENTER FOR
CONSTRUCTION
EDUCATION AND
RESEARCH

GAS TUNGSTEN ARC WELDING (GTAW) - CARBON STEEL PIPE

OBJECTIVES

Upon completion of this module, the trainee will be able to:

1. Make GTAW open-root V-groove welds on carbon steel pipe in the 1G position using carbon steel filler rod and argon gas.
2. Make GTAW open-root V-groove welds on carbon steel pipe in the 2G position using carbon steel filler rod and argon gas.
3. Make GTAW open-root V-groove welds on carbon steel pipe in the 5G position using carbon steel filler rod and argon gas.
4. Make GTAW open-root V-groove welds on carbon steel pipe in the 6G position using carbon steel filler rod and argon gas.

Prerequisites

Successful completion of the following module is required before beginning study of this module:

- *Gas Tungsten Arc Welding - (GTAW) - Plate # 09305*

Required Student Materials

Each trainee will need:

1. Personal protective equipment
2. Leather welding gloves
3. Leather jacket or sleeves
4. Welding shield
5. Welding table with positioning arm
6. GTAW welding equipment
   - Power source
   - Torch
   - Torch nozzles
7. Tungsten electrodes (EWTh-2, 3/32- and 1/8-inch)
8. Carbon steel filler rod, 3/32- and 1/8-inch diameter

9. Welding grade argon shielding gas
10. Welding grade argon backing gas (if used)
11. Cutting goggles
12. Chipping hammer
13. Wire brush
14. Pliers
15. Diagonal cutting pliers
16. Tape measure
17. Soapstone
18. Pipe for weld coupons: 6-, 8-, 10- or 12-inch, Schedule 40 or Schedule 80

Each trainee will need access to:

1. Oxyfuel cutting equipment
2. Portable angle-head grinders
3. Bench grinder (for tungstens)
4. Flat mill files
5. Pipe alignment clamps
6. Pipe beveling equipment (optional)

this work may be

Course Map Information

This course map shows all of the *Wheels of Learning* task modules in the third level of the Welding curricula. The suggested training order begins at the bottom and proceeds up. Skill levels increase as a trainee advances on the course map. The training order may be adjusted by the local Training Program Sponsor.

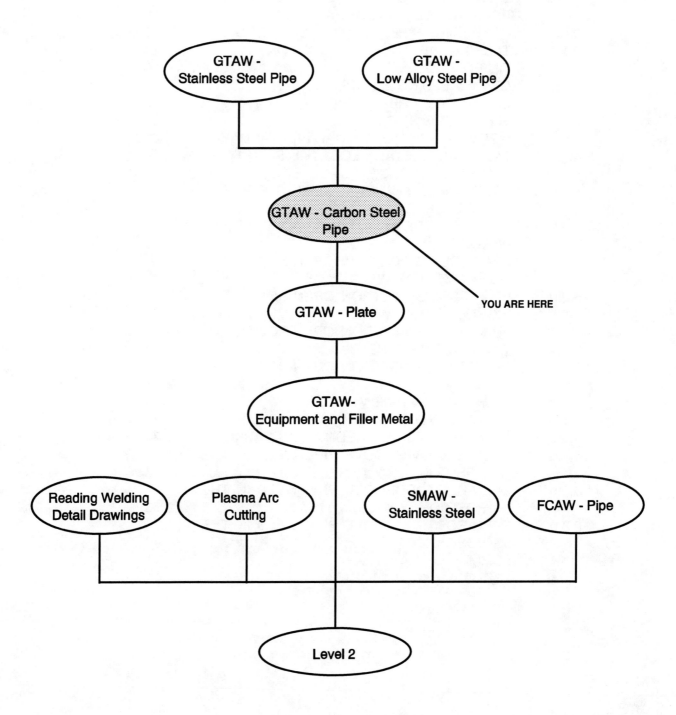

TABLE OF CONTENTS

Trade Terms Introduced in this Module

Backing Gas: An inert gas used on the back side of a joint to protect it from oxidation and atmospheric contamination during welding.

Critical Piping: High pressure piping typically used for high pressure and power systems which usually must meet stringent code requirements.

cfh: Abbreviation for cubic feet per hour.

Direct Current Electrode Negative (DCEN): Also called Direct Current Straight Polarity (DCSP). The DC welding lead arrangement where the base metal (workpiece) lead is connected to the positive (+) welding machine post and the torch lead is connected to the negative (-) welding machine post.

Direct Current Electrode Positive (DCEP): Also called Direct Current Reverse Polarity (DCRP). The DC welding lead arrangement where the base metal (workpiece) lead is connected to the negative (-) welding machine post and the torch lead is connected to the positive (+) welding machine post.

Hot Pass: The first filler pass after the root pass.

Out-Of-Position Weld: Any of the welding positions except the flat welding position where both the axis and the face of the weld are in the horizontal plane.

Penetration: The distance that the weld fusion line extends below the surface of the base metal.

Subcritical Piping: Low pressure piping typically used for heating and air conditioning, simple water supply systems and assorted service installations. Welds in subcritical piping are evaluated against less stringent code requirements than critical piping.

Torch Gas: The shielding gas flowing through the GTAW torch during welding.

Weld Coupon: The metal to be welded as a test or practice.

Welding Variables: Items that affect the quality of a weld such as weld voltage, weld current, weld travel rate, shielding gas type, torch angle and electrode stickout.

1.0.0 INTRODUCTION

Gas Tungsten Arc Welding (GTAW), is used to make high quality welds in metallic piping. The process allows the operator greater control of penetration and fill than with almost any other process. Also, because the process uses inert gas for weld shielding, there usually is no flux used. This eliminates most post weld cleaning and problems of slag inclusions. The resulting welds are typically of higher quality with more uniform weld deposit. Because the process does produce such high quality uniform weld beads, it is often used to make the root pass on pipe even when the filler and cover passes are to be made with a process with a higher deposition rate such as SMAW or GMAW.

This module explains how to set up GTAW equipment, prepare carbon steel pipe for open-root V-groove welds and techniques for welding pipe with carbon steel filler rod in the 1G, 2G, 5G and 6G pipe welding positions. *Figure 1-1* shows the 1G, 2G, 5G and 6G welding positions.

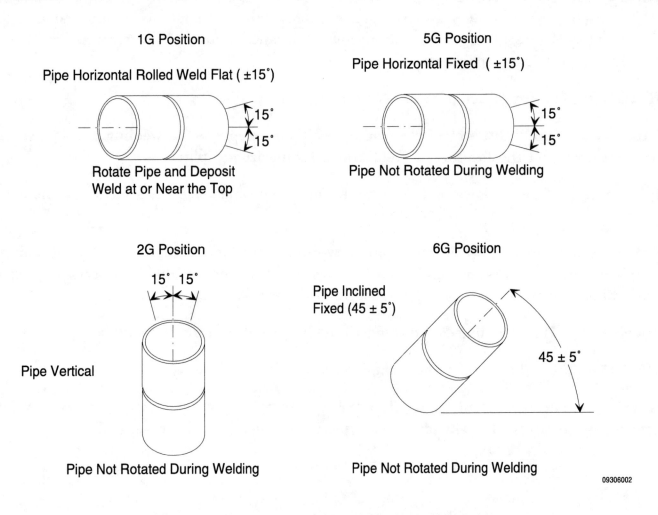

Figure 1-1. 1G, 2G, 5G and 6G Welding Positions

Performance demonstrations will require that the welds meet the qualification standards in ASME Boiler and Pressure Vessel Code, Section IX, *Welding and Brazing Qualifications.*

WELDING THREE TRAINEE TASK MODULE 09306

2.0.0   ROOT BACKSIDE PROTECTION

Welding processes which use torch gas to protect the molten puddle (on the face of the weld) from the atmosphere during welding, do not provide protection for the backside of the weld. Although backside protection is usually not required for carbon steel and some low alloys, it becomes essential as the alloy content increases. This is because alloying elements are more susceptible to combining with the oxygen in the atmosphere to form oxides. The oxides are objectionable because they can decrease the corrosion resistance to an unacceptable level and can be pulled into the weld puddle during subsequent welding causing defects. The most common methods of providing backside protection are:

- Inert Gas Backing
- Backup Flux
- ER7OS

2.1.0   INERT GAS BACKING

Inert gas backing provides the best and highest quality protection when welding open root joints. Any inert gas can be used, but almost always, nitrogen or argon are used because of their properties and cost. Argon provides better protection than nitrogen, but nitrogen typically cost 1/4 to 1/3 less. Because it costs less, nitrogen is used where it will not adversely affect the weld. To be effective, the inert gas has to:

- Replace the atmospheric gases to an acceptable level
- Be contained to provide the protection where it is needed
- Be continuous replenished to compensate for the gas which is lost during welding

For short and small diameter piping (small volume), the openings (ends) can be plugged or capped to provide containment and reduce the amount (cost) of the gas used. This method of containment is not cost effective for large piping or long sections (large volume). To decrease the volume of piping requiring inert gas replacement, the piping is plugged near the weld, on both sides, with removable devices called dams. Dams may include inflatable bladders, foam rubber, inflated plastic bags or water-soluble dams which are typically made of water-soluble paper. When dams are used, provisions have to be made to remove the dams after welding is completed. Another procedure to reduce the cost when argon is required as the backing gas, is to purge with nitrogen before purging with the argon.

When setting up inert gas backing, the containment is established and then the weld joint is sealed (typically with masking tape). The inert gas (purge gas) is then introduced through a tube or hose which passes through or past one of the containment devices or through a nozzle (needle) which directs the inert gas through the root opening of the weld. Purge gas should always enter at the bottom and there should be a vent located near the top as an exit for the atmospheric gases. This is because the purge gas is typically heavier than the air it is displacing. If the vent hole was in the bottom, the purge gas would vent out without displacing the atmospheric air in the top section of the pipe. The WPS will give the backing gas flow rate range. This range is for maintaining the inert gas atmosphere during welding after the pipe has been purged. Higher flow rates may be used for the initial purge to reduce the time it takes to initially purge the pipe. Verification that the atmosphere inside the pipe is correct for welding is often performed with an oxygen analyzer. When the oxygen analyzer is used, one percent maximum oxygen is the typical criteria for an atmosphere acceptable for welding.

Figure 2-1 shows typical pipe purging methods.

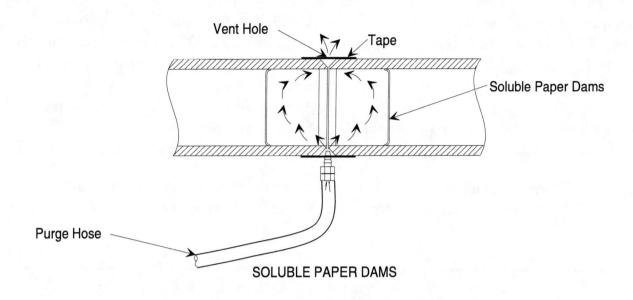

SOLUBLE PAPER DAMS

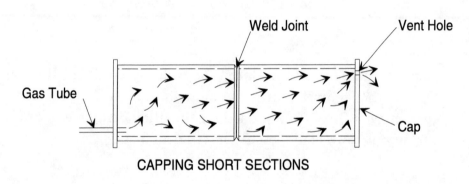

CAPPING SHORT SECTIONS

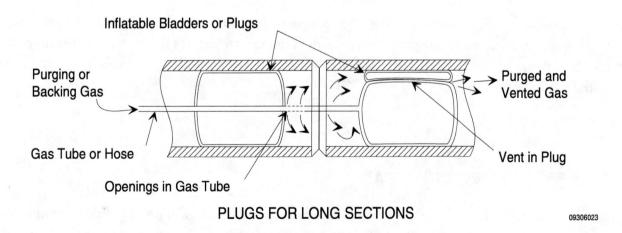

PLUGS FOR LONG SECTIONS

09306023

Figure 2-1. Typical Pipe Purging Methods

2.2.0 BACKUP FLUX

Backup flux is another common method of providing root protection for open root welds when weld quality requirements are less critical. Typically the flux comes as a dry powder which is mixed with alcohol (methanol) or acetone to a consistency of thick cream. After the mixture has set for several minutes to allow the alcohol or acetone to react with the flux, an even coat of the mixture is brushed onto the back side of the joint. The mixture dries rapidly leaving a coating of flux. The joint is then welded in the normal manner. During welding the flux powder melts forming a barrier that prevents air from coming into contact with the molten metal. To remove the flux after welding, a wire brush and hot water can be used.

In addition to dry powder flux, special flux coated and flux cored rods are also available to provide root backside protection when using the GTAW process.

WARNING!   Backup flux may contain silica, fluorides or other toxic materials. Weld only in well ventilated spaces.

CAUTION   Backup fluxes are formulated for welding specific alloy types. Check the manufactures recommendations to be sure the correct type flux is being used for the alloy being welded. Using the wrong flux could result in weld defects.

3.0.0   GTAW WELDING EQUIPMENT SET-UP

Before welding can take place the work area has to be made ready, the welding equipment set up and the metal to be welded prepared. The following sections will explain how to prepare the area, set up the equipment and perform GTAW welding of carbon or low-carbon steel pipe.

3.1.0   PREPARE WELDING AREA

To practice welding, a welding table, bench or stand is needed. The welding surface must be steel and provisions must be made for mounting practice welding coupons out of position. A simple mounting for out of position coupons is a pipe stand that can be welded vertically to the steel table top. To make a pipe stand, weld a three- or four-foot length of pipe vertically to the table top. Then cut a short section pipe to slide over the vertical pipe. Drill a hole in the slide. Weld a nut over the hole so a bolt can be used to lock the slide in place on the

WELDING THREE TRAINEE TASK MODULE 09306

vertical pipe. Weld a piece of pipe or angle horizontally to the slide. The slide can be rotated and adjusted vertically to position the horizontal arm. The weld coupon can be tack-welded to the horizontal arm.

WARNING!   The table must be heavy enough to support the weight of the welding coupon extended on the horizontal arm without falling over. It must also support the coupon during chipping or grinding. Serious injury will result if the table falls onto someone in the area.

Figure 3-1 shows a welding table with out-of-position support arm.

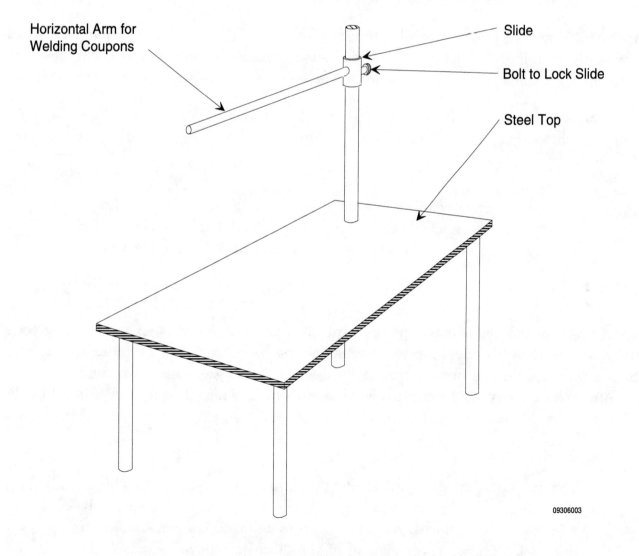

Figure 3-1. Welding Table With Out-Of-Position Support Arm

To set up the area for welding follow these steps.

Step 1   Check to be sure the area is properly ventilated. Make use of doors, windows and fans.

Step 2   Check the area for fire hazards. Remove any flammable materials before proceeding.

Step 3   Check the location of the nearest fire extinguisher. Do not proceed unless the extinguisher is charged and you know how to use it.

Step 4   Position a welding table within convenient distance to the GTAW welding machine.

Step 5   Set up flash shields around the welding area.

3.2.0   PREPARING PIPE WELDING COUPONS

The open-root V-groove butt weld joint is the most common weld joint used for joining critical and non-critical piping. The joint is easy to prepare because it only requires the beveling of the matching members. However, welding the root pass does take practice to perfect.

Welding coupons should be cut from 6-, 8-, 10- or 12-inch schedule 40 or schedule 80 carbon steel pipe. Each welded joint will require two coupons of the same size and schedule pipe.

Note   Pipe for practice welding is expensive and difficult to obtain. Every effort should be made to conserve and not waste the material that is available. Re-use weld coupons until all surfaces have been welded on by cutting the coupon apart and reusing the pieces.

The bevel angle of the open-root V-groove joint used on pipe depends on the welding code used at your site. The two most common codes are the API (American Petroleum Institute) which generally specifies a bevel angle of 30 degrees (60-degree included angle) and the ASME (American Society of Mechanical Engineers) which generally specifies a bevel angle of 37-1/2 degrees (75-degree included angle).

Note   The examples used in this student manual show a 30-degree bevel. Depending on your site requirements, the 37-1/2 degree bevel can be substituted. Check with your instructor to determine if you should use the 37-1/2 degree bevel instead of the 30-degree bevel.

To prepare pipe welding coupons for open-root V-groove weld joints follow these steps:

Step 1   Clean heavy rust or mill scale from the pipe with a grinder or wire brush.

Step 2   Bevel the end of the pipe to 30 degrees (plus 5 or minus 0 degrees) or 37-1/2 degrees (plus or minus 2-1/2 degrees) by any acceptable beveling method such as flame cutting or grinding.

Step 3   Cut off approximately 3 inches of the beveled pipe end.

Step 4   Continue cutting coupons until enough pairs are cut for practice welding.

Step 5   Check the bevel. It should be free of slag and notches more than 1/16-inch deep. The bevel for ASME specifications should be 37-1/2 degrees (plus or minus 2-1/2 degrees). The bevel angle for API specifications should be 30 degrees (plus 5 or minus 0 degrees).

Step 6   Grind a 1/16-inch root face (land) on the bevel.

Note   The most common land used is 1/16-inch although welding procedures generally allow the land on open-root welds to be from 1/32 inch to 1/8 inch wide. Adjust the land size as needed when you start the welding practices.

Figure 3-2 shows typical ASME and API pipe bevel specifications for bevel angles, root lands and root openings.

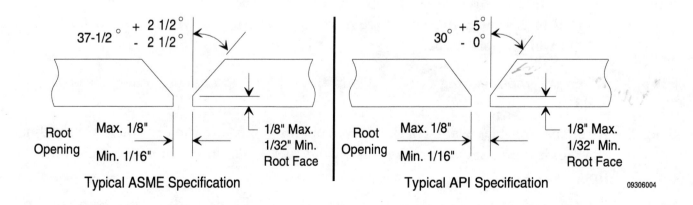

Figure 3-2. ASME and API Pipe Bevel Specifications

3.3.0 TACKING PIPE WELDING COUPONS

When tack welding coupons, the same root backing required for welding must be used. This is because the tack welds become part of the finished root pass.

CAUTION   Failure to provide root backing for tack welds when required will result in weld defects.

Step 1   Provide for root backing using backup flux or gas backing.

CAUTION   Check your site quality standards and/or WPS to determine the type of backing required. Do not use backup flux unless it is specifically approved at your site. Check with your supervisor if you are unsure of the backing requirements.

Step 2   Align the two pipe sections so that the ID (inside diameter) is even all around. Small diameter pipe can be aligned by clamping both pieces to a piece of angle iron. Larger diameter pipe can be aligned with the aid of a pipe alignment jig or by holding a straight edge across the joint, parallel to the pipe axis. The straight edge must be used all around the inside of the pipe, in case one or both sections is out of round.

Step 3   Gap the root opening at 1/8-inch with pieces of 1/8-inch filler rod or metal shims.

Note   The welding procedures generally allow the root opening to be from 1/16 to 1/8 inch. Adjust the root opening as needed when you start the welding practices.

Step 4   When the root opening is correct and the pipe ID is aligned, make the first tack weld.

Step 5   After the first tack weld, check the root opening opposite the first tack weld. Adjust the gap if necessary and make the second tack weld on the opposite side from the first tack.

Step 6   Check the root opening again and weld the third tack midway between the first two tacks.

Step 7   Weld the fourth tack opposite the third tack and midway between the first and second tacks. There should now be four tack welds evenly spaced every 90 degrees around the pipe coupon. Use a minimum of four tack welds on pipe from 4 inches to 8 inches in diameter.

Note   Larger pipe sizes or materials other than carbon steel pipe may require additional tack welds to maintain the root opening during welding.

Step 8   After all tack welds are completed, use a file or grinder to feather the tack welds as shown in *Figure 3-3*.

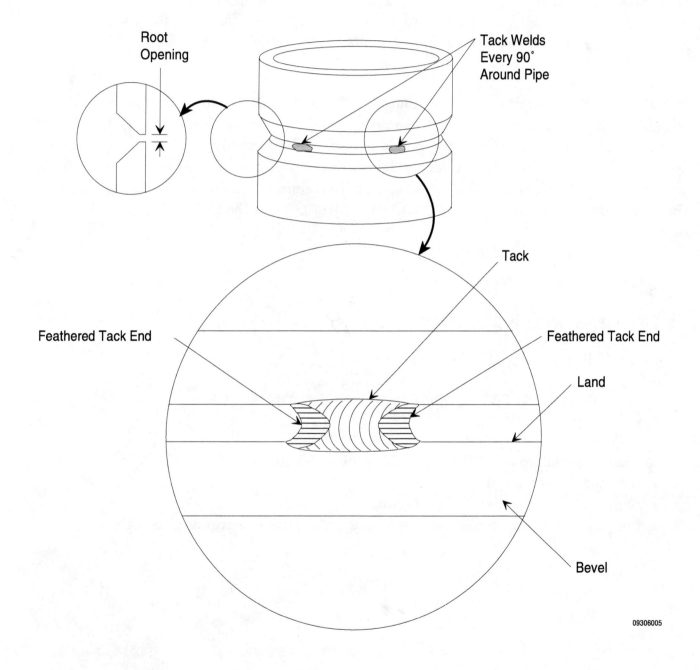

Figure 3-3. Tacked Open-Root V-Groove Pipe Weld Coupon

3.4.0   FILLER RODS

Filler rods are selected to be compatible with the base metal to be welded. The WPS or job site standards will specify the filler rod type and size to be used.

CAUTION   When the WPS or Site Quality Standards specify a filler rod it must be used. Severe penalties will result from not following a WPS or Site Quality Standard if available.

For the welding exercises in this module, 3/32- and/or 1/8-inch carbon steel filler rods classification ER70S-2 or equivalent should be used to weld the carbon steel pipe coupons. Remove only a small number of filler rods at a time. Keep the remainder in its package to keep it clean. Before using the filler rod, check it for contamination such as corrosion, dirt, oil or grease, or burnt ends. All of these forms of contamination will cause weld defects. Clean the filler rod with a clean, oil free rag and snip burnt ends. If the filler rod cannot be cleaned, do not use it.

3.5.0   GTAW WELDING EQUIPMENT

Locate the following GTAW equipment:

- Constant current power supply (specifically designed for GTAW if possible)
- GTAW torch and cable assembly
- Collet assembly (3/32- and 1/8-inch)
- Tungsten electrodes, EWTh-2 (3/32- and/or 1/8-inch)
- Gas nozzles for torch
- Shielding gas cylinder and regulator-flow meter
- Purging gas cylinder and regulator/flowmeter (if gas purging is to be used)

Figure 3-4 shows a complete GTAW system.

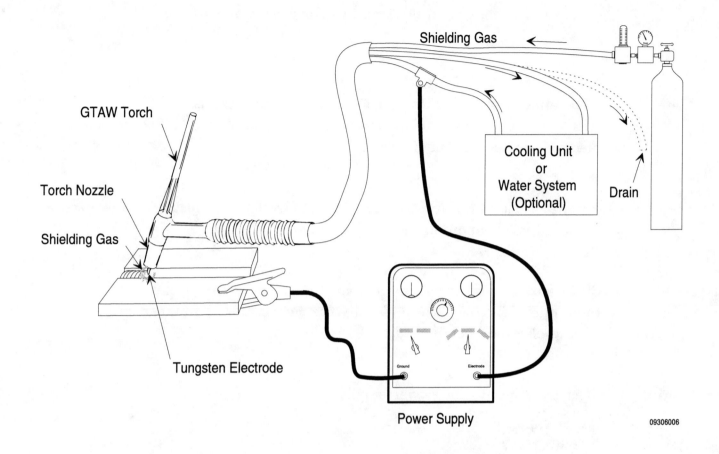

Figure 3-4. Complete GTAW System

Perform the following steps to set up the GTAW equipment.

Step 1 Verify that the welding machine is a constant current type DC or AC/DC power supply.

Step 2 Check to be sure that the welding machine is properly grounded through the primary current receptacle.

Step 3 Verify the location of the primary current disconnect.

Step 4 Plug the power source into the power receptacle.

Step 5 Locate a GTAW torch, and if the power supply is not a dedicated GTAW machine, make sure the GTAW torch is adapted to connect to the power supply (special power cable lug).

Step 6 Install the 1/8-inch collet and tungsten electrode in the torch. Tighten the collet nut (hand tighten only).

CAUTION   If the collet nut is over-tightened the threads will be damaged when the torch heats when being used. If the collet is left too loose, the collet and electrode will overheat and damage the torch.

Note   Depending on site conditions, a 3/32-inch electrode can be used. Check with your instructor to determine if you should use a 3/32-inch electrode instead of the 1/8-inch.

Step 7   Check the nozzle to be sure it is the correct size for the electrode to be used and that it is in good condition (no cracks or chips).

Note   Since nozzle numbering systems vary by manufacture, refer to the torch manufactures data to identify the correct nozzle number (size) for the electrode being used.

Step 8   Connect the torch to the power supply.

- Power cable
- Shielding gas
- Cooling water (if used)

Step 9   If used, connect the remote foot control cable to the power supply and set the remote/local switch to remote.

Step 10   Set up the welding machine for direct current straight polarity (DCSP).

Step 11   Set the high frequency (if equipped) to "START".

Step 12   Locate one cylinder (two cylinders if gas backing is to be used) of welding grade argon gas and secure the cylinder(s) in an upright position. Locate the shielding gas cylinder near the power supply. The purging gas cylinder can be located nearer the weld.

Step 13   Remove the protective cap(s) and clear the valve(s) of visible debris. Momentarily crack the cylinder valve(s) to blow out any remaining dirt.

Step 14   Install the gas regulator/flowmeter to the shielding gas cylinder and connect the gas hose from the flowmeter to the "Gas In" fitting on the welding machine or to the torch shielding gas supply line (power lug) if using a non-GTAW power supply.

Step 15   Turn on the welding machine.

Step 16 Fully open the shielding gas cylinder valve and then open the flow meter and activate the purge control on the welding machine to purge the torch line and torch for a few seconds. After purging, adjust the gas flowmeter to the specified flow rate (15-20 cfh).

Step 17 Install the gas regulator/flowmeter to the purging gas cylinder and connect the gas hose. Do not open the cylinder vale or adjust the flow rate until the purging hose is connected to the weldment.

Step 18 Set the welding amperage as follows:

- Root pass   90-95 amperes
- Filler passes   100-110 amperes

Note If a remote current control is being used, the amperage set on the power supply will be the maximum amperage available when the remote control is fully depressed. Set the amperage slightly higher so additional current will be available if required for the start of the weld, misalignment or at tacks.

Step 19 Set any other applicable variables. Refer to the manufacturer's specifications and instructions for setting:

- Hot start current
- Gas preflow (prepurge) duration
- Gas postflow (post purge) duration
- High frequency intensity
- Pulse current
- Current slope up
- Current slope down

3.6.0 ELECTRODE PREPARATION

The EWTh-2 tungsten electrodes used with DCSP should be taper ground with a dulled point. The taper length should be between two and three times the electrode diameter. The electrode should be ground so that the grind marks are along the direction of the electrode axis. Do not grind around the point. Also, do not grind the color-coded end of the tungsten or the identification code will be lost.

Note   Either EWTh-1 or EWTh-3 electrodes can be used for carbon steel welding if EWTh-2 electrodes are not available.

Figure 3-5 shows a GTAW tungsten electrode properly ground for DCSP.

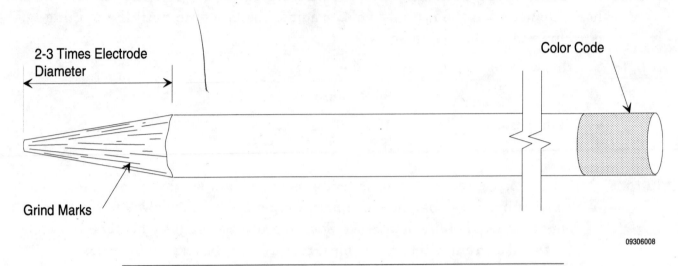

09306008

Figure 3-5. GTAW Tungsten Electrode Properly Ground For DCSP

3.6.1 Electrode Stickout

Electrode stickout is measured from the end of the gas nozzle to the end of the electrode. It is normally two to three times the electrode diameter. For example, normal stickout for a 1/8 inch diameter electrode would be 1/4 to 3/8 inch (2 x 1/8 inch = 1/4 inch, 3 x 1/8 inch = 3/8 inch).

Figure 3-6 shows GTAW tungsten electrode stickout.

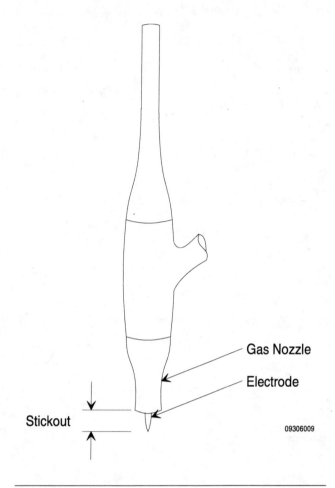

Figure 3-6. GTAW Tungsten Electrode Stickout

4.0.0   PIPE GROOVE WELD QUALITY STANDARDS

Groove welds may be made in all positions on pipe. The weld position is determined by the axis of the pipe. Four standard weld positions are used with pipe:

1G Position - Pipe axis is horizontal and pipe is slowly rotated while being welded on the top; weld beads are flat

2G Position - Pipe axis is vertical and pipe is fixed; weld beads are horizontal

5G Position - Pipe axis is horizontal and pipe is fixed; weld beads are flat, vertical and overhead

6G Position - Pipe axis is inclined (nominally at 45 degrees) and pipe is fixed; weld beads are horizontal, vertical and overhead

Pipe positions 1G, 2G and 5G can vary plus or minus 15 degrees from the basic position.

Position 6G can vary plus or minus 5 degrees from the nominal 45 degree angle. *Figure 4-1* shows the four basic pipe weld test positions.

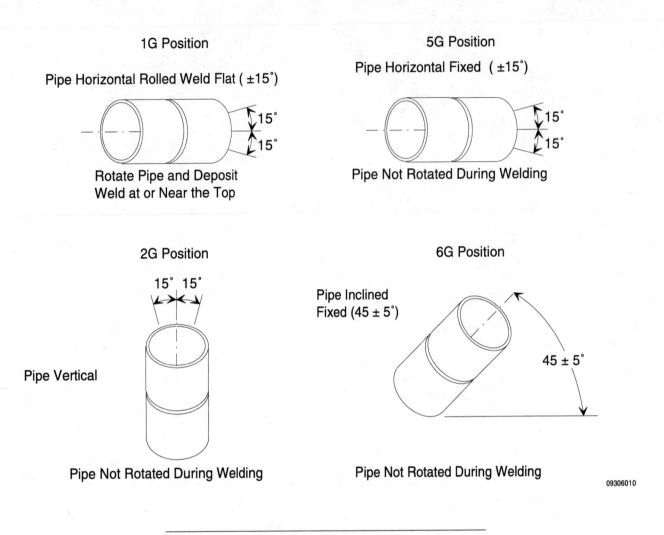

Figure 4-1. Four Basic Pipe Weld Test Positions

4.1.0   TACK AND RESTART LOCATIONS ON 5G AND 6G COUPONS.

When 5G or 6G welds are to be destructively tested, the test specimens will be cut from the four regions midway between the 12-o'clock and 3-o'clock positions, the 3-o'clock and 6-o'clock positions, the 6-o'clock and 9-o'clock positions and the 9-o'clock and 12-o'clock positions. (The 12-o'clock position is the top of the weld groove; the 6-o'clock position is the bottom of the weld groove.) Use special care when tieing into tack welds and when making restarts since these areas are more susceptible to weld defects.

Figure 4-2 shows the test specimen regions of a 5G or 6G position pipe.

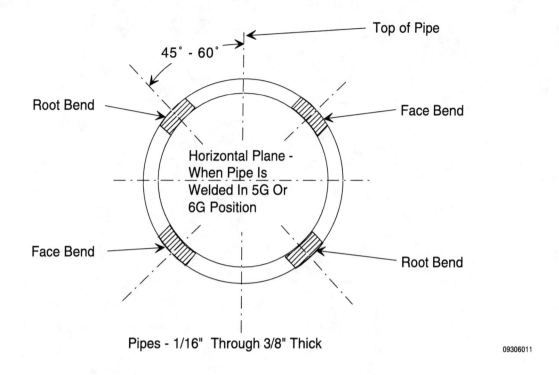

Figure 4-2. Test Specimen Regions Of 5G Or 6G Position Pipe

4.2.0   ACCEPTABLE AND UNACCEPTABLE PIPE WELD PROFILES

Pipe groove welds should be made with slight reinforcement (not exceeding 1/8 inch for base metals greater than 1/4-inch thickness) and a gradual transition to the base metal at each toe. The root pass should have complete penetration with the protrusion on the inside of the pipe being flush to a maximum of 1/8-inch for base metals greater than 1/4-inch thickness. Pipe groove welds must not have excess convexity, insufficient throat, excessive undercut or overlap. If a weld has any of these defects it must be repaired.

CAUTION   Refer to your site's WPS (Welding Procedure Specifications) for specific requirements on pipe groove welds. The information in this manual is provided as a general guideline only. The site WPS or Quality Specifications must be followed for all welds. Check with your supervisor if you are unsure of the specifications for your application.

Figure 4-3 shows acceptable and unacceptable pipe groove weld profiles.

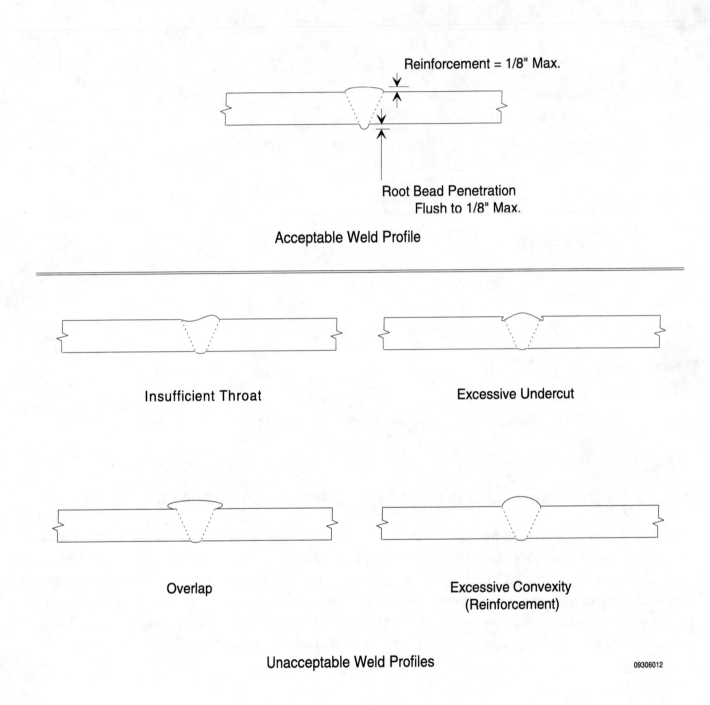

Reinforcement = 1/8" Max.

Root Bead Penetration
Flush to 1/8" Max.

Acceptable Weld Profile

Insufficient Throat

Excessive Undercut

Overlap

Excessive Convexity
(Reinforcement)

Unacceptable Weld Profiles

09306012

Figure 4-3. Acceptable and Unacceptable Pipe Groove Weld Profiles

5.0.0 PRACTICE OPEN-ROOT V-GROOVE PIPE WELDS

Practice open-root V-groove pipe welds in the four standard welding positions: 1G, 2G, 5G and 6G.

5.1.0 PRACTICE 1G POSITION OPEN-ROOT V-GROOVE PIPE WELD

Practice the 1G (horizontal and rolled) position open-root V-groove pipe weld using GTAW and stringer beads for all passes. Use either the walking-the-cup or the freehand torch handling technique and a 15- to 20-degree longitudinal torch push angle. For the root pass, either use the on-the-wire filler rod technique or feed the filler rod into the root opening from the face side of the weld at an angle of about 15 degrees. Pay particular attention at the termination of the weld to fill the crater.

Follow these steps to practice open-root V-groove GTAW pipe welds in the 1G position.

Step 1 Tack-weld the practice coupon together following the example given in section 3.3.0, "Tacking Pipe Welding Coupons."

Step 2 Position the pipe weld coupon horizontally on two sets of rollers at a comfortable welding height. If rollers are not available, lay the coupon on a smooth welding bench so that it is free to roll as it is welded.

Step 3 If gas backing is to be used, attach the gas retaining device, backing gas hose and set the backing gas flowmeter for about 8-cfh.

Figure 5-1 shows three styles of roller supports commonly found in pipe welding shops.

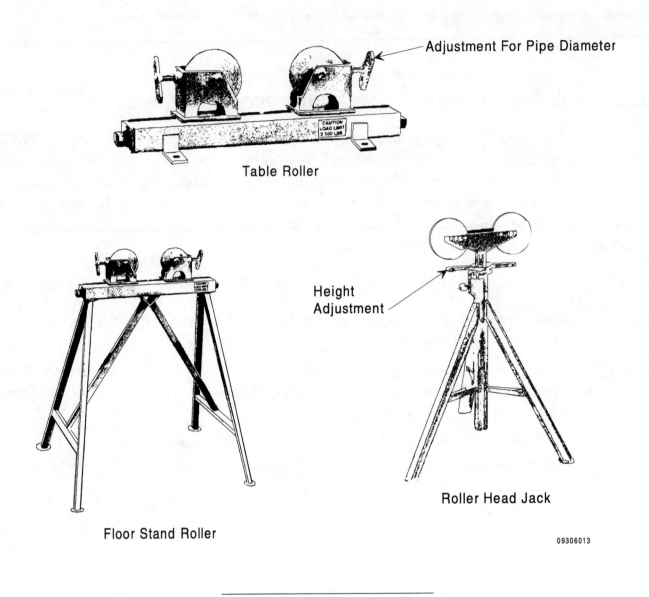

Adjustment For Pipe Diameter

Table Roller

Height Adjustment

Roller Head Jack

Floor Stand Roller

09306013

Figure 5-1. Pipe Roller Supports

Step 4   Attach the ground clamp directly onto the pipe coupon. This will prevent the welding current from passing through the roller bearings or arcing between the rollers and the pipe coupon.

CAUTION   Failure to attach the ground clamp to the pipe coupon can result in variations in welding current, damage to the roller bearings and arcing on the rollers and pipe coupon.

Step 5   Run the root pass by starting the weld bead on a tack weld positioned at the 11 o'clock position and advance toward the 12 o'clock position.

Step 6   Roll the pipe as necessary to keep the weld at the top of the coupon (in the flat position).

Step 7   Use the same rolling procedure to make the filler and cover passes. Pay particular attention at the tie-ins to prevent excess buildup.

Figure 5-2 shows the 1G pipe coupon position and bead sequence.

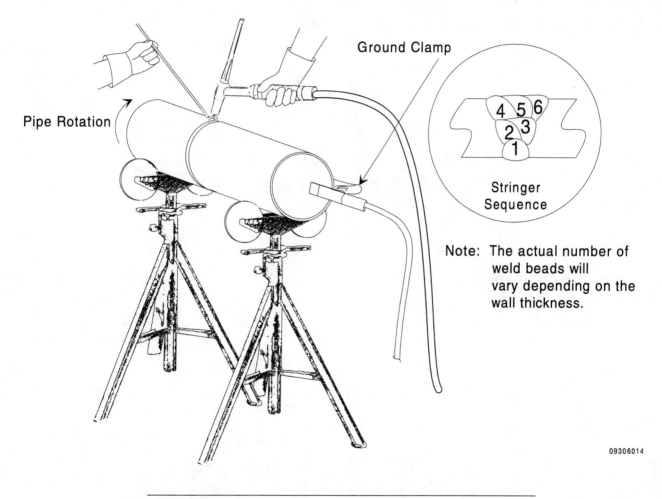

Figure 5-2. 1G Pipe Coupon Position And Bead Sequence

5.2.0   PRACTICE 2G POSITION OPEN-ROOT V-GROOVE PIPE WELD

Practice the 2G position open-root V-groove pipe welds using GTAW and stringer beads for all passes. Use a 15- to 20-degree longitudinal torch push angle all around the joint. For the root pass, use either the on-the-wire rod technique or feed the rod into the root from the face side of the weld at an angel of about 15 degrees. When making the filler and cover passes use either the walking-the-cup or the freehand torch handling technique. Pay particular attention at the termination of the weld to fill the crater.

Follow these steps to practice open-root V-groove pipe welds in the 2G position.

Step 1 Tack-weld the practice coupon together following the example given in section 3.3.0, "Tacking Pipe Welding Coupons."

Step 2 Tack-weld the pipe coupon to the positioning arm on the welding table with the axis of the pipe vertical.

Step 3 If gas backing is to be used, attach the gas retaining device, backing gas hose and set the backing gas flowmeter for about 8-cfh.

Figure 5-3 shows the 2G pipe coupon position and bead sequence.

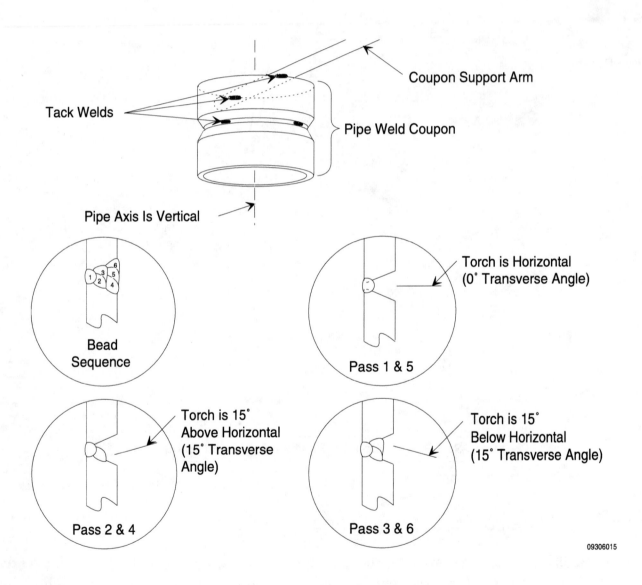

Figure 5-3. 2G Pipe Coupon Position and Bead Sequence

Step 4   Position the arm to allow comfortable welding access all around the coupon.

Step 5   Run the root pass by starting on a tack weld and continuing until it is completed.

Step 6   Run the filler and cover passes.

5.3.0   PRACTICE 5G POSITION OPEN-ROOT V-GROOVE PIPE WELD

Practice the 5G position open-root V-groove pipe welds using stringer beads. Weld uphill by starting on a tack positioned at the bottom of the pipe and welding uphill toward the top. Start with an overhead torch push angle of 15 to 20 degrees from a normal to the pipe surface. As the overhead position gradually alters to the vertical position, gradually increase the torch push angle to 30 degrees from the normal and the rod angle to 45 degrees from a tangent to the pipe. As the vertical position gradually alters to the flat position, gradually decrease the torch and rod angles until the torch push angle is 15 to 20 degrees from the normal and the rod is again 15 to 20 degrees from the tangent to the pipe surface.

Figure 5-4 shows the 5G pipe position torch and rod angles.

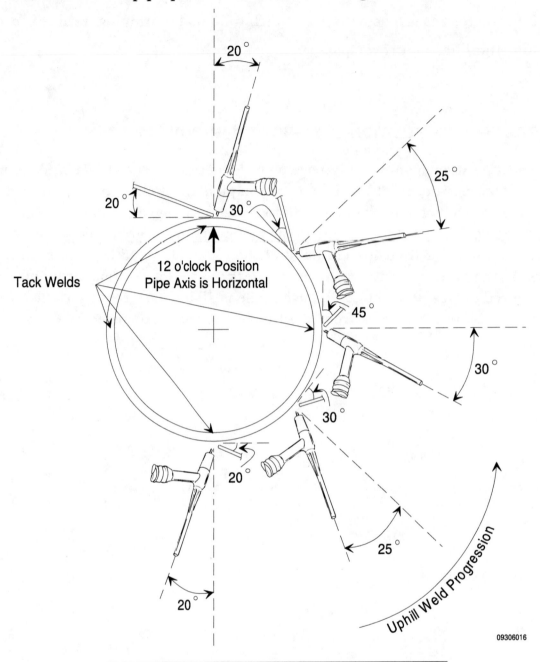

Figure 5-4. 5G Pipe Position Torch And Rod Angles

Pay particular attention at the termination of the weld to fill the crater. Follow these steps to practice open-root V-groove pipe welds in the 5G position.

Step 1 Tack-weld the practice coupon together following the example given in section 3.3.0, "Tacking Pipe Welding Coupons."

Step 2 Tack weld the pipe weld coupon to the positioning arm with the pipe axis horizontal. Be sure to position the coupon so that the tack welds are not in the regions from which the test coupons will be cut.

WELDING THREE TRAINEE TASK MODULE 09306

Step 3 If gas backing is to be used, attach the gas retaining device, backing gas hose and set the backing gas flowmeter for about 8cfh.

Figure 5-5 shows the 5G pipe coupon position and bead sequence.

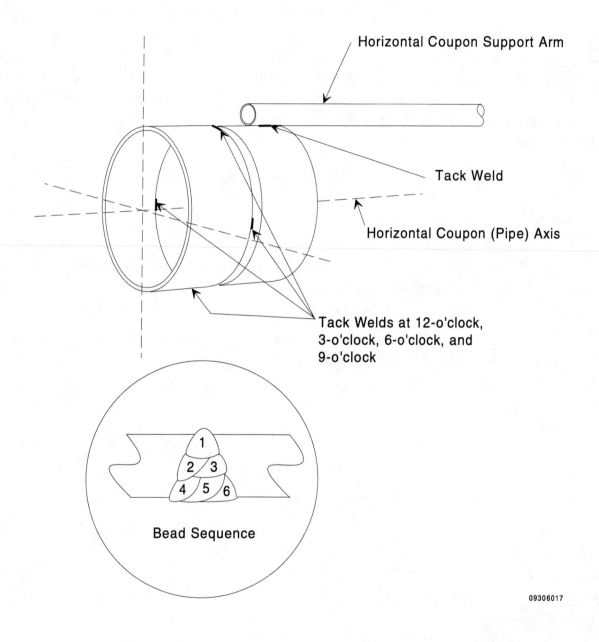

Horizontal Coupon Support Arm

Tack Weld

Horizontal Coupon (Pipe) Axis

Tack Welds at 12-o'clock, 3-o'clock, 6-o'clock, and 9-o'clock

Bead Sequence

09306017

Figure 5-5. 5G Pipe Coupon Position And Bead Sequence

Step 4 Position the arm and coupon to allow welding access all around the coupon.

Step 5 Run the root pass by starting on the tack weld positioned at the bottom of the coupon.

Step 6 Continue to run the filler and cover passes.

5.4.0 PRACTICE 6G POSITION OPEN-ROOT V-GROOVE PIPE WELD

Practice the 6G position open-root V-butt pipe welds using stringer beads. Weld uphill by starting at the bottom of the pipe and welding toward the top. Start with an overhead torch push angle of 15 to 20 degrees from normal to the pipe surface. As the overhead position gradually alters to the vertical position, gradually increase the torch push angle to 30 degrees from the normal and increase the rod angle to 45 degrees from a tangent to the pipe. As the vertical position gradually alters to a horizontal position, gradually decrease the torch and rod angles until the torch is 15 to 20 degrees from the normal and the rod is 15 to 20 degrees from the tangent to the pipe surface. Constantly adjust the transverse angle to keep the arc on the surfaces requiring the heat. Pay particular attention at the termination of the weld to fill the crater.

Follow these steps to practice open-root V-butt pipe welds in the 6G position.

Step 1   Tack weld the practice coupon together following the example given in section 3.3.0, "Tacking Pipe Welding Coupons."

Step 2   Tack weld the pipe weld coupon to the positioning arm with the pipe axis inclined 45 degrees to the horizontal. Be sure to position the coupon so that the tack welds are not in the regions from which the test coupons will be cut.

Step 3   If gas backing is to be used, attach the gas retaining device, backing gas hose and set the backing gas flowmeter for about 8cfh.

Figure 5-6 shows the 6G pipe coupon position and bead sequence.

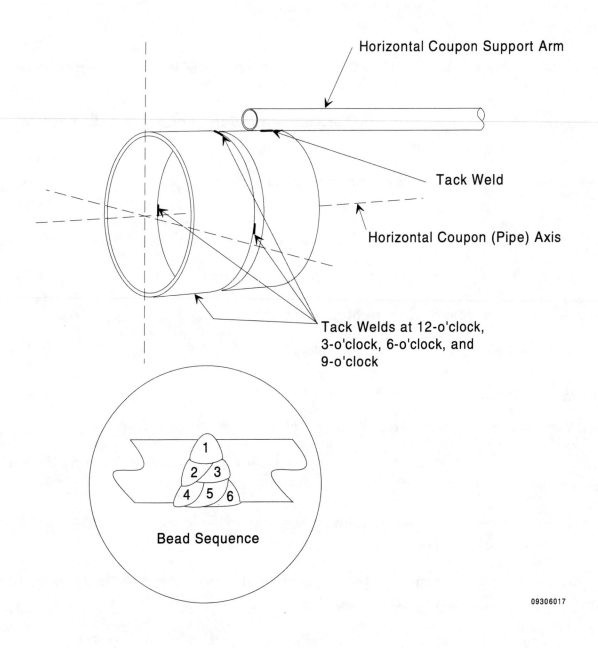

Figure 5-6. 6G Pipe Coupon Position And Bead Sequence

Step 4 Position the arm and coupon to allow welding access all around the coupon.

Step 5 Run the root pass by starting the root pass on the tack weld positioned at the bottom of the coupon.

Step 6 Run the filler and cover passes using stringer beads.

SELF-CHECK REVIEW

1. What is the recommendation for the use of root protection (backing) when performing GTAW?

2. Which type of backing, (gas or flux) provides the best and highest quality protection when welding open root joints?

3. What is the most common weld joint for joining medium- and thick-walled critical and non-critical piping.

4. When using GTAW, what are the usual pipe-end bevel angles for open-root V-groove pipe joints?

5. If backing is required for welding, does backing have to be used for tacking?

6. For GTAW, what is the land size range for open-root pipe weld?

7. What care must be taken when tightening the torch collet?

8. What is the correct GTAW tungsten tip shape for welding V-grooves in carbon steel pipe?

9. What is the usual GTAW electrode stickout length (range)?

10. What V-groove pipe positions can be acceptably welded with GTAW?

11. From what regions will the destructive test specimens be removed on 5G and 6G position V-groove pipe welds?

12. What is the maximum root penetration on the inside of a pipe, for material thickness greater than 1/4-inch, allowed by common welding codes?

13. Why must the ground clamp be attached to the pipe (as opposed to the table or roller stand) when welding in the 1G position?

14. When welding the root pass in the 5G or 6G positions, where should the root pass start?

ANSWERS TO SELF-CHECK REVIEW

1. Root protection (backing) is generally recommended when tacking and welding open root joints using the GTAW process. (2.0.0)

2. Gas backing provides the best and highest quality protection when welding open root joints. (2.1.0)

3. The most common pipe weld joint is the open-root V-groove butt joint. (3.2.0)

4. Pipe end bevel angles of either 30 or 37-1/2 degrees. (3.2.0)

5. Yes. (3.3.0)

6. Land widths may range between 1/32 and 3/32 inch. (3.3.0)

7. If the collet nut is over-tightened the threads will be damaged when the torch heats when being used. If the collet is left too loose, the collet and electrode will overheat and damage the torch. (3.5.0)

8. Tapered two to three times its diameter, with a dull point. (3.6.0)

9. The usual GTAW stickout is from two to three times the electrode diameter extending beyond the cup end. (3.6.1)

10. All pipe welding positions. (4.0.0)

11. When 5G or 6G welds are to be destructively tested, the test specimens will be cut from the four regions midway between the 12-o'clock and 3-o'clock positions, the 3-o'clock and 6-o'clock positions, the 6-o'clock and 9-o'clock positions and the 9-o'clock and 12-o'clock positions. (The 12-o'clock position is the top of the weld groove; the 6-o'clock position is the bottom of the weld groove.) (4.1.0)

12. 1/8 inch maximum. (4.2.0)

13. Failure to attach the ground clamp to the pipe coupon can result in variations in welding current, damage to the roller bearings and arcing on the rollers and pipe coupon. (5.1.0)

14. On a tack weld at the bottom of the pipe. (5.3.0, 5.4.0)

6.0.0 PERFORMANCE ACCREDITATION TASKS

The following tasks are designed to evaluate your ability to run open-root V-groove pipe welds with SMAW equipment in the standard test positions using carbon steel filler rod. Perform each task when you are instructed to do so by your instructor. As you complete each task, take it to your instructor for evaluation. Do not proceed to the next task until instructed to do so by your instructor.

6.1.0 OPEN-ROOT V-GROOVE PIPE WELD IN THE 2G POSITION

Using 1/8 inch carbon steel filler rod for the root pass and either 3/32 or 1/8 inch carbon steel filler rod for the filler and cover passes, make an open-root V-groove weld on pipe in the 2G position as shown in *Figure 6-1*.

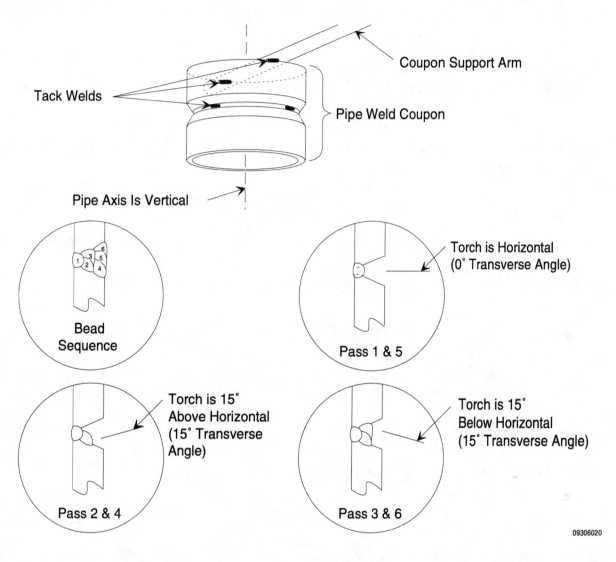

Figure 6-1. Open-Root V-groove Pipe Weld in the 2G Position

Criteria For Acceptance:

- Uniform rippled appearance on the bead face
- Craters and restarts filled to the full cross section of the weld
- Uniform weld size plus or minus 1/16 inch
- Acceptable weld profile in accordance with the *ASME Boiler and Pressure Vessel Code, Section IX*.
- Smooth transition with complete fusion at the toes of the weld
- Complete, uniform root penetration at least flush with the inside of the pipe to a maximum protrusion of 1/8 inch.

- No porosity
- No oxidation
- No undercut
- No tungsten inclusions
- No pinholes (fisheyes)
- Acceptable guided bend test results

6.2.0   OPEN-ROOT V-GROOVE PIPE WELD IN THE 5G POSITION

Using 1/8 inch carbon steel filler rod for the root pass and either 3/32 or 1/8 inch carbon steel filler rod for the filler and cover passes, make an open-root V-groove weld on pipe in the 5G position as shown in *Figure 6-2*.

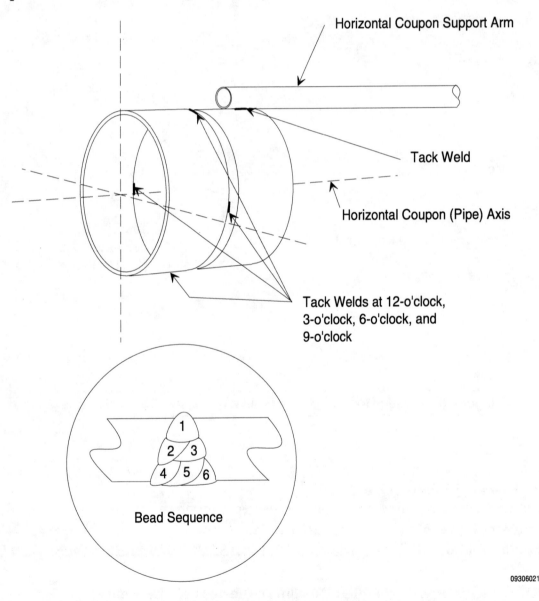

09306021

Figure 6-2. Open-Root V-groove Pipe Weld in the 5G Position

Criteria For Acceptance:

- Uniform rippled appearance on the bead face
- Craters and restarts filled to the full cross section of the weld
- Uniform weld size plus or minus 1/16 inch
- Acceptable weld profile in accordance with the *ASME Boiler and Pressure Vessel Code, Section IX.*
- Smooth transition with complete fusion at the toes of the weld
- Complete, uniform root penetration at least flush with the inside of the pipe to a maximum protrusion of 1/8 inch.
- No porosity
- No oxidation
- No undercut
- No tungsten inclusions
- No pinholes (fisheyes)
- Acceptable guided bend test results

6.3.0   OPEN-ROOT V-GROOVE PIPE WELD IN THE 6G POSITION

Using 1/8 inch carbon steel filler rod for the root pass and either 3/32 or 1/8 inch carbon steel filler rod for the filler and cover passes, make an open-root V-groove weld on pipe in the 6G position as shown in *Figure 6-3*.

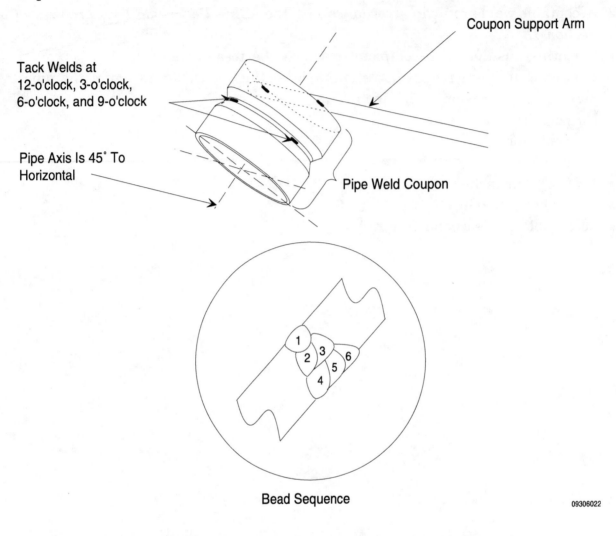

Figure 6-3. Open-Root V-groove Pipe Weld in the 6G Position.

Criteria For Acceptance:

- Uniform rippled appearance on the bead face
- Craters and restarts filled to the full cross section of the weld
- Uniform weld size plus or minus 1/16 inch
- Acceptable weld profile in accordance with the *ASME Boiler and Pressure Vessel Code, Section IX.*
- Smooth transition with complete fusion at the toes of the weld
- Complete, uniform root penetration at least flush with the inside of the pipe to a maximum protrusion of 1/8 inch.

- No porosity
- No oxidation
- No undercut
- No tungsten inclusions
- No pinholes (fisheyes)
- Acceptable guided bend test results

SUMMARY

Being able to make open-root V-groove welds on pipe in all positions is one of the more difficult skills a welder must develop. Practice these welds until you can consistently produce acceptable welds as defined in the "Criteria for Acceptance" sections of the Performance Qualification Tasks.

References

For advanced study of topics covered in this module, the following works are suggested:

Pipe Welding Techniques (1985), Griffin, Roden, Jeffus and Briggs, Delmar Publishers Inc., 2 Computer Drive West, Box 15-015, Albany, N.Y. 12212. Phone 1-800-347-7707.

Welding Skills, Giachino and Weeks, American Technical Publishers Inc, Homewood, IL, 60430. Phone 1-800-323-3471.

Welding Principles and Applications (1985), Jeffus and Johnson, Delmar Publishers, Inc., 2 Computer Drive West, Box 15-015, Albany, N.Y. 12212. Phone 1-800-347-7707.

Modern Welding (1988), Althouse, Turnquist, Bowditch, and Bowditch, Goodheart-Willcox Publishers, South Holland, IL. Phone 1-800-323-0440.

GTAW Handbook (1988), William H. Minnick, Goodheart-Willcox Publishers, South Holland, IL. Phone 1-800-323-0440.

PERFORMANCE/LABORATORY EXERCISES

1. Weld pipe, open-root V-groove joint in the 1G position, using GTAW and carbon steel filler rod.
2. Weld pipe, open-root V-groove joint in the 2G position, using GTAW and carbon steel filler rod.
3. Weld pipe, open-root V-groove joint in the 5G position, using GTAW and carbon steel filler rod.
4. Weld pipe, open-root V-groove joint in the 6G position, using GTAW and carbon steel filler rod.

The NCCER makes every effort to keep these manuals up-to-date and free of technical errors. We appreciate your help in this process. If you have an idea for improving this manual, or if you find an error, a typographical mistake, or an inaccuracy in the *Wheels of Learning*, please write us, using this form or a photocopy. Be sure to include the exact module number, page number, a description of the problem, and the correction, if possible. We'll do our best to correct it in later editions. Thank you for your assistance.

Write: *Wheels of Learning*
National Center for Construction Education and Research
P.O. Box 141104
Gainesville, FL 32614-1104

Fax: 352-334-0932

WHEELS OF LEARNING USER UPDATE

Please let us know if you have found an inaccuracy, error, or other problem in a *Wheels of Learning* manual. Use this form or write us a letter. Please be sure to tell us the exact module name and module number, the page number, and the problem. Thanks for your help.

Craft _____ Module Name _____

Module Number _____ Page Number(s) _____

Description of Problem _____

_____

_____

(Optional) Correction of Problem _____

_____

_____

(Optional) Your Name and Address _____

_____

Oxyfuel Cutting

Module 09101

Welding Trainee Task Module 09101

NATIONAL
CENTER FOR
CONSTRUCTION
EDUCATION AND
RESEARCH

OXYFUEL CUTTING

OBJECTIVES

Upon completion of this module, the trainee will be able to:

1. Explain oxyfuel cutting safety.
2. Identify and explain oxyfuel cutting equipment.
3. Identify and explain oxyfuel flames.
4. Identify and explain backfire and flashbacks.
5. Set up oxyfuel equipment.
6. Light and adjust an oxyfuel torch.
7. Shut down oxyfuel cutting equipment.
8. Disassemble oxyfuel equipment.
9. Change empty cylinders.
10. Perform housekeeping tasks.
11. Perform oxyfuel cutting:
    - Straight line and square shapes
    - Piercing and slot cutting
    - Bevels
    - Washing
    - Gouging

Prerequisites

Successful completion of the following modules is required before beginning study of this module:

Common Core Curricula

Required Student Materials

Each trainee will need:

1. Personal protective equipment
2. Leather gauntlet-type gloves
3. Oxyfuel cutting equipment
4. Tip cleaners
5. Cutting goggles
6. Chipping hammer
7. Pliers
8. Tape measure
9. Soapstone
10. Friction lighter

Each trainee will need access to:

1. Framing squares
2. Combination squares with protractor
3. Various sizes of cutting tips for 14-gauge to 1-inch plate
4. Washing tips
5. Gouging tips
6. Steel plate
   - Thin (16 to 10 gauge)
   - Thick (1/4 to 1 inch)
7. 10" crescent wrench

Course Map Information

This course map shows all of the Wheels of Learning modules in the first level of the Welding curricula. The suggested training order begins at the bottom and proceeds up. Skill levels increase as a trainee advances on the course map. The training order may be adjusted by the site Training Program Sponsor.

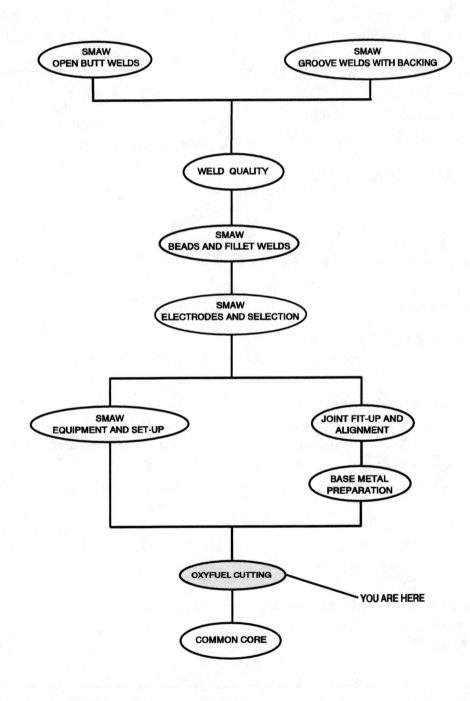

TABLE OF CONTENTS

Trade Terms Used in this Module

Backfire: A loud snap or pop as the torch flame is extinguished.

Carburizing Flame: A flame burning with an excess amount of fuel.

Drag Lines: The lines on the kerf that result from the travel of the cutting oxygen stream into, through and out of the metal.

Ferrous Metals: Metals containing iron.

Flashback: The flame burning back into the tip, torch, hose or regulator, causing a high-pitched whistling or hissing sound.

Kerf: The edge of the cut.

Neutral Flame: A flame burning with equal amounts of fuel gas and oxygen.

Oxidizing Flame: A flame burning with an excess amount of oxygen.

Pierce: To penetrate through steel plate with an oxyfuel cutting torch.

Slag: The material that is expelled from the kerf when oxyfuel cutting.

Soapstone: Soft white stone used to mark metal.

Weldment: An assembly whose component parts are joined by welding

1.0.0 INTRODUCTION

Oxyfuel cutting is a process that uses a flame to cut ferrous metals. The flame is produced by burning a fuel gas mixed with pure oxygen. The flame heats the metal to be cut to the kindling temperature, then a stream of pure oxygen is directed at the metal's surface. This causes the metal to instantaneously oxidize or burn. The cutting process results in oxides which mix with molten iron and produce slag which is blown from the cut by the jet of cutting oxygen. The oxidation process which takes place during the cutting operation is similar to a greatly speeded-up rusting process. *Figure 1-1* shows oxyfuel cutting.

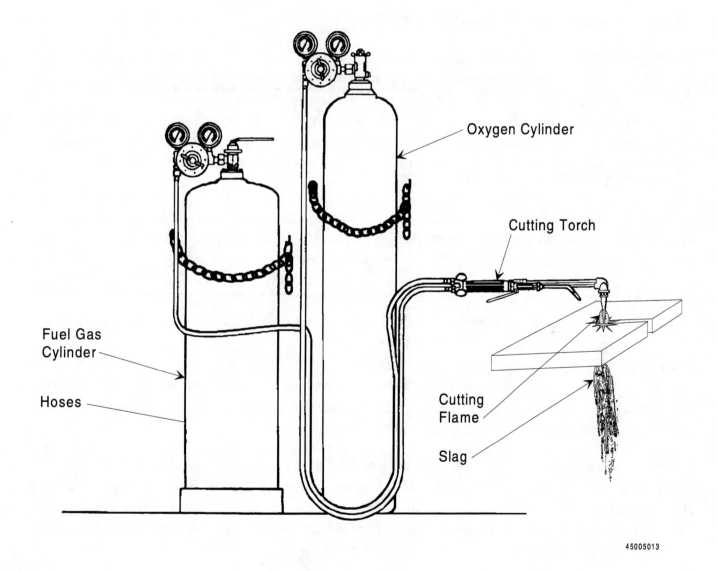

45005013

Figure 1-1. Oxyfuel Cutting

The oxyfuel cutting process can be used to quickly cut, trim and shape metals. It can be used to easily cut the hardest metals, but can only be used on ferrous metals such as straight carbon steels which oxidize rapidly. Oxyfuel cutting does not work with stainless steels and non-ferrous metals because they do not readily oxidize.

2.0.0   OXYFUEL SAFETY

The proper safety equipment and precautions must be used when working with oxyfuel equipment because of the potential danger from the high pressure flammable gases and high temperatures used. The following sections will explain the safety equipment and procedures to use.

2.1.0   PERSONAL PROTECTIVE EQUIPMENT

When performing oxyfuel cutting wear clothing that will protect you from flying sparks and heat. Shirts should be long-sleeved, have pocket flaps and be worn with the collar buttoned. Pants should not have cuffs and should fit so they hang straight down the leg, touching the shoe-tops without creases. Cuffs and creases can catch sparks, which can cause fires. Never wear polyester or other synthetic fibers. Sparks will melt these materials, causing serious burns. Wool and cotton are more resistant to sparks and should be worn instead of synthetic fibers.

WARNING!   Never carry matches or gas-filled lighters in your pockets. Sparks could cause the matches to ignite or the lighter to explode, causing serious injury.

Never wear low-cut shoes when oxyfuel cutting. Sparks and slag can enter low-cut shoes, causing burns to the feet. Wear leather work boots with high tops (at least eight inches). They should fit tightly around the leg to prevent sparks from going down into the boot. A cap, worn with the bill pointing to the back, will protect the head and prevent sparks from going down the back of your shirt collar. To protect the hands and wrists, always wear all-leather gauntlet-type welding gloves when cutting.

Snug-fitting cutting goggles must also be worn to protect the eyes from glare, heat, flying sparks and slag. Cutting goggles should have an approved colored filter lens covered by a clear protective cover lens. Filter lenses with shade five or six are generally recommended for most oxyfuel cutting. The clear lens is placed outside the colored filter plate to protect it from spatter.

Ear plugs should also be worn when cutting. The ear plugs prevent flying sparks from entering the ear and burning the ear canal or perforating the eardrum.

Figure 2-1 shows personal protective equipment.

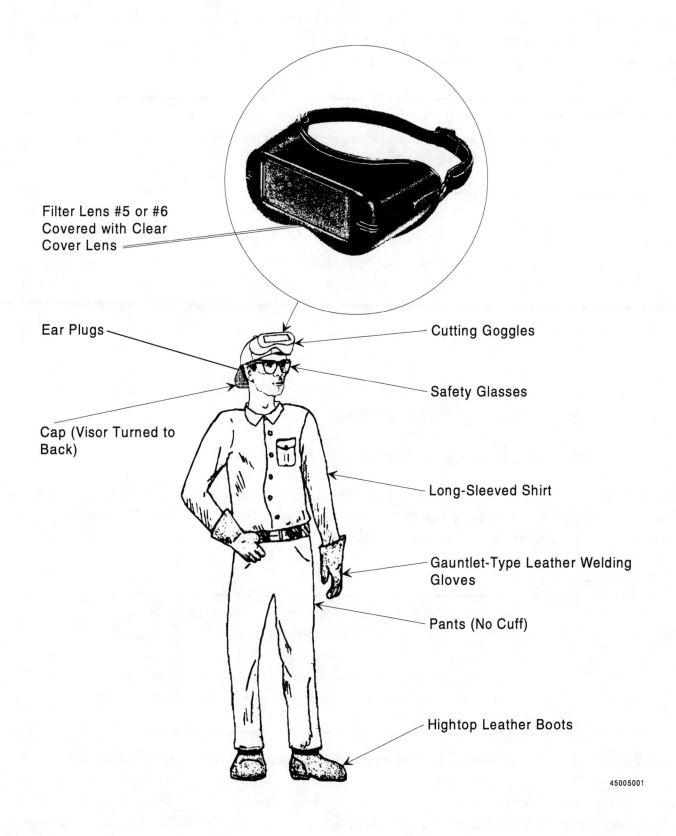

Filter Lens #5 or #6 Covered with Clear Cover Lens

Ear Plugs

Cap (Visor Turned to Back)

Cutting Goggles

Safety Glasses

Long-Sleeved Shirt

Gauntlet-Type Leather Welding Gloves

Pants (No Cuff)

Hightop Leather Boots

45005001

Figure 2-1. Personal Protective Equipment

2.2.0 VENTILATION

Always perform cutting operations in a well-ventilated area. Heating and cutting metals with an oxyfuel torch can create toxic fumes. The most common hazardous material is zinc. Zinc is present in galvanized coatings and brass.

WARNING!	Never cut galvanized material without proper ventilation. The zinc oxide fumes given off as the galvanized material is cut are hazardous. Also use a respirator when cutting galvanized material.

2.3.0 AREA SAFETY

Before beginning a cutting operation, you must check the area for fire hazards. Cutting sparks can fly 30 feet or more and can drop several floors. Any flammable material in the area must be moved or covered. In addition, have an approved fire extinguisher available before starting any cutting operation.

Note	Most sites require a fire watch and/or hot work permit. Check your site requirements regarding fire watches and hot work permits before performing any cutting operations.

2.4.0 HOT WORK PERMITS AND FIRE WATCHES

Whenever oxyfuel cutting equipment is used there is a greater danger of fire. Hot work permits and fire watches have been developed to minimize this danger. Most sites require the use of hot work permits and fire watches and when they are violated, there are severe penalties imposed.

WARNING!	Never perform any type of heating, cutting or welding until you have obtained a hot work permit and established a fire watch. If you are unsure of the procedure, check with your supervisor. Violation of the hot work permit and fire watch can result in serious injury or death.

A hot work permit is an official authorization from the site manager to perform work which may pose a fire hazard. The permit will include information such as the time, location and type of work being done. The hot work permit system promotes the development of standard fire safety guidelines. Permits also help managers to keep records of who is working where and at what time. This information is essential in the event of an emergency or other times when personnel need to be evacuated.

During a fire watch, one person other than the cutting operator constantly scans the work area for fires. Fire watch personnel should have ready access to fire extinguishers and alarms and know how to use them. Cutting operations should never be performed without a fire watch.

2.5.0 CUTTING CONTAINERS

Before welding or cutting containers such as tanks or barrels, check to see if they have contained combustible and/or hazardous materials or residues of these materials. Such materials include:

- Petroleum products
- Chemicals that give off toxic fumes when heated
- Acids that could produce hydrogen gas as a result of a chemical reaction
- Any explosive or flammable residue

To identify the contents of containers, check the label and then refer to the MSDS (Material Safety Data Sheet). The MSDS provides information about the chemical that will help you determine if the material is hazardous. If the container label is missing, or if you suspect the container has been used to hold materials other than what is on the label, do not proceed until the material is identified.

If a container has held any hazardous materials, it must be cleaned before cutting or welding takes place. Clean containers by steam cleaning, flushing with water or washing with detergent until all traces of the material have been removed.

WARNING!   Clean containers only in well-ventilated areas. Vapors
                 can accumulate during cleaning, causing explosions or
                 injury.

After cleaning the container, fill it with water or an inert gas such as argon or carbon dioxide (CO_2) for additional safety. Air, which contains oxygen, is displaced from inside the container by the water or inert gas. Without oxygen, combustion cannot take place. When using water, position the container to minimize the air space. When using an inert gas, provide a vent hole so the inert gas can purge out the air containing oxygen.

Figure 2-2 shows how to use water in a container to minimize free air space.

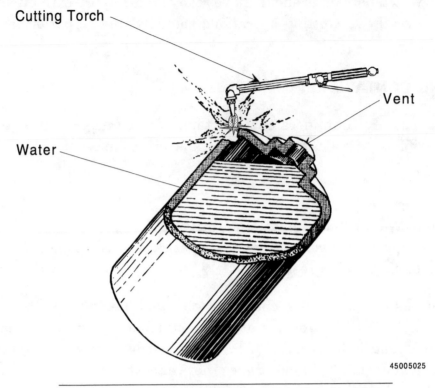

Figure 2-2. Using Water to Minimize the Air Space

2.6.0. OXYGEN

Oxygen supports combustion. Its presence around flammable materials or sparks can create conditions that encourage flames to burn out of control. Therefore, keep oxygen away from combustible materials and sources of flame, especially substances containing oil, grease or other petroleum products. In addition, the oxygen in cylinders should never be used as compressed air. The oxygen will react with the oil residue in air tools, causing explosions. Do not use oxygen to blow dust or debris. Even a small spark will cause an explosion.

WARNING!   Never use oxygen to blow off clothing. If a spark hits the clothing, the oxygen trapped in the clothing will cause it to burn rapidly and out of control.

2.7.0 PERFORMING HOUSEKEEPING TASKS

Maintaining a clean and neat work area promotes safety and efficiency. Be sure to:

- Pick up cutting scraps at the completion of work.
- Sweep up any scraps or debris around the work area at the completion of work.
- Ensure that cylinders and equipment are returned to the proper places at the completion of work.
- Ensure cut metals and slag are cooled before disposing of them to provent fire.

3.0.0   OXYFUEL CUTTING EQUIPMENT

The equipment used to perform oxyfuel cutting includes oxygen and fuel gas cylinders, oxygen and fuel gas regulators, hoses and a cutting torch. Oxyfuel cutting equipment is shown in *Figure 3-1*.

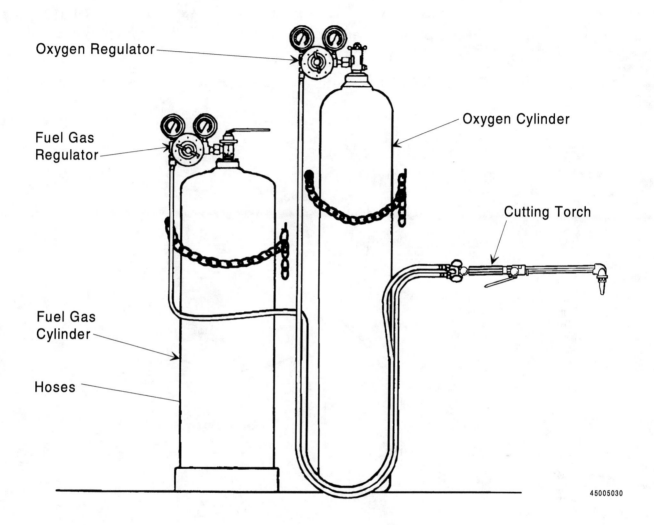

Figure 3-1. Oxyfuel Cutting Equipment

3.1.0   OXYGEN

Oxygen is a colorless, odorless, tasteless gas that supports combustion. Combined with burning material, pure oxygen causes the fire to flare and burn out of control. When mixed with fuel gases, oxygen produces the high-temperature flame required for flame cutting metals.

3.1.1 Oxygen Cylinders

Oxygen is stored at over 2000 psi in hollow steel cylinders. The cylinders come in a variety of sizes based on the cubic feet of oxygen they hold. The smallest cylinder holds about 40 cubic feet of oxygen and the largest holds about 300 cubic feet. The most common size oxygen cylinder used for welding and cutting operations is the 244-cubic-foot cylinder. It is nearly five feet tall and nine inches in diameter. The shoulder of the oxygen cylinder has the name of the gas stamped, labelled or stencilled on it. *Figure 3-2* shows oxygen cylinders.

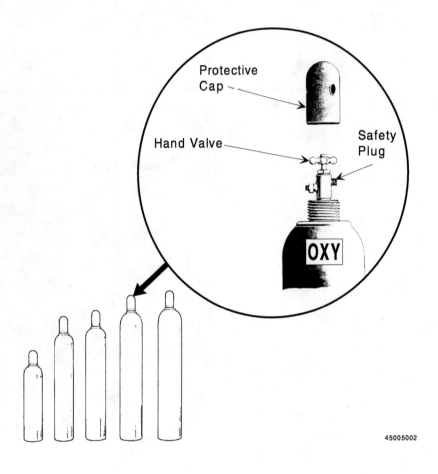

45005002

Figure 3-2. Oxygen Cylinders

The oxygen cylinder has a bronze cylinder valve on top. Turning the cylinder valve's handwheel controls the flow of oxygen out of the cylinder. A safety plug on the side of the bronze cylinder valve allows oxygen in the cylinder to escape if the pressure in the cylinder rises to high. Care must be used when handling oxygen cylinders because oxygen is stored at such high pressures. When not in use, the cylinder valve must always be covered with the protective steel cap.

WARNING!　　Do not remove the protective cap unless the cylinder is secured. If the cylinder fell over and the nozzle broke off, the cylinder would shoot like a rocket, causing severe injury or death to anyone in its way.

3.2.0　ACETYLENE

Acetylene gas is a compound of carbon and hydrogen that is formed by dissolving calcium carbide in water. It has a strong, distinctive odor. In its gaseous form acetylene is extremely unstable and explodes easily. Because of this instability, it must remain at pressures less than 15 psi when in its gaseous form. At higher pressures, acetylene gas breaks down chemically, producing heat and pressure that could result in a violent explosion. When combined with oxygen, acetylene creates a flame that burns hotter than 5500 degrees Fahrenheit, one of the hottest gas flames. Acetylene can be used for flame cutting, welding, heating, flame hardening and stress relieving.

3.2.1　Acetylene Cylinders

Because of the explosive nature of acetylene gas, it cannot be stored in its gaseous form above 15 psi. Acetylene cylinders are specially constructed to store acetylene at higher pressures. The acetylene cylinder is filled with a porous material that is soaked with acetone to absorb the acetylene. Because of the liquid acetone inside the cylinder, acetylene cylinders must always be used in an upright position. If the cylinder is tipped over, stand the cylinder upright and wait at least 30 minutes before using it. If liquid acetone is withdrawn from a cylinder it will gum up the safety check valves and regulators.

Acetylene cylinders have safety fuse plugs in the top and bottom of the cylinder that melt at 220 degrees Fahrenheit. In the event of a fire the fuse plugs will release the acetylene gas, preventing the cylinder from exploding. There are different styles of valves for the tops of acetylene cylinders. Some have a handwheel, while others require a special square key wrench to operate.

Figure 3-3 shows an acetylene cylinder.

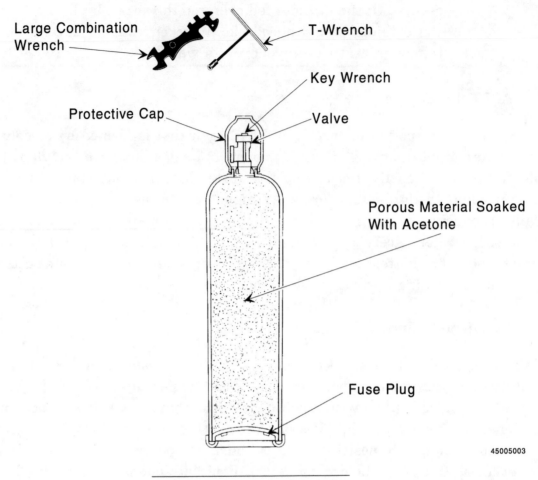

Figure 3-3. Acetylene Cylinder

3.3.0 LIQUEFIED FUEL GASES

Many fuel gases other than acetylene are used for cutting. They include natural gas and liquefied fuel gases such as MAPP, propylene and propane. Their flames are not as hot as acetylene but they are cheaper and safer to use. The supervisor at your job site will determine which fuel gas to use.

The following chart compares the temperatures of oxygen and fuel gas flames.

Acetylene	Over 5500°F
MAPP	5300°F
Propylene	5190°F
Natural Gas	4600°F
Propane	4580°F

"MAPP" stands for methylacetylene propadiene and is a Dow Chemical Company product that is a chemical combination of acetylene and propane gases. MAPP gas burns at temperatures almost as high as acetylene and has the stability of propane. Because of this stability it can be used at pressures over 15 psi and is not as likely as acetylene to backfire or flashback. MAPP also has an offensive odor which can be detected easily. MAPP gas can be used for flame cutting, scarfing, heating, stress relieving, brazing and soldering.

Propylene mixtures are hydrocarbon-based gases that are stable and shock-resistant. They are purchased under trade names such as High Purity Gas (HPG), APACHI, and PRESTOLENE. HPG and these other gases have distinctive odors to make leak-detection easier. They burn at temperatures around 5193 degrees Fahrenheit, hotter than natural gas and propane. These gases are stable and resistant to shock, making them relatively safe to use. Propylene gases are used for flame cutting, scarfing, heating, stress relieving, brazing and soldering.

Propane is also known as LP, or liquefied petroleum, gas. It is stable and shock-resistant, and it has a distinctive odor for easy leak detection. It burns at 4580 degrees Fahrenheit, which is the lowest temperature of any fuel gas. It has a slight tendency toward backfire and flashback and is used quite commonly for cutting procedures.

Natural gas is delivered by pipeline rather than by cylinders. It burns at about 4600 degrees Fahrenheit. Natural gas is relatively stable and shock resistant, with a slight tendency toward backfire and flashback. Because of its recognizable odor, leaks are easily detectable. Natural gas is used primarily for cutting on job sites where permanent cutting stations are set up.

3.3.1   Liquefied Fuel Gas Cylinders

Liquefied fuel gases are shipped in hollow steel cylinders. When empty, they are much lighter than acetylene cylinders.

The liquefied fuel gas is stored in hollow steel cylinders of various sizes. They can hold from 30 to 225 pounds of fuel gas. As the cylinder valve is opened, the vaporized gas is withdrawn from the cylinder. The remaining liquefied gas absorbs heat and releases additional vaporized gas. The pressure of the vaporized gas varies with the outside temperature. The colder the outside temperature, the lower the vaporized gas pressure will be. If high volumes of gas are removed from a liquefied fuel gas cylinder, the pressure drops and the temperature of the cylinder will also drop. A ring of frost can form around the base of the cylinder. If high withdrawal rates continue, the regulator may also start to ice up. If high withdrawal rates are required, special regulators with electric heaters should be used.

WARNING!   Never apply heat directly to a cylinder or regulator. This
                        can cause excessive pressures, resulting in an explosion.

The pressure inside a liquefied fuel gas cylinder is not an indicator of how full or empty the cylinder is. The weight of a cylinder determines how much liquefied gas is left.

Figure 3-4 shows a liquefied fuel gas cylinder.

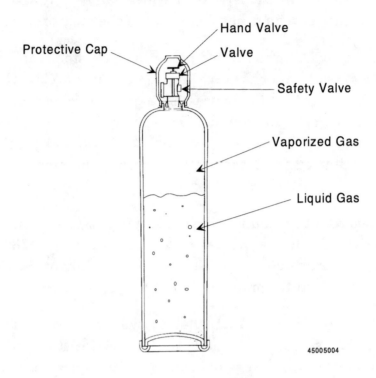

Figure 3-4. Liquefied Fuel Gas Cylinder

Liquefied fuel gas cylinders have a safety valve built into the valve at the top of the cylinder. The safety valve releases gas if the pressure begins to rise. Care must be used when handling fuel gas cylinders because the gas in cylinders is stored at such high pressures. Cylinders should never be dropped or hit with heavy objects, and should always be stored in an upright position. When not in use, the cylinder valve must always be covered with the protective steel cap.

WARNING!   Do not remove the protective cap unless the cylinder is secured. If the cylinder falls over and the nozzle breaks off, the cylinder will release highly explosive gas.

3.4.0   REGULATORS

Regulators are attached to the cylinder valve. They reduce the high cylinder pressures to the required lower working pressures and maintain a steady flow of gas from the cylinder.

WELDING ONE TRAINEE TASK MODULE 09101

Figure 3-5 shows typical regulators.

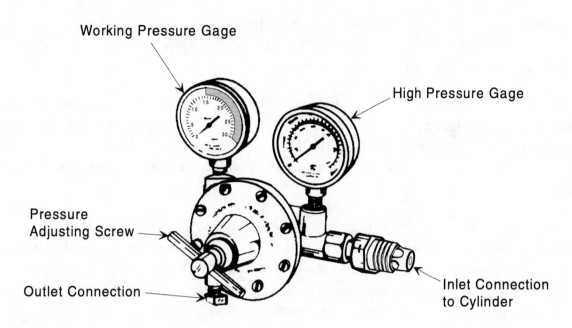

Working Pressure Gage

High Pressure Gage

Pressure Adjusting Screw

Outlet Connection

Inlet Connection to Cylinder

Fuel Gas Regulator

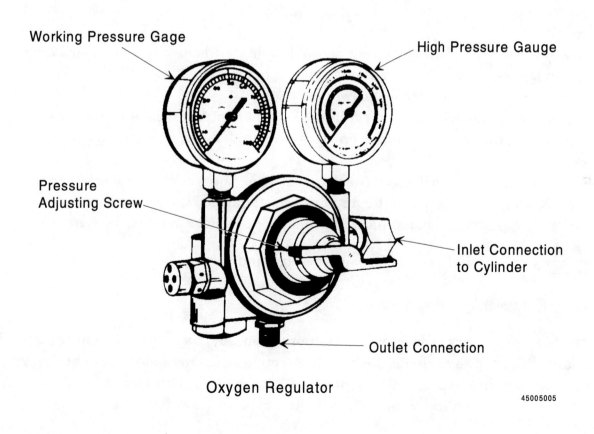

Working Pressure Gage

High Pressure Gauge

Pressure Adjusting Screw

Inlet Connection to Cylinder

Outlet Connection

Oxygen Regulator

45005005

Figure 3-5. Regulators

The pressure adjusting screw controls the gas pressure. Turned clockwise it increases the flow of gas. Turned counterclockwise it reduces or stops the flow of gas.

Most regulators contain two gauges. The high pressure or cylinder pressure gauge indicates the actual cylinder pressure, while the low pressure, or working pressure, gauge indicates the pressure of the gas leaving the regulator.

Oxygen and fuel gas regulators are different. Oxygen regulators are often painted green and always have right-hand threads on all connections. The oxygen regulator's high pressure gauge generally reads up to 3000 psi and includes a second scale which shows the amount of oxygen in the cylinder in terms of cubic feet. The low pressure or working pressure gauge may read to 100 psi or higher.

Fuel gas regulators are often painted red and always have left-hand threads on all the connections. As a reminder that the regulator has left-hand threads a V-notch is cut around the nut. The fuel gas high pressure gauge usually reads up 400 psi. The low pressure or working pressure gauge may read up to 40 psi but is always red-lined at 15 psi as a reminder that acetylene pressure should not be increased over 15 psi.

To prevent damage to regulators, always follow these guidelines:

- Never submit regulators to jarring or shaking, as this can damage the equipment beyond repair.
- Always check that the adjusting screw is released before the cylinder valve is turned on and released when the welding has been completed.
- Always open cylinder valves slowly.
- Never use oil to lubricate a regulator, as this can result in an explosion.
- Never use fuel gas regulators on oxygen cylinders or oxygen regulators on fuel gas cylinders.
- Never work with a defective regulator. If it is not working properly, shut off the gas supply and have the regulator repaired by someone who is qualified to work on it.
- Never use pliers or channel-locks to install or remove regulators.

There are two types of regulators, single-stage and two-stage.

3.4.1 Single-Stage Regulators

Single-stage regulators reduce pressure in one step. As gas is drawn from the cylinder, the internal pressure of the cylinder decreases. A single-stage regulator is unable to compensate for this decrease in internal cylinder pressure. Therefore, it becomes necessary to adjust the output pressure periodically as the gas in the cylinder is consumed.

Figure 3-6 shows a single-stage regulator.

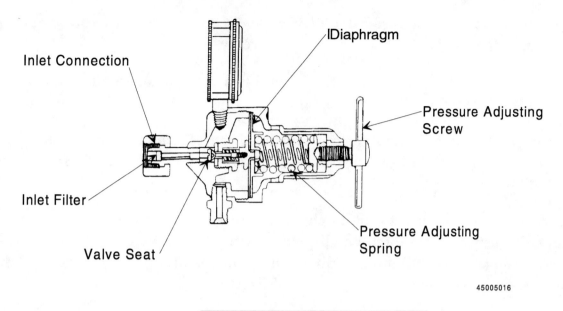

Figure 3-6. Single-Stage Regulator

3.4.2   Two-Stage Regulators

The two-stage regulator reduces pressure in two steps. It first reduces the input pressure from the cylinder to a predetermined intermediate pressure. The intermediate pressure is then adjusted by the pressure adjusting screw. With this type of regulator the delivery pressure to the torch remains constant and no readjustment is necessary as the gas in the cylinder is consumed. *Figure 3-7* shows a two-stage regulator.

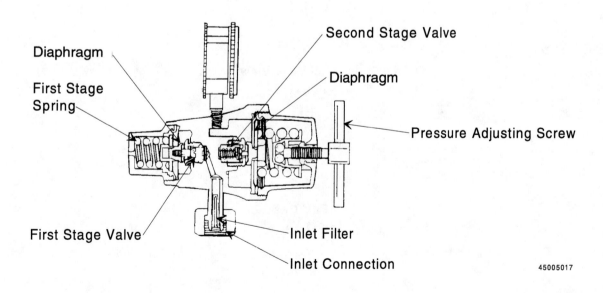

Figure 3-7. Two-Stage Regulator

3.4.3 Check Valves and Flash Arrestors

Check valves and flash arrestors are safety devices for regulators. Check valves allow gas to flow in one direction only. Flash arrestors stop fire.

Check valves consist of a ball and spring that open inside a cylinder to allow gas to move in one direction, but that close if the gas attempts to flow in the opposite direction.

Flashback arrestors prevent flashbacks from reaching the regulator. They have a flame-retarding filter that will allow heat but not flames to pass through. Flashback arrestors often contain a check valve.

Flashback arrestors are attached to the torch handle connections. Check valves and flash arrestors can be attached to the torch handle connections or to the outlet of the regulator. They have arrows on them to indicate flow direction. When installing check valves and flash arrestors be sure the arrow matches the gas flow direction.

Note   Occasionally test reverse-flow check valves by blowing into them to ensure that they work correctly.

Figure 3-8 shows check valves.

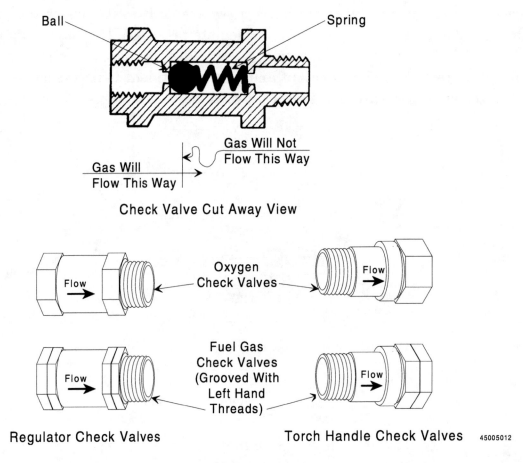

Figure 3-8. Check Valves

3.5.0   HOSES

Hoses transport gases from the regulators to the torch. Oxygen hoses are usually green or black with right-hand threaded connections, while hoses for fuel gas are usually red and have left-hand threaded connections. The fuel gas connections are also grooved as a reminder that they have left-hand threads.

Proper care and maintenance of the hose is important for maintaining a safe, efficient work area. Remember the following guidelines for hoses:

- Protect the hose from molten slag or sparks which will burn the exterior. Although some hoses are flame retardant, they will burn.
- Remove the hoses from under the metal being cut. If the metal falls on the hose it will cut it.
- Frequently inspect and replace hoses that show signs of cuts, burns, worn areas, cracks or damaged fittings.
- Never use pipe-fitting compounds or lubricants around hose connections. These compounds often contain oil or grease, which ignite and burn or explode around oxygen.

3.6.0   CUTTING TORCHES

Cutting torches mix oxygen and fuel gas for the torch flame and control the stream of oxygen necessary for the cutting jet. Depending on the job site, you may use either a one-piece or combination cutting torch.

3.6.1   One-Piece Cutting Torch

The one-piece cutting torch contains the fuel gas and oxygen valves that allow the gases to enter the chambers and then flow into the tip, where they are mixed. The main body of the torch is called the handle. The torch valves control the fuel gas and oxygen. The cutting oxygen lever, which is spring-loaded, controls the jet of cutting oxygen. Hose connections are located at the end of the torch body behind the valves. *Figure 3-9* shows a one-piece cutting torch.

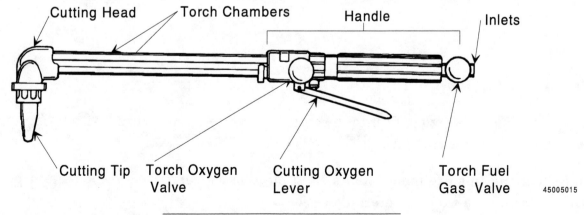

Figure 3-9. One-Piece Cutting Torch

3.6.2 Combination Cutting Torch

The combination cutting torch consists of a cutting torch attachment that fits onto a welding torch handle. There are fuel gas and oxygen valves on the torch handle. The cutting attachment has a cutting oxygen lever and another oxygen valve to control the preheat flame. When the cutting attachment is screwed onto the torch handle, the torch handle oxygen valve is opened all the way and the preheat oxygen is controlled by a oxygen valve on the cutting attachment. When the cutting attachment is removed welding and heating tips can be screwed onto the torch handle. *Figure 3-10* shows a combination cutting torch.

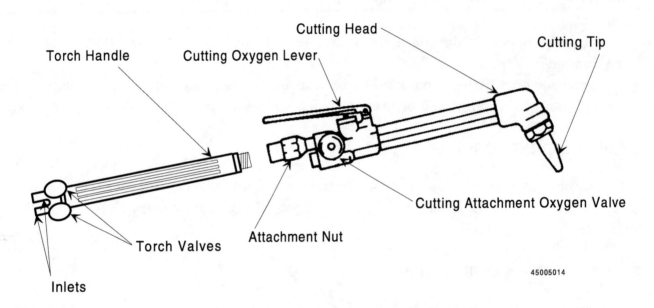

45005014

Figure 3-10. Combination Cutting Torch

3.7.0 CUTTING TORCH TIPS

Cutting torch tips, or nozzles, fit into the cutting torch and are secured with a tip nut. There are one- and two-piece cutting tips.

One-piece cutting tips are made from a solid piece of copper. Two-piece cutting tips have a separate external sleeve and internal section.

Figure 3-11 shows one- and two-piece torch tips.

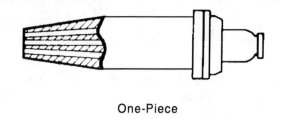

One-Piece

Two-Piece

45005018

Figure 3-11. One- and Two-Piece Cutting Tips

Most tip manufacturers supply literature explaining the appropriate torch tips and gas pressures to be used with various applications. *Figure 3-12* shows a cutting tip chart supplied by Victor Equipment Company that lists recommended tip sizes and gas pressures for use with acetylene fuel gas.

ACETYLENE CUTTING TIP CHART
CUTTING TIP SERIES 1-101, 3-101 & 5-101

Metal Thickness	Tip Size	Cutting Oxygen Pressure (PSIG)	Pre-Heat Oxygen (PSIG)	Acetylene Pressure (PSIG)	Speed I.P.M.	Kerf Width
1/8"	000	20/25	3/5	3/5	20/30	.04
1/4"	00	20/25	3/5	3/5	20/28	.05
3/8"	0	25/30	3/5	3/5	18/26	.06
1/2"	0	30/35	3/6	3/5	16/22	.06
3/4"	1	30/35	4/7	3/5	15/20	.07
1"	2	35/40	4/8	3/6	13/18	.09
2"	3	40/45	5/10	4/8	10/12	.11
3"	4	40/50	5/10	5/11	8/10	.12
4"	5	45/55	6/12	6/13	6/9	.15
6"	6	45/55	6/15	8/14	4/7	.15
10"	7	45/55	6/20	10/15	3/5	.34
12"	8	45/55	7/25	10/15	3/4	.41

Figure 3-12. Victor Equipment Company Cutting Tip Chart

45005019

The cutting torch tip used depends on the base metal thickness and fuel gas being used. Special purpose tips are also available for use on such operations as gouging and grooving.

3.7.1 Cutting Tips for Acetylene

One-piece torch tips are always used with acetylene cutting because of the high temperatures involved. They can have four, six or eight preheat holes in addition to the single cutting hole. *Figure 3-13* shows acetylene cutting torch tips.

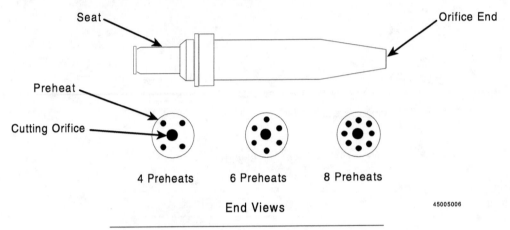

Figure 3-13. Acetylene Cutting Torch Tips

3.7.2 Cutting Tips for Liquefied Fuel Gases

Tips used with liquefied fuel gases must have at least six preheat holes. Since fuel gases burn at lower temperatures than acetylene, more holes are necessary for preheating. Tips used with liquefied fuel gases can be one- or two-piece cutting tips. *Figure 3-14* shows tips used with liquefied fuel gases.

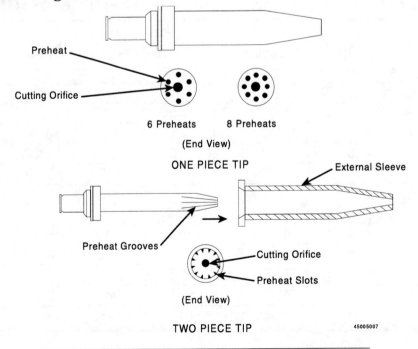

Figure 3-14. Cutting Tips for Liquefied Fuel Gases

WELDING ONE TRAINEE TASK MODULE 09101

3.7.3 Special Purpose Cutting Tips

Special cutting tips are available for special cutting jobs. These include cutting sheet metal, rivets, risers and flues, as well as washing and gouging. *Figure 3-15* illustrates special purpose cutting torch tips.

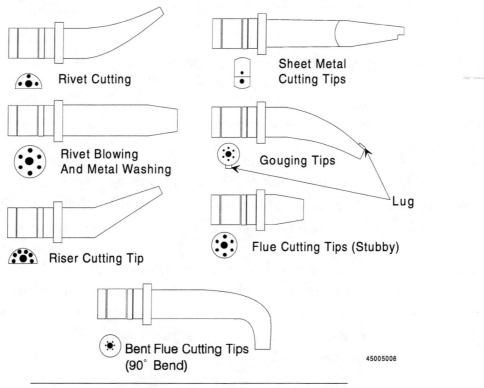

Figure 3-15. Special Purpose Cutting Torch Tips

The sheet metal cutting tip has only one preheat hole. This minimizes the heat and prevents distortion in the sheet metal. These tips are normally used with a motorized carriage but can be used for hand-cutting.

Rivet cutting tips are used to cut off rivet heads, bolt heads and nuts.

Riser cutting tips are similar to rivet cutting tips and can also be used to cut off rivet heads, bolt heads and nuts. They have extra preheat holes to cut risers, flanges or angle legs faster. They can be used for any operation which requires a cut close and parallel to another surface, such as removing a metal backing.

Rivet blowing and metal washing tips are heavy duty tips designed to withstand high heat. They are used for coarse cutting and for removing such items as clips, angles and brackets.

Gouging tips are used to groove metal in preparation for welding.

Flue cutting tips are designed to cut flues inside boilers. They also can be used for any cutting operation in tight quarters where it is difficult to get a conventional tip into position.

3.8.0   TIP CLEANERS AND TIP DRILLS

With use, cutting tips become dirty. Carbon and other impurities build up inside the holes, and molten metal often sprays and sticks onto the surface of the tip. A dirty tip will result in a poor-quality cut with an uneven kerf and excessive slag buildup. To ensure good cuts with straight kerfs and minimal slag buildup, cutting tips must be cleaned with tip cleaners or tip drills.

Tip cleaners are small round files. They usually come in a set, with files to match the diameters of the various tip holes. In addition, each set usually includes a file that can be used to recondition the face of the cutting tip. Tip cleaners are inserted into the tip hole and moved back and forth a few times to remove deposits from the hole.

Tip drills are used for major cleaning or for holes that are overly worn. They are more brittle than tip cleaners, making them more difficult to use. Tip drills are tiny drill bits sized to match the diameters of tip holes. The drill fits into a drill handle for use. The handle is held and the drill bit is turned carefully inside the hole to remove debris.

CAUTION   Tip cleaners and tip drills are brittle. Unless care is taken, they
                may break off inside a hole. Broken tip cleaners are difficult to
                remove. Improper use of tip cleaners or tip drills can enlarge
                the tip, causing improper burning of gases. If this occurs, tips
                must be discarded.

Figure 3-16 shows tip cleaners and tip drills.

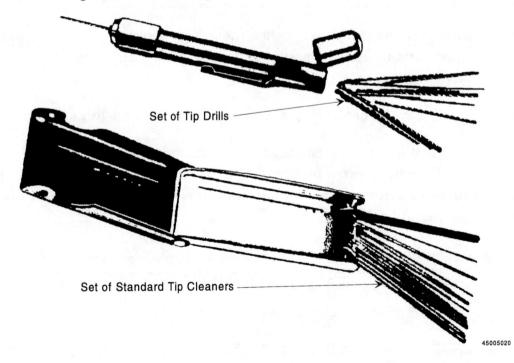

Set of Tip Drills

Set of Standard Tip Cleaners

45005020

Figure 3-16. Tip Cleaners and Tip Drills

WELDING ONE TRAINEE TASK MODULE 09101

3.9.0 FRICTION LIGHTERS

A friction lighter, also known as a striker or spark-lighter, should always be used to ignite the cutting torch. The friction lighter works by rubbing a piece of flint on a steel surface to create sparks.

WARNING!   Do not use a match or gas-filled lighter to light a torch. This could result in severe burns or could cause the lighter to explode.

Figure 3-17 shows a friction lighter.

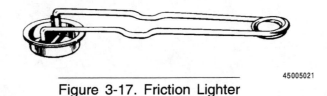

45005021

Figure 3-17. Friction Lighter

3.10.0 CYLINDER CART

The cylinder cart, or bottle cart, is a modified hand truck that has been equipped with seats and chains to hold cylinders firmly in place. Bottle carts help ensure the safe transportation of gas cylinders. *Figure 3-18* shows a cutting rig in a cylinder cart.

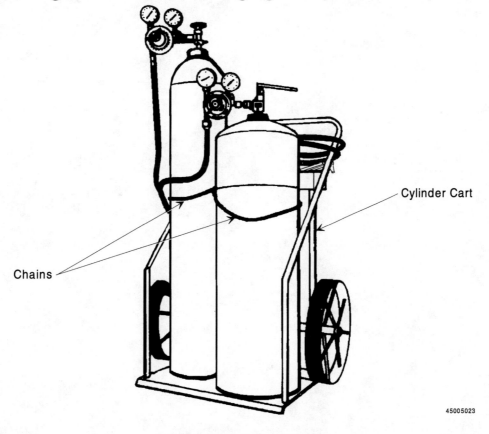

Cylinder Cart

Chains

45005023

Figure 3-18. Cylinder Cart

notes

SELF-CHECK REVIEW 1

1. Explain how the oxyfuel cutting process works.
2. List the personal protective equipment needed during oxyfuel cutting.
3. What common material gives off hazardous fumes when cut?
4. What is a hot work permit?
5. What is the proper method of cleaning containers before welding?
6. Why must oxygen be used with great caution?
7. What happens to acetylene gas at pressures higher than 15 psi?
8. Why are MAPP and other fuel gases used more frequently than acetylene gas for cutting purposes?
9. What is the purpose of regulators?
10. What are check valves?
11. How can you tell the difference between oxygen and fuel gas hoses?
12. Can two-piece cutting tips be used with acetylene?
13. Why must special care be used when using tip cleaners and tip drills?

ANSWERS TO SELF-CHECK REVIEW 1

1. Fuel gas is mixed with oxygen to produce a flame that heats a base metal to the kindling point. A stream of pure oxygen is then directed at the metal's surface. This causes the base metal to instantaneously oxidize or burn. (1.0.0)

2. Long-sleeved, buttoned-collar shirts with pocket flaps. Cuffless pants that hang straight down the leg. Clothing should only be wool or cotton. High-top, heavy leather work boots with steel toes. A cap worn with the bill pointing backward. All-leather gauntlet-type welding gloves long enough to cover the ends of the shirt sleeves. Snug-fitting cutting goggles with an approved colored filter lens covered by a clear protective cover lens. Ear plugs. (2.1.0)

3. Zinc in galvanized coatings. (2.2.0)

4. Approval given by the site manager for work involving fire hazards. (2.4.0)

5. Clean containers by steam cleaning, flushing with water or washing with detergent until all traces of the material have been removed. (2.5.0)

6. It supports combustion and can create conditions that encourage flames to burn out of control when it is near combustible or flammable materials. It can cause fires to burn rapidly and out of control. (2.6.0)

7. It explodes easily. At higher pressures, acetylene gas breaks down chemically. This quickly produces great amounts of heat and pressure and the explosion that often results is violent. (3.2.0)

8. Even though their flames are not as hot as acetylene, they are cheaper and safer to use. (3.3.0)

9. They reduce the high cylinder pressures of gases to the desired low working pressures. They maintain constant working pressures to produce a steady flow of gas from a cylinder. (3.4.0)

10. Check valves are safety devices that allow gas to flow in one direction only. (3.4.3)

11. Oxygen hoses are usually green or black with right-hand threaded connections, while hoses used with fuel gas are usually red and have grooved left-hand threaded connections. (3.5.0)

12. No. (3.7.1)

13. Tip cleaners and tip drills are brittle and unless care is taken, they may break off inside a hole. Improper use can also enlarge tip holes. (3.8.0)

4.0.0 SETTING UP OXYFUEL EQUIPMENT

When setting up oxyfuel equipment, you must follow certain procedures to ensure that the equipment operates properly and safely. The following sections explain the procedure and sequence for setting up oxyfuel equipment.

4.1.0 TRANSPORT AND SECURE CYLINDERS

Follow these steps to transport and secure cylinders.

WARNING!   Always handle cylinders with care. They are under high pressure and should never be dropped, knocked over or exposed to excessive heat. When moving cylinders, always be certain that the valve caps are in place. Use a cylinder cage to lift cylinders. Never use a sling or electromagnet.

Step 1 Transport cylinders in the upright position to the work station on a hand truck or bottle cart.

Step 2 Secure the cylinders at the work station.

Step 3 Remove the protective cap from each cylinder and inspect the outlet nozzles to make sure that the seat and threads are not damaged. Place the protective caps where they will not be lost and where they will be available when the cylinders are empty.

CAUTION   Do not transport or immediately use an acetylene cylinder found resting on its side. Stand it upright and wait at least thirty minutes to allow the acetone to settle before using it.

Figure 4-1 shows transporting cylinders.

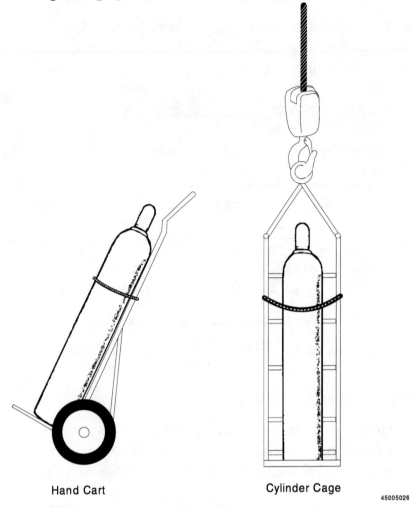

Hand Cart   Cylinder Cage

45005026

Figure 4-1. Transporting Cylinders

4.2.0 CRACK CYLINDER VALVES

Follow these steps to crack cylinder valves.

Step 1 Crack both the oxygen and fuel gas cylinder valves momentarily to remove any dirt from the valves.

WARNING! Always stand to one side of the valves when opening them to avoid injury from dirt which may be lodged in the valve.

Step 2 Wipe out the connection seat with a clean cloth. (Dirt frequently collects in the outlet nozzle of the cylinder valve and must be cleaned out to keep dirt from getting into the regulator when pressure is turned on.)

WELDING ONE TRAINEE TASK MODULE 09101

WARNING!	Be sure the cloth used does not have any oil or grease on it. Oil or grease compressed with oxygen will explode.

Figure 4-2 shows the cylinder valves.

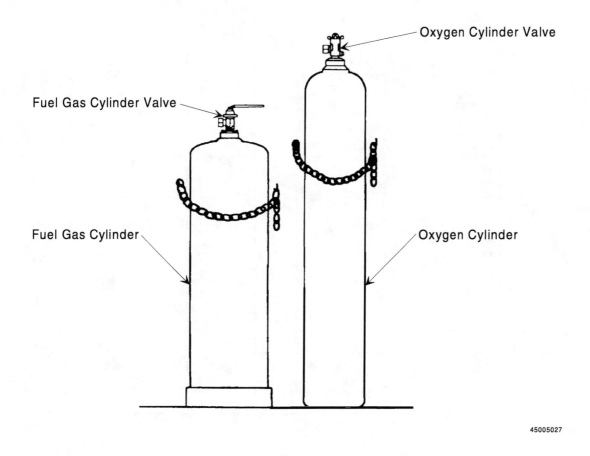

Figure 4-2. Cylinder Valves

4.3.0 ATTACH REGULATORS

Follow these steps to attach the regulators.

Step 1 Check the regulator fittings to ensure that they are free of oil or grease.

Step 2 Connect and tighten the oxygen regulator to the oxygen cylinder.

CAUTION	Do not work with a regulator that shows signs of damage such as cracked gauges, bent thumbscrews or worn threads. Set it aside for repairs. Never attempt to repair regulators yourself.

Step 3 Connect and tighten the fuel gas regulator to the fuel gas cylinder. Remember that all fuel gas fittings have left-hand threads.

Figure 4-3 shows attaching regulators.

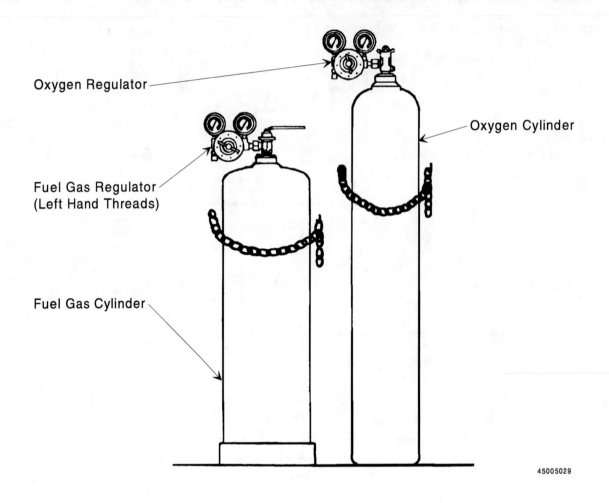

Oxygen Regulator

Oxygen Cylinder

Fuel Gas Regulator
(Left Hand Threads)

Fuel Gas Cylinder

45005029

Figure 4-3. Attaching Regulators

4.4.0   INSTALL CHECK VALVES

Follow these steps to install check valves.

Note   Check valves can be attached to either the regulator or the torch.

Step 1   Attach a reverse-flow check valve or flash arrestor (if available) to the hose connection on the oxygen regulator.

Step 2   Attach a reverse-flow check valve or flash arrestor (if available) to the hose connection on the fuel gas regulator. Keep in mind that all fuel gas fittings have left-hand threads.

Figure 4-4 shows a check valve attached to the oxygen regulator.

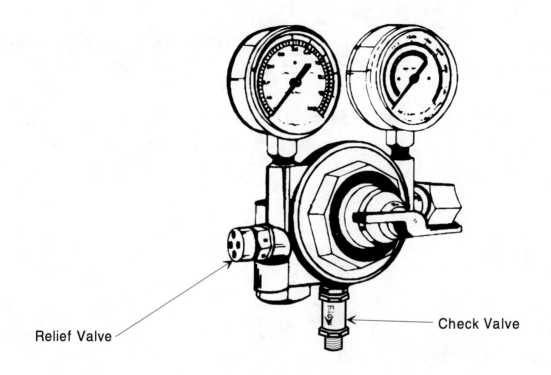

Relief Valve

Check Valve

Oxygen Regulator

45005028

Figure 4-4. Check Valve

4.5.0   CONNECT HOSES TO REGULATORS

New hoses contain talc and loose bits of rubber. These materials must be blown out of the hoses before the torch is connected. If they are not blown out they will clog the torch needle valves.

WARNING!   Never blow out hoses with compressed air. Compressed
air often contains some oil which could explode or cause
a fire when compressed in the hose with oxygen.

Follow these steps to connect the hoses to the regulators.

Step 1   Inspect both the oxygen and fuel gas hoses for any damage, burns, cuts or fraying.

Step 2   Repair any damaged hoses.

Step 3   Connect the oxygen hose to the oxygen regulator.

Step 4   Connect the fuel gas hose to the fuel gas regulator. Keep in mind that all fuel gas fittings have left-hand threads.

4.6.0   ATTACH HOSES TO THE TORCH

Follow these steps to attach the hoses to the torch:

Step 1   Attach the reverse-flow check valve (if available) to the oxygen and fuel gas hose connection on the torch body. Keep in mind that all fuel gas fittings have left-hand threads.

Step 2   Attach the oxygen hose to the oxygen fitting on the cutting torch.

Step 3   Attach the red hose to the fuel gas fitting on the cutting torch. Remember that all fuel gas fittings have left-hand threads.

Figure 4-5 shows attaching hoses.

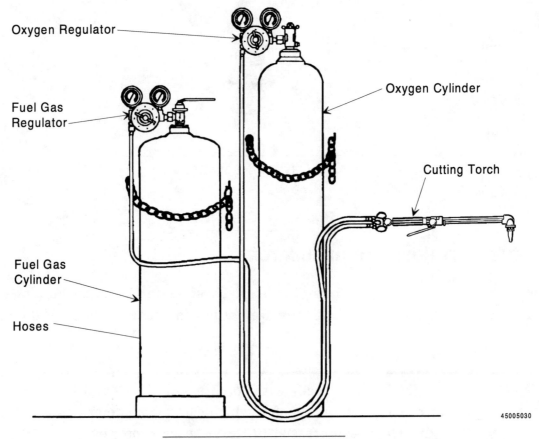

Figure 4-5. Attaching Hoses

4.7.0   INSTALL CUTTING TIP

Follow these steps to install a cutting tip in the cutting torch.

Step 1   Identify the thickness of the material to be cut.

Step 2   Identify the proper size cutting tip from the manufacturer's recommended tip size chart.

Step 3   Install the cutting tip in the cutting torch.

WELDING ONE TRAINEE TASK MODULE 09101

4.8.0 CLOSE TORCH VALVES AND RELEASE REGULATOR ADJUSTING SCREWS

Follow these steps to close the torch valves and release the regulator adjusting screws.

Step 1 Check the torch oxygen valve on the torch to be sure it is closed.

CAUTION Closing the torch gas valves prevents gases from backing up inside the torch.

Step 2 Check the oxygen regulator adjusting screw to be sure it is released (backed out) and loose.

Step 3 Check the fuel gas regulator adjusting screw to be sure it is released and loose.

CAUTION Releasing the regulator adjusting screws prevents damage to the regulator diaphragm when the cylinder valves are opened.

4.9.0 OPEN CYLINDER VALVES

Follow these steps to open the cylinder valves.

Step 1 Standing to one side of the oxygen regulator, slowly open the oxygen cylinder valve all the way, allowing the pressure in the gauge to rise gradually. The oxygen cylinder valve has a seat at the top. Opening this valve all the way prevents leakage by engaging this seat.

WARNING! Never stand directly in front of or behind the regulator. It can blow apart, causing serious injury. Always open the cylinder valve gradually. Quick openings can damage the regulator or gauge or even cause the gauge to explode.

Step 2 Standing to one side of the fuel gas regulator, slowly open the fuel gas cylinder valve a quarter turn or until the cylinder pressure gauge indicates the cylinder pressure. Opening the cylinder valve a quarter turn allows it to be quickly closed in case of of a fire.

Figure 4-6 shows opening cylinder valves.

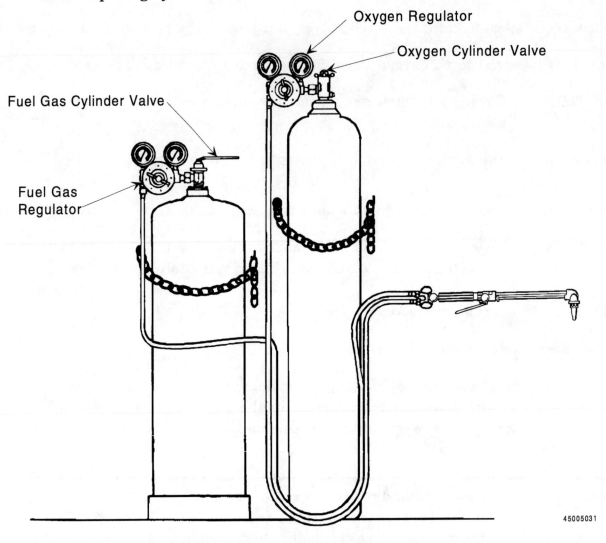

Figure 4-6. Opening Cylinder Valves

4.10.0   ADJUST THE WORKING PRESSURE

Follow these steps to adjust the working pressure.

Step 1   Slowly turn the oxygen regulator adjusting screw until the working pressure on the gauge is set as recommended for the cutting tip to be used.

Step 2   Slowly turn the fuel gas regulator adjusting screw until the working pressure is set as recommended for the cutting tip to be used.

Figure 4-7 shows adjusting the working pressure.

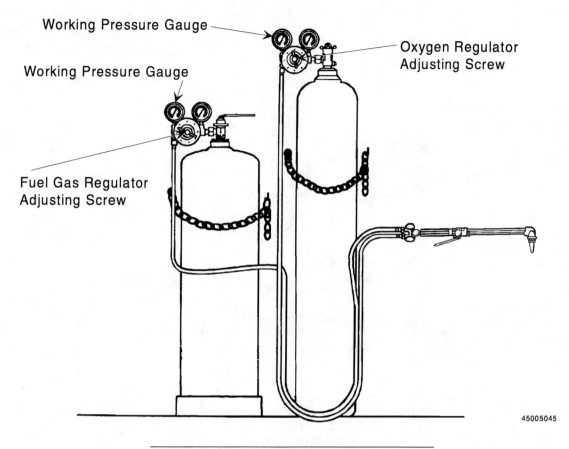

Figure 4-7. Adjusting the Working Pressure

4.11.0   TEST FOR LEAKS

Equipment must be tested for leaks immediately after it is set up. Leaks could cause a fire or explosion if undetected. To test for leaks brush a commercially prepared leak-testing formula or a solution of detergent and water on the following points. If bubbles form, a leak is present.

WARNING!   Use only a non-oil-based detergent. In the presence of
                    oxygen, oil can cause fires or explosions.

Leak points include:

- Oxygen cylinder valve
- Fuel gas cylinder valve
- Oxygen regulator inlet connection
- Fuel gas regulator inlet connection
- Hose connections at the regulators and torch

Figure 4-8 shows these leak points.

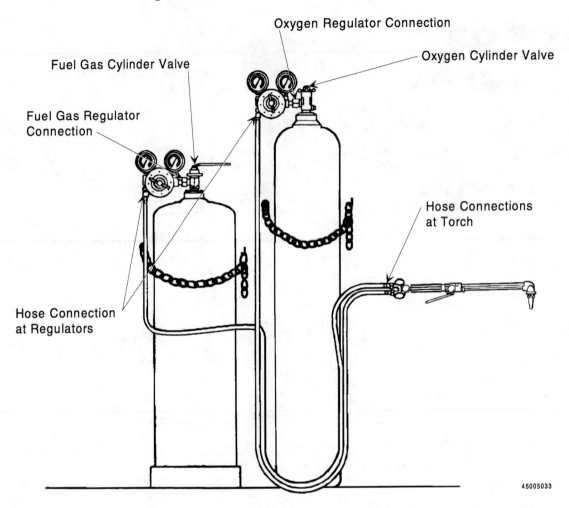

Figure 4-8. Leak Points

If there is a leak at the fuel gas cylinder valve stem, attempt to stop it by tightening the packing gland. If this does not stop the leak, mark and remove the cylinder and notify the supplier. For other leaks, tighten the connections a bit with a wrench. If this does not stop the leak, turn off the gas pressure, open all connections and inspect the screw threads.

CAUTION   Use care not to over-tighten connections. The brass connections used will strip if over-tightened. Repair or replace equipment that does not seal properly.

5.0.0 CONTROLLING THE OXYFUEL TORCH FLAME

To be able to safely use a cutting torch the operator must understand the flame and be able to adjust it and react to unsafe conditions. The following sections will explain the oxyfuel flame and how to control it safely.

5.1.0 OXYFUEL FLAMES

There are three types of flames: neutral, carburizing and oxidizing.

A neutral flame burns equal amounts of oxygen and fuel gas. The inner cone will be light blue in color, surrounded by a darker-blue outer flame envelope that results when the oxygen in the air combines with the super-heated gases from the inner cone. A neutral flame is used for all but special cutting applications.

A carburizing flame has a white feather created by excess fuel. The length of the feather depends on the amount of excess fuel present in the flame. The outer flame envelope is longer than that of the neutral flame, and it is much brighter in color. The excess fuel in the carburizing flame (especially acetylene) produces large amounts of carbon. The carbon will combine with red-hot or molten metal, making the metal hard and brittle. The carburizing flame is cooler than a neutral flame and is never used for cutting. It is used for some special heating applications.

An oxidizing flame has an excess of oxygen. Its inner cone is shorter, much bluer in color and more pointed than a neutral flame. The outer flame envelope is very short and often fans out at the ends. An oxidizing flame is the hottest flame. Some special fuel gases will recommend a slightly oxidizing flame but in most cases it is not used. The excess oxygen in the flame can combine with many metals, forming a hard, brittle, low-strength oxide.

Figure 5-1 shows the three types of oxyfuel flames.

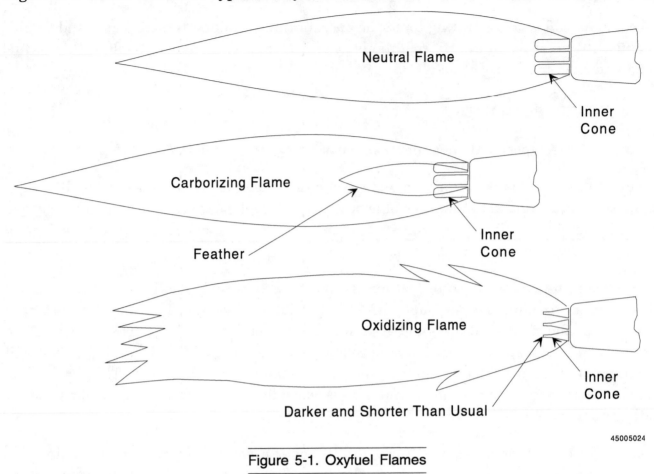

Figure 5-1. Oxyfuel Flames

5.2.0   BACKFIRE AND FLASHBACKS

When the torch flame goes out with a loud pop or snap a backfire has occurred. Backfires are usually caused when the tip or nozzle touches the work surface or when a bit of hot slag briefly interrupts the flame. When a backfire occurs, you can relight the torch immediately. Sometimes the torch even relights itself. If a backfire recurs without the tip making contact with the base metal shut off the torch and find the cause. Possible causes are:

- Improper operating pressures
- A loose torch tip
- Dirt in the torch tip seat or a bad seat

When the flame goes out and burns back inside the torch with a hissing or whistling sound, a flashback is occurring. Immediately shut off the torch. The flame is burning inside the torch. If the flame is not extinguished quickly the end of the torch will melt off. The flashback will stop as soon as the oxygen valve is closed. Therefore, quick action is crucial. Flashbacks can cause fires and explosions within the cutting rig and therefore are very dangerous. Flashbacks can be caused by:

WELDING ONE TRAINEE TASK MODULE 09101

- Equipment failure
- Overheated torch tip
- Slag or spatter hitting and sticking to the torch tip

After a flashback has occurred, wait until the torch has cooled before relighting it. Before you relight it, blow oxygen (not fuel gas) through the torch for several seconds to remove soot that may have built up in the torch during the flashback. If you hear the hissing or whistling, or if the flame does not appear normal, shut off the torch immediately and have the torch serviced by a qualified technician.

5.3.0   LIGHTING THE TORCH AND ADJUSTING THE FLAME

After the cutting equipment has been properly set up the torch can be lit and the flame adjusted for cutting. Follow these steps to light the torch.

Step 1   Choose the appropriate cutting torch tip according to the base metal thickness you will be cutting and fuel gas you are using.

Note   Refer to manufacturer's charts. You may have to readjust the oxygen and fuel gas pressure depending on the tip selected.

Step 2   Attach the tip to the cutting torch or cutting attachment by placing it on the end of the torch and tightening the nut.

Note   Some manufacturers recommend tightening the nut with a wrench. Others recommend tightening the nut by hand. Check the tip manual to see if the manufacturer recommends that the nut be tightened manually or with a wrench.

Step 3   Put on cutting goggles. Position the cutting goggles on top of the head ready to be lowered over the eyes when needed.

Step 4   Put on leather gauntlet-type welding gloves.

Figure 5-2 shows a cutting operator wearing the proper protective equipment.

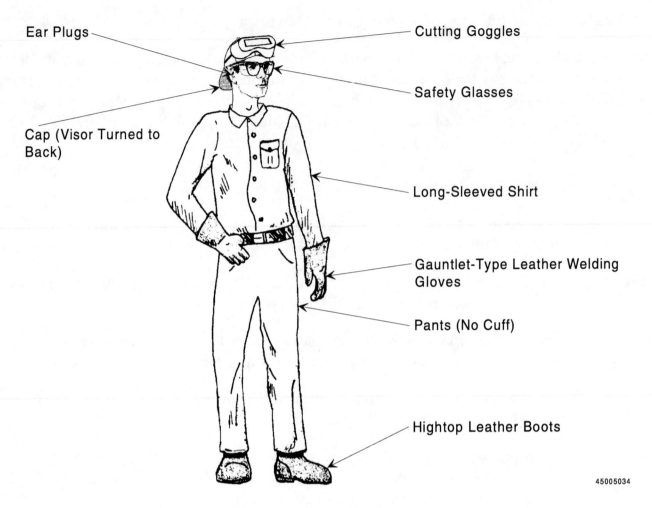

Ear Plugs

Cutting Goggles

Safety Glasses

Cap (Visor Turned to Back)

Long-Sleeved Shirt

Gauntlet-Type Leather Welding Gloves

Pants (No Cuff)

Hightop Leather Boots

45005034

Figure 5-2. Protective Equipment

Step 5   Open the fuel gas needle valve on the torch handle about one-quarter turn.

WARNING!	Hold the friction lighter near the end of the tip when lighting the torch. Do not cover the tip with the lighter. Always use a friction lighter. Never use matches or cigarette lighters to light the torch, as this could result in severe burns or could cause the lighter to explode.

Step 6   Holding the friction lighter near the front of the torch tip, light the torch.

Step 7   Adjust the torch fuel gas flame once the torch is lit by adjusting the flow of fuel gas with the fuel gas valve. Increase the flow of fuel gas until the flame stops smoking.

Step 8   Turn on the oxygen torch valve slowly and adjust the torch flame to a neutral flame.

Step 9   Press the cutting oxygen lever all the way down and observe the flame. It should have a long center high-pressure oxygen cutting jet up to eight inches long extending from the cutting oxygen hole. If it does not:

• Check that the working pressures are as recommended on the manufacturer's chart.

• Clean the cutting tip. If this does not clear up the problem, change the cutting tip.

Figure 5-3 shows a proper flame.

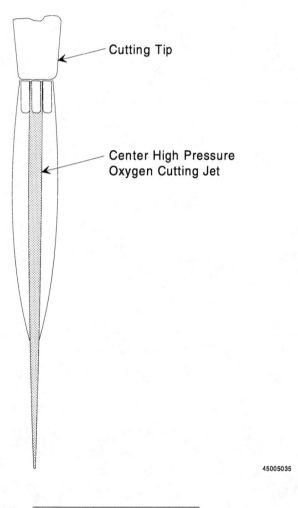

Figure 5-3. Proper Flame

5.4.0   SHUTTING OFF THE TORCH

Follow these steps to shut off the torch.

Step 1   Close the torch fuel gas valve quickly to extinguish the flame.

Step 2   Close the torch oxygen valve.

WELDING ONE TRAINEE TASK MODULE 09101

SELF-CHECK REVIEW 2

1. How should cylinders be transported?
2. Why should you always stand to one side of the valve when opening it?
3. Why should regulator fittings be checked before use to ensure that they are free of oil or grease?
4. Where are check valves attached?
5. Why should new hoses be blown out before they are used?
6. Why should torch needle valves be closed before lighting the torch?
7. Why should you always open the cylinder valve gradually?
8. How is a cutting rig tested for leaks?
9. What is a neutral flame?
10. What is a flashback and what are its effects?
11. When lighting the torch which torch valve is opened first?
12. Why should a friction lighter be used to light the torch?
13. When shutting off the torch which torch valve is closed first?

ANSWERS TO SELF-CHECK REVIEW 2

1. Upright in bottle carts or cylinder cages only. (4.1.0)
2. To avoid injury from dirt which may be lodged in the valve and that will spray out when it is opened. (4.2.0)
3. When oil or grease come in contact with oxygen they can explode or cause a fire. (4.2.0)
4. To the regulator or to the torch. (4.4.0)
5. They contain talc and bits of rubber that will clog the torch needle valves if they are not blown out. (4.5.0)
6. To keep gases from backing up inside the torch. (4.8.0)
7. Quick openings can damage the regulator or gauge or even cause the gauge to explode. (4.9.0)
8. Brush a commercially prepared leak-testing formula or a solution of detergent and water on the connections to test for leaks. If bubbles form, a leak is present. (4.11.0)
9. It burns equal amounts of oxygen and fuel gas. The inner cone will be light blue in color, surrounded by a darker-blue outer flame envelope that results when the oxygen in the air combines with the super-heated gases from the inner cone. (5.1.0)
10. When the flame goes out and burns back inside the torch with a hissing or whistling sound. If the flame is not extinguished quickly the end of the torch will melt off. Flashbacks can cause fires and explosions within the cutting rig and are very dangerous. (5.2.0)
11. Fuel gas. (5.3.0)
12. Using matches or cigarette lighters can result in severe burns or could cause the lighter to explode. (5.3.0)
13. Fuel gas. (5.4.0)

6.0.0 SHUTTING DOWN OXYFUEL CUTTING EQUIPMENT

When a cutting job is completed and the oxyfuel equipment is no longer needed it must be shut down. Follow these steps to shut down the oxyfuel cutting equipment.

Step 1 Close the fuel gas cylinder valve.

Step 2 Close the oxygen cylinder valve.

Step 3 Open the fuel gas torch valve to allow gas to escape. Do not proceed until all pressure is released and the fuel gas working pressure gauge reads "0."

Step 4 Back out the fuel gas regulator adjusting screw until it is loose.

Step 5 Close the fuel gas torch valve.

Step 6 Open the oxygen torch valve to allow gas to escape. Do not proceed until all pressure is released and the oxygen working pressure gauge reads "0."

Step 7 Back out the oxygen regulator adjusting screw until it is loose.

Step 8 Close the oxygen torch valve.

Step 9 Coil up the hose and secure the torch to prevent damage.

Figure 6-1 shows shutting down oxyfuel cutting equipment.

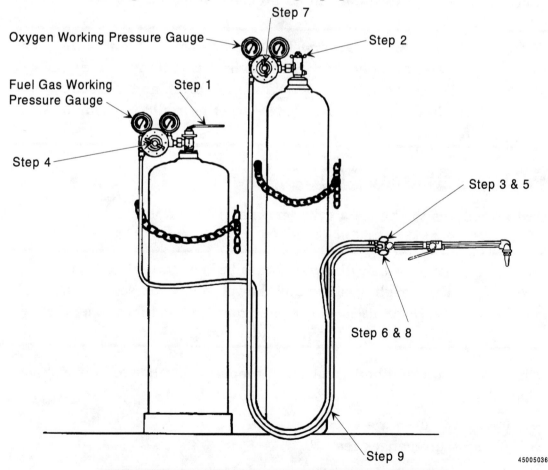

45005036

Figure 6-1. Shutting Down Oxyfuel Cutting Equipment

7.0.0 DISASSEMBLING OXYFUEL EQUIPMENT

Follow these steps if the oxyfuel equipment must be disassembled after use.

Step 1 Check to be sure equipment has been properly shut down. This includes checking that:

- Cylinder valves are closed.
- All pressure gauges read "0."

Step 2 Remove both hoses from the torch.

Step 3 Remove both hoses from the regulators.

Step 4 Remove both regulators from the cylinder valves.

Step 5 Replace the protective caps on the cylinders.

Step 6 Return the oxygen cylinder to its proper storage place.

CAUTION Always transport and store gas cylinders in the upright position. Be sure they are properly secured (chained) and capped.

Step 7 Return the fuel gas cylinder to its proper storage place.

WARNING! Regardless if the cylinders are empty or full, never store fuel gas cylinders and oxygen cylinders together. Storing cylinders together is a violation of OSHA and local fire regulations. Storing cylinders together could result in a fire and explosion.

8.0.0 CHANGING EMPTY CYLINDERS

Follow these procedures to change a cylinder when it is empty.

WARNING! When moving cylinders, always be certain that they are in the upright position and the valve caps are secured in place. Never use a sling or electromagnet to lift cylinders. To lift cylinders, use a cylinder cage.

Step 1 Check to be sure equipment has been properly shut down. This includes checking that:

- Cylinder valves are closed.
- All pressure gauges read "0."

Step 2 Remove the regulator from the empty cylinder.

Step 3   Replace the protective cap on the empty cylinder.

Step 4   Transport the empty cylinder from the work station to the storage area.

Step 5   Mark "MT" (or the accepted site notation for indicating an empty cylinder) on the shoulder of the cylinder.

Step 6   Place the empty cylinder in the empty cylinder section of the cylinder storage area for the type of gas in the cylinder.

WARNING!   Regardless if the cylinders are empty or full, never store fuel gas cylinders and oxygen cylinders together. Storing cylinders together is a violation of OSHA and local fire regulations. Storing cylinders together could result in a fire and explosion.

Figure 8-1 shows a properly marked empty cylinder.

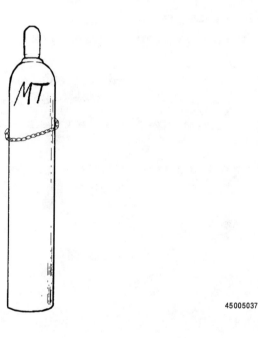

45005037

Figure 8-1. Empty Cylinder

9.0.0   PERFORMING CUTTING PROCEDURES

The following sections explain how to recognize good and bad cuts, prepare for cutting operations and perform straight-line cutting, piercing, bevel cutting, washing and gouging.

9.1.0 INSPECTING THE CUT

Before attempting to make a cut you must be able to recognize good and bad cuts and know what causes bad cuts. A good kerf, or cut edge, has several characteristics. These are explained in the following sections.

- A good cut features a square top edge that is sharp and straight, not ragged. The bottom edge can have some slag adhering to it but not an excessive amount. What slag there is should be easily removed with a chipping hammer. The drag lines should be near vertical and not very pronounced.
- When preheat is insufficient, bad gouging results at the bottom of the cut because of too slow travel speed.
- Too much preheat will result in the top surface melting over the cut, an irregular cut edge and an excessive amount of slag.
- When the cutting oxygen pressure is too low, the top edge will melt over because of the resulting slow cutting speed.
- Using cutting oxygen pressure that is too high will cause the operator to lose control of the cut, resulting in an uneven kerf.
- A travel speed that is too slow results in bad gouging at the bottom of the cut and irregular drag lines.
- When the travel speed is too fast there will be gouging at the bottom of the cut, a pronounced break in the dragline and an irregular kerf.
- A torch that is held or moved unsteadily across the metal being cut can result in a wavy and irregular kerf.
- When a cut is lost and then not restarted carefully, bad gouges at the point where the cut is restarted will result.

Figure 9-1 shows examples of good and bad cuts.

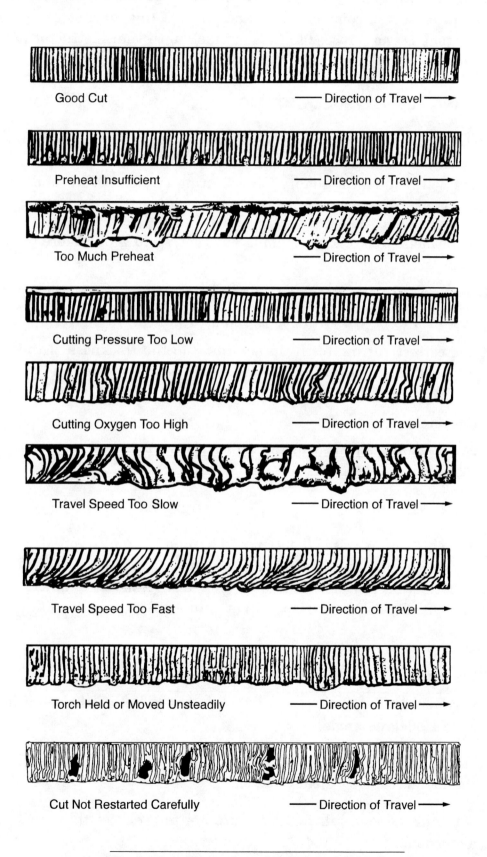

Figure 9-1. Examples of Good and Bad Cuts

Note   The following tasks are designed to develop your skills with a cutting torch. Practice each task until you are thoroughly familiar with the procedure. As you complete each task, take it to your instructor for evaluation. Do not proceed to the next task until your instructor tells you to.

9.2.0   PREPARING FOR OXYFUEL CUTTING

Before metal can be cut the equipment must be set up and the metal prepared. One important step is to properly lay out the cut by marking it with soapstone or punch marks. The few minutes this takes will result in a quality job reflecting craftsmanship and pride in your work. Follow these steps to prepare to make a cut.

Step 1   Prepare the metal to be cut by cleaning any rust, scale or other foreign matter from the surface.

Step 2   If possible, position the work so you will be comfortable when cutting.

Step 3   Mark the lines to be cut with soapstone or a punch.

Step 4   Select the correct cutting torch tip according to the thickness of the metal to be cut, the type of cut to be made, the amount of preheat needed and the type of fuel gas used.

Step 5   Light the torch.

Step 6   Use the procedures outlined in the following sections for performing particular types of cutting operations.

9.3.0   CUTTING THIN STEEL

Thin steel is 3/16-inch thick or less. When cutting by hand use a tip one size larger than is recommended. A major concern when cutting thin steel is distortion caused by the heat of the torch and the cutting process. To minimize distortion move as quickly as you can without loosing the cut. Follow these steps to cut thin steel.

Step 1   Prepare the metal surface.

Step 2   Light the torch.

Step 3   Hold the torch so that the tip is pointing in the direction the torch is travelling at a 15- to 20-degree angle.

CAUTION   Holding the tip upright when cutting thin steel will over-heat the metal, causing distortion.

Step 4   Preheat the metal to a dull red. Use care not to overheat thin steel as this will cause distortion.

Note   The edge of the tip can be rested on the surface of the metal being cut and then slid along the surface when making the cut.

Step 5   Press the cutting oxygen lever to start the cut and then move quickly along the line to be cut. Move as quickly as you can without losing the cut to minimize distortion.

Figure 9-2 shows cutting thin steel.

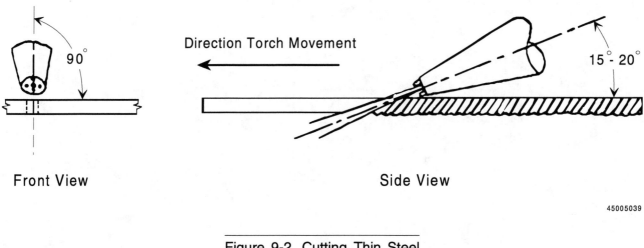

Front View   Side View

45005039

Figure 9-2. Cutting Thin Steel

9.4.0   CUTTING THICK STEEL

Most oxyfuel cutting will be on steels over 3/16-inch thick. Whenever heat is applied to metal, distortion is a problem, but as the steel gets thicker it becomes less of a problem. Follow these steps to cut thick steel.

Step 1   Prepare the metal surface and torch. Light the torch.

Step 2   Preheat the top edge by angling the torch tip toward the corner. As the edge approaches the preheat temperature of a bright cherry red rotate the torch tip to an upright position parallel with the surface to be cut.

Step 3   Press the cutting oxygen lever to start the cut.

Step 4   Move the torch tip along the line to be cut, keeping the torch parallel to the surface. Move at a steady pace so the torch makes a steady ripping sound.

Note   The torch can be moved from either right to left or left to right. Choose the direction that is the most comfortable for you.

Step 5   Continue moving the torch until the tip has passed the edge of the metal.

Figure 9-3 shows cutting thick steel.

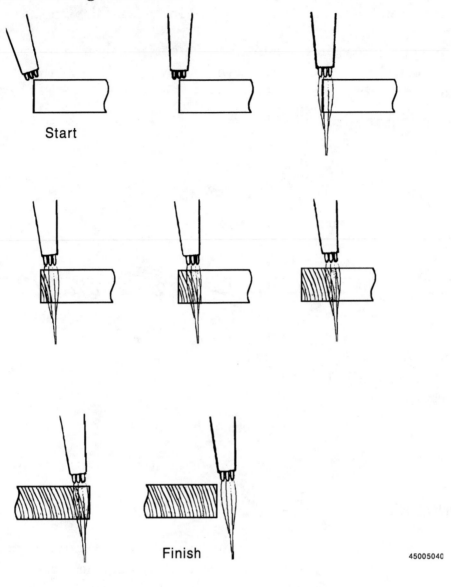

Start

Finish

45005040

Figure 9-3. Cutting Thick Steel

9.5.0 PIERCING PLATE

Before holes or slots can be cut in a plate, the plate must be pierced. Piercing places a small hole through metal where the cut can be started. Because more preheat is necessary on the surface of a plate than at the edge, choose the next-larger cutting tip than is recommended for the thickness to be pierced. When piercing steel over three inches thick it may help to preheat the bottom side of the plate directly under the spot to be pierced. Follow these steps to pierce plate for cutting.

Step 1 Prepare the metal surface and torch. Light the torch.

Step 2 Hold the torch tip 1/4 inch to 5/16 inch above the spot to be pierced until the surface is a bright cherry red.

Step 3   Slowly press the cutting oxygen lever. As the cut starts, raise the tip about 1/2 inch above the metal surface and tilt the torch slightly so that molten metal does not blow back onto the tip. The tip should be raised and tipped before the cutting oxygen lever is fully depressed.

Step 4   Maintain this position until a hole burns through the plate.

Step 5   Lower the torch tip to about 3/16inch above the metal surface and continue to cut outward from the original hole to the edge of the line to be cut.

Figure 9-4 shows piercing steel.

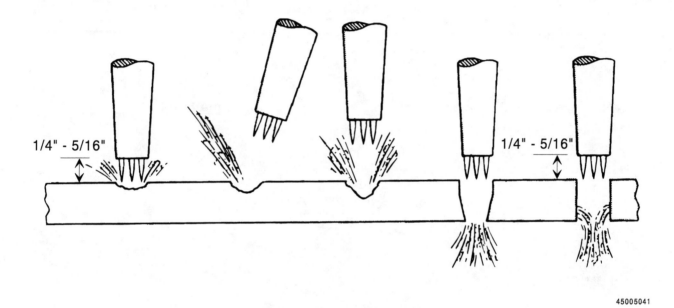

1/4" - 5/16"   1/4" - 5/16"

45005041

Figure 9-4. Piercing Steel

9.6.0   CUTTING BEVELS

Bevel cutting is often performed to prepare the edge of steel plate for welding. Follow these steps to perform bevel cutting.

Step 1   Prepare the metal surface and the torch. Light the torch.

Step 2   Hold the torch so that the tip faces the metal at the desired bevel angle.

Note   An angle iron can be used as a guide.

Step 3   Preheat the edge to a bright cherry red.

Step 4   Press the cutting oxygen lever to start the cut.

Step 5   As cutting begins, move the torch tip at a steady rate along the line to be cut. Pay particular attention to the torch angle to ensure it is uniform along the entire length of the cut.

Figure 9-5 shows cutting a bevel.

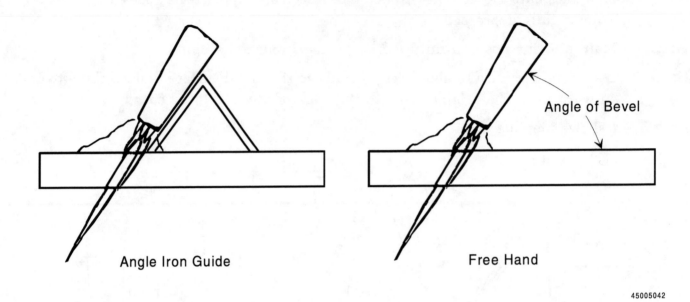

Angle of Bevel

Angle Iron Guide

Free Hand

45005042

Figure 9-5. Cutting A Bevel

9.7.0   WASHING

Washing is a term used to describe the process of cutting out bolts or rivets. Washing operations use a special tip that has a large cutting hole which produces a low-velocity stream of oxygen. The low-velocity oxygen stream helps prevent cutting into the surrounding base metal. Washing tips can also be used to remove items such as blocks, angles or channels that are welded onto a surface. Follow these steps to perform washing.

Step 1   Prepare the metal surface and torch. Light the torch.

Step 2   Preheat the metal to be cut until it is a bright cherry red.

Step 3   Move the cutting torch at a 55-degree angle to the metal surface.

Step 4   Press the cutting oxygen lever to cut the material to be removed. Use care not to cut into the surrounding metal.

CAUTION   As the surrounding metal heats up there is a greater danger of cutting into it. Try to complete the washing operation as quickly as possible. If the surrounding metal gets to hot, stop and let it cool down.

Figure 9-6 shows washing.

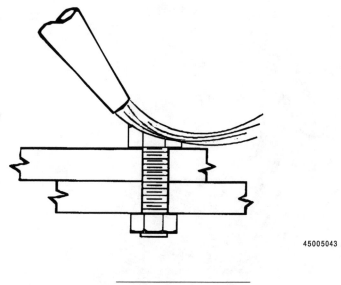

45005043

Figure 9-6. Washing

9.8.0 GOUGING

Gouging is a process of cutting a groove into a surface. Gouging operations use a special curved tip that produces a low-velocity stream of oxygen which curves up, allowing the operator to control the depth and width of the groove. It is an effective means to gouge out cracks or weld defects for welding. Gouging tips can also be used to remove steel backing from welds or to wash off bolt or rivet heads. Gouging tips are not as effective as washing tips for removing the shank of the bolt or rivet. Follow these steps to gouge.

Step 1 Prepare the metal surface and torch. Light the torch.

Step 2 Holding the torch so that the preheat holes are pointed directly at the metal, preheat the surface until it becomes a bright cherry red.

Step 3 When the steel has been heated to a bright cherry red, slowly roll the torch away from the metal so that the holes are at an angle that will enable you to cut the correct-depth gouge. As you roll the torch away, press the cutting oxygen lever gradually.

Step 4 Move the cutting torch along the line of the gouge. As you move the torch, rock it back and forth as necessary to create a gouge of the required depth and width.

Note The travel speed and torch angle are very important when gouging. If the travel speed or torch angle are incorrect, the gouge will be irregular and there will be a build-up of slag inside the gouge. Practice until the gouge is clean and even.

Figure 9-7 shows gouging.

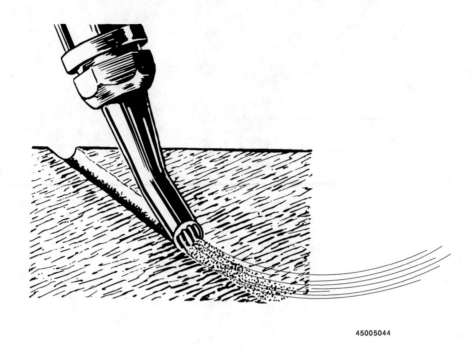

45005044

Figure 9-7. Gouging

10.0.0 PERFORMANCE ACCREDITATION TASKS

The following tasks are designed to test your competency with a cutting torch. Do not perform a cutting task until directed to do so by your instructor.

10.1.0   SET UP, LIGHT AND ADJUST, AND SHUT DOWN OXYFUEL EQUIPMENT

Using oxyfuel equipment that has been completely disassembled demonstrate how to:

- Set up oxyfuel equipment
- Light and adjust the flame
  - Carbonizing
  - Neutral
  - Oxidizing
- Shut off the torch
- Shut down the oxyfuel equipment

Criteria for Acceptance

- Set up the oxyfuel equipment in the correct sequence as specified in Section 4.0.0
- No leaks
- Properly adjust all three flames as specified in Section 5.1.0
- Shut off the torch in the correct sequence as specified in Section 5.4.0
- Shut down the oxyfuel equipment as specified in Section 6.0.0

10.2.0   CUT A SHAPE FROM THIN STEEL

Using mild steel plate, lay out and cut the shape shown in *Figure 10-1*.

Note: Material MS 1/8 to 1/4 inch Thickness
Holes 3/4" Diameter
Slots   3/4' X 1 1/2"

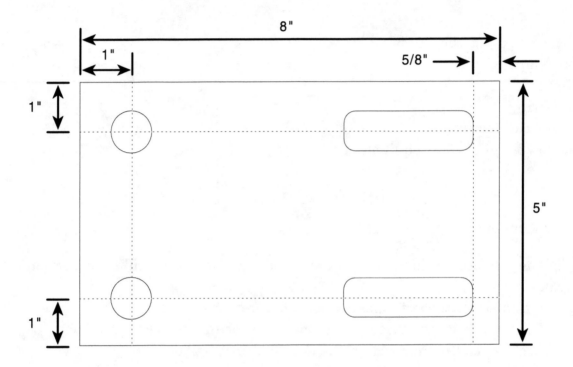

45005010

Figure 10-1. Shape From Thin Steel

Criteria For Acceptance:

- Outside dimensions plus or minus 1/16 inch
- Inside dimensions (holes and slots) plus or minus 1/8 inch
- Square plus or minus five degrees
- Minimal amount of slag sticking to plate which can be easily removed
- Square kerf face with minimal notching not exceeding 1/16-inch deep

10.3.0 CUT A SHAPE FROM THICK STEEL

Using mild steel plate, lay out and cut the shape shown in *Figure 10-2*.

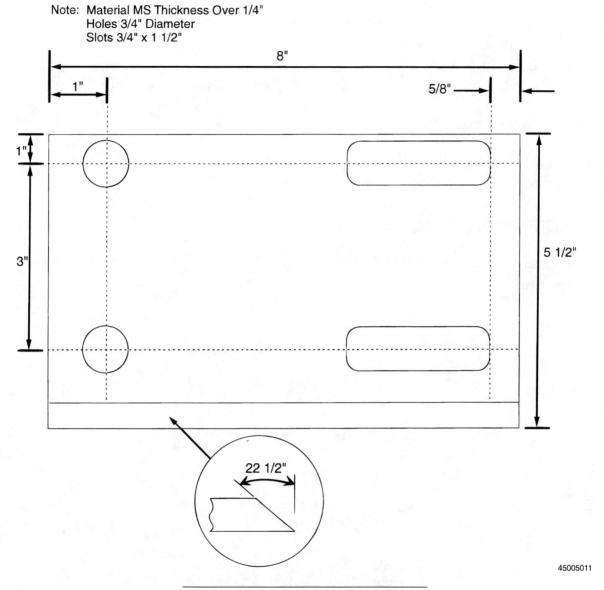

Note: Material MS Thickness Over 1/4"
Holes 3/4" Diameter
Slots 3/4" x 1 1/2"

45005011

Figure 10-2. Shape From Thick Steel

Criteria For Acceptance:

- Outside dimensions plus or minus 1/16 inch
- Inside (holes and slots) dimensions plus or minus 1/8 inch
- Square plus or minus five degrees
- Bevel plus or minus two degrees
- Minimal amount of slag sticking to plate which can be easily removed
- Square kerf face with minimal notching not exceeding 1/16-inch deep

10.4.0 PERFORM WASHING

Using any of the material identified below, perform oxyfuel washing to remove the portion identified by the instructor. Materials that can be used for this task include:

- Steel backing strip on a butt weld
- Excess buildup on the face of a weld
- Rivets or bolts in a plate
- Blocks, angles, clips, eyes, D-rings or other debris welded to a plate

Criteria For Acceptance:

- Material removed flush with the base metal surface
- No notching in the surface of the base metal

10.5.0 PERFORM OXYFUEL GOUGING

Using mild steel plate at least 1/2-inch thick or thicker, gouge a U-groove at least eight inches long as shown in *Figure 10-3*.

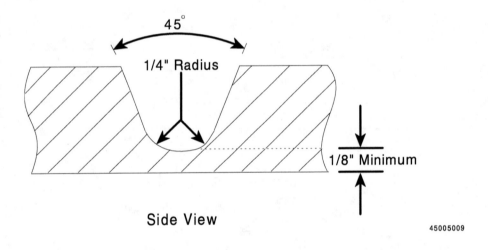

Figure 10-3. Oxyfuel Gouging

Criteria For Acceptance:

- Groove width and depth uniform along its length to plus or minus 1/16 inch
- Root face minimum 1/8 inch
- Radius dimension plus 1/4 inch, minus 0
- Included groove angle plus 10 degrees, minus 0
- Uniform and smooth groove walls
- Minimum to no slag in groove as cut

SUMMARY

Oxyfuel cutting has many uses on job sites. It can be used to cut plate and shapes to size, prepare joints for welding, clean metals or welds, and disassemble structures. Because of the high pressures and flammable gases there is a danger of fire and explosion when using oxyfuel equipment. However, these risks can be minimized when the oxyfuel cutting operator is well-trained and knowledgeable. Be sure you know and understand the safety precautions and equipment presented in this module before using oxyfuel equipment.

SELF-CHECK REVIEW 3

1. What are the first two steps when shutting down oxyfuel equipment after the torch has been shut off?
2. When changing an empty cylinder, how do you indicate that the cylinder is empty?
3. Describe a quality oxyfuel cut.
4. What should be done to the steel to be cut before lighting the torch?
5. Explain how the torch is held for cutting thin steel.
6. Explain how the torch is held for cutting thick steel.
7. Explain how to pierce a hole with a torch.
8. When cutting bevels what must you pay particular attention to?
9. What kind of tip is used for washing?
10. What is a gouging tip used for?

ANSWERS TO SELF-CHECK REVIEW 3

1. Close the fuel gas and oxygen cylinder valves. (6.0.0)
2. Mark "MT" (or the accepted site notation for indicating an empty cylinder) on the shoulder of the cylinder. Place the empty cylinder in the empty cylinder section of the cylinder storage area for the type of gas in the cylinder. (8.0.0)
3. The top edge should be sharp and straight, not ragged. The bottom edge can have some slag adhering to it but not an excessive amount. What slag there is should be easily removable with a chipping hammer. The drag lines should be near vertical and not very pronounced. (9.1.0)
4. Clean the surface, position the work if possible, and mark the lines to be cut with soapstone or a punch. (9.2.0)
5. Hold the torch so that the tip is pointing in the direction the torch is travelling at a 15- to 20-degree angle. (9.3.0)
6. Hold the torch parallel to the surface. (9.4.0)
7. Hold the torch tip 1/8 inch to 1/4 inch above the spot to be pierced until the surface is a bright cherry red. Slowly press the cutting oxygen lever. As the cut starts, raise the tip about 1/2 inch above the metal surface and tilt the torch slightly so that molten metal does not blow back onto the tip. The tip should be raised and tipped before the cutting oxygen lever is fully depressed. (9.5.0)
8. Pay particular attention to the torch angle to ensure it is uniform along the entire length of the cut. (9.6.0)
9. A special washing tip that has a large cutting hole which produces a low-velocity stream of oxygen. (9.7.0)
10. Cutting a groove into a surface to gouge out cracks or weld defects for welding. Gouging tips can also be used to remove steel backing from welds or wash off bolt or rivet heads. (9.8.0)

References

For advanced study of topics covered in this module, the following works are suggested:

Modern Welding, Althouse, Turnquist, Bowditch and Bowditch, Goodheart-Willcox Company, Inc., South Holland, Illinois, 1988. Phone 1-800-323-0440.

Welding Principles and Practices, Jeffus and Johnson, Delmar Publishers, Inc., Albany, NY 1988. Phone 1-800-347-7707.

PERFORMANCE/LABORATORY EXERCISES

1. Set up oxyfuel cutting equipment.
2. Light and adjust the cutting torch.
3. Shut down oxyfuel equipment.
4. Cut thin steel.
5. Cut thick steel.
6. Pierce plate.
7. Cut bevels.
8. Perform washing.
9. Perform oxyfuel gouging.

The NCCER makes every effort to keep these manuals up-to-date and free of technical errors. We appreciate your help in this process. If you have an idea for improving this manual, or if you find an error, a typographical mistake, or an inaccuracy in the *Wheels of Learning*, please write us, using this form or a photocopy. Be sure to include the exact module number, page number, a description of the problem, and the correction, if possible. We'll do our best to correct it in later editions. Thank you for your assistance.

Write: *Wheels of Learning*
National Center for Construction Education and Research
P.O. Box 141104
Gainesville, FL 32614-1104
Fax: 352-334-0932

WHEELS OF LEARNING USER UPDATE

Please let us know if you have found an inaccuracy, error, or other problem in a *Wheels of Learning* manual. Use this form or write us a letter. Please be sure to tell us the exact module name and module number, the page number, and the problem. Thanks for your help.

Craft _____ Module Name _____

Module Number _____ Page Number(s) _____

Description of Problem _____

_____

_____

_____

(Optional) Correction of Problem _____

_____

_____

(Optional) Your Name and Address _____

_____

Plasma Arc Cutting (PAC)

Module 09302

Welder Trainee Task Module 09302

NATIONAL
CENTER FOR
CONSTRUCTION
EDUCATION AND
RESEARCH

PLASMA-ARC CUTTING (PAC)

OBJECTIVES

Upon completion of this module, the trainee will be able to:

1. Set up plasma-arc cutting equipment.
2. Prepare the work area to safely perform plasma-arc cutting.
3. Select the correct amperage and gas pressures or flow rates for the type and thickness of metal to be cut.
4. Use plasma-arc cutting equipment to pierce and cut slots in metal.
5. Use plasma-arc cutting equipment to square cut metal.
6. Use plasma-arc equipment to bevel cut metal.
7. Dismantle and store the equipment and clean the work area.

Prerequisites

Successful completion of the following module(s) is required before beginning study of this module:

- Common Core (Safety)

Required Student Materials

Each trainee will need:

1. Personal protective equipment
2. Leather welding gloves
3. Welding shield
4. Ear plugs
5. Chipping hammer
6. Wire brush
7. Pliers
8. Soapstone
9. Scrap steel sheet or plate, 12 gauge to 1/2 inch thick
10. Scrap stainless steel sheet or plate, 12 gauge to 1/2 inch thick (if available)
11. Scrap aluminum plate, 3/16 to 1/2 inch thick (if available)

Each trainee will need access to:

1. Plasma-arc cutting unit complete with cutting torch and appropriate gas sources

Course Map Information

This course map shows all of the *Wheels of Learning* task modules in the third level of the Welding curricula. The suggested training order begins at the bottom and proceeds up. Skill levels increase as a trainee advances on the course map. The training order may be adjusted by the local Training Program Sponsor.

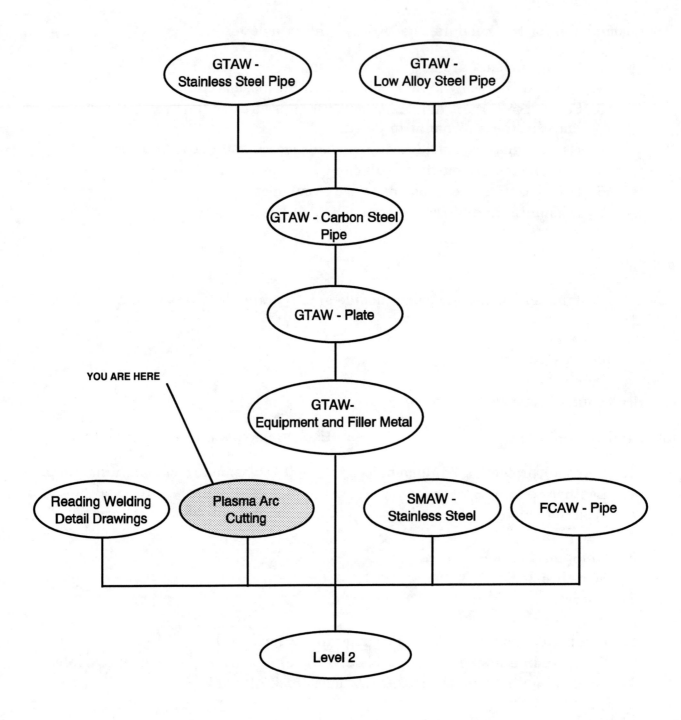

WELDING THREE TRAINEE TASK MODULE 09302

TABLE OF CONTENTS

Trade Terms Introduced In This Module

DCSP (Direct Current Straight Polarity): Another term for DCEN (Direct Current Electrode Negative). The DC current in the plasma cutting arc flows from the electrode tip to nozzle and work piece. To achieve this flow, the torch is connected to the power supply's negative (-) side, and the work lead clamp is connected to the power supply's positive (+) side.

Dross: Oxidized metal expelled from the cut, or which forms on the cut face during plasma-arc cutting.

Electrode: The cathode in the plasma-arc torch. It is normally made of tungsten and can be straight or specially configured depending on the manufacturer of the torch.

Oxidize: To combine with oxygen such as in burning (rapid oxidation) or rusting (slow oxidation).

Plasma: A fourth state of matter (not solid, liquid or gas) created by heating a gas to such a high temperature that it boils electrons off the gas molecules (ionization).

Potential: The relative electrical voltage between two points of reference.

1.0.0 INTRODUCTION

Plasma-arc cutting (PAC) uses a jet of plasma to pierce, cut and gouge metal. The plasma is created by superheating gas in an electric arc. The temperature of the plasma jet is hot enough to melt any metal. The processes can be set up for manual or mechanized operation and in most cases is faster and more efficient than any other cutting methods. Among the PAC advantages are:

- Cuts both ferrous and nonferrous metals
- Minimal slag
- Very thin heat-affected zone
- Very little or no distortion
- High cutting speeds

The plasma-arc cutting process requires an electrical power supply and a cutting gas. Depending on the equipment, a separate shielding or cooling gas may also be required. PAC torches convert the cutting gas into plasma by passing the gas through an electric arc inside the torch. The torch has a relativity small opening (orifice) which constricts the arc and high pressure gas flow. This results in the plasma stream exiting the torch as a supersonic jet hotter than any flame. Depending on the current flow, the plasma jet may reach temperatures of 30,000 degrees Fahrenheit. When the jet contacts metal, it transfers its tremendous heat causing the metal to instantly melt. The high velocity jet then blasts the molten metal away forming a hole, groove or gouge.

Figure 1-1 shows plasma-arc cutting.

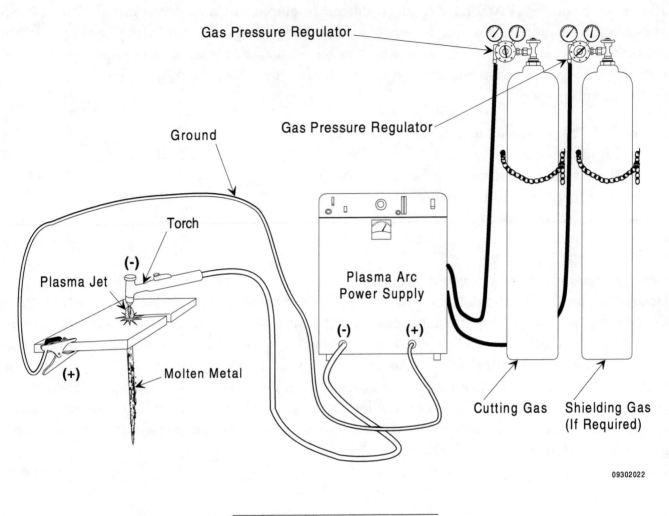

Figure 1-1, Plasma-Arc Cutting.

2.0.0   PLASMA-ARC CUTTING (PAC) PROCESS

There are two types of plasma cutting processes: transferred arc and nontransferred arc. The transferred arc process is the most common and is used to cut materials that are electrically conductive such as metals. The nontransferred arc process is less common and is used to cut materials that are not electrically conductive such as ceramics and concrete.

2.1.0   TRANSFERRED ARC PROCESS

In the transferred arc process the work piece is part of the electrical circuit as in other arc processes. The arc is established between the electrode and the work as well as a smaller arc between the electrode and the torch nozzle inside the torch. The smaller arc between the torch electrode and torch nozzle is called an idle arc. The idle arc is used to start the ionization process. The process also has a high frequency pilot arc. When the torch is brought near the base metal the high frequency pilot arc is established between the torch and the work. The heavier cutting arc and plasma jet is then established between the torch and the work following the high frequency pilot arc's path. For the transferred arc process to work, one lead of the power supply (the ground) must be connected to the metal to be cut. *Figure 2-1* shows the transferred arc process.

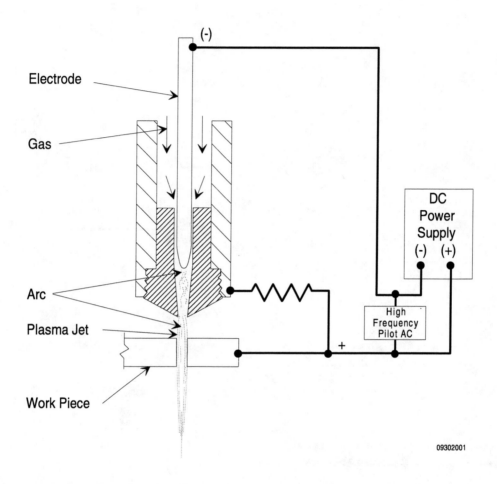

Figure 2-1, Transferred Arc Process.

2.2.0   NONTRANSFERRED ARC PROCESS

In the nontransferred arc process, the entire arc is established within the torch between the torch electrode and the torch nozzle. The material being cut is not connected electrically to the arc circuit. Nontransferred arc is not normally found in facilities that cut only metal.

Figure 2-2 shows the nontransferred arc process.

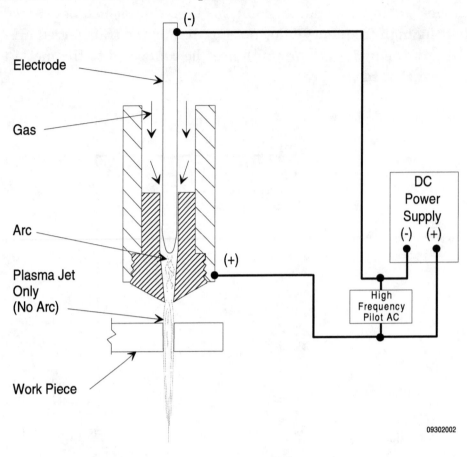

Figure 2-2, Nontransferred Arc Process.

3.0.0   PLASMA-ARC CUTTING SAFETY

The three principal safety hazards associated with plasma-arc cutting are intense heat, arc flash and metal splash. The intense heat of the process can cause severe burns to unprotected skin. The arc generates damaging ultraviolet and infrared rays which will burn unprotected eyes and skin much as a sunburn does, but much quicker and more severely. Arc burn to the eyes can cause permanent eye damage. Metal splash can ignite any combustible material within its range. In out-of-position cutting, metal splash or sparks can enter an unprotected ear canal and burn the ear and puncture the eardrum. The following sections explain how plasma-arc cutting can be safely performed.

3.1.0   PERSONAL PROTECTIVE EQUIPMENT

Because of the potential dangers of PAC, the proper personal protective equipment must be used. This equipment includes the proper work clothing, boots, leathers, gloves, safety glasses and special welding helmets. This personal protective equipment is explained in the following sections.

3.1.1 Work Clothing

To avoid burns from metal splash and ultraviolet rays, persons performing PAC must wear the appropriate clothing. Never wear polyester or other synthetic fibers. Sparks or intense heat will melt these materials, causing severe burns. Wool or cotton is more resistant to sparks and so should be worn. Dark clothing is also preferred because dark clothes minimize the reflection of arc rays which could be deflected under the welding helmet.

To prevent sparks from lodging in clothing and causing a fire or burns, collars should be kept buttoned and pockets should have flaps that can be buttoned to keep out the sparks. A soft cotton cap worn with the visor reversed will protect the top of your head and keep sparks from going down the back of your collar. Pants should be cuffless and hang straight down the leg. Pant cuffs will catch sparks and catch on fire. Never wear frayed or fuzzy materials. These materials will trap sparks which will catch the clothing on fire. Low-top shoes should never be worn while cutting. Sparks will fall into the shoes, causing severe burns. Leather work boots at least eight inches high should be used to protect against sparks and arc flash. *Figure 3-1* shows proper work clothing.

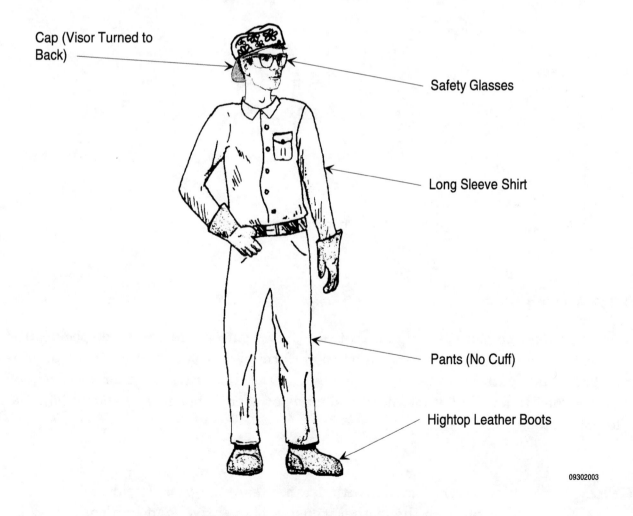

Cap (Visor Turned to Back)

Safety Glasses

Long Sleeve Shirt

Pants (No Cuff)

Hightop Leather Boots

09302003

Figure 3-1. Proper Work Clothing

3.1.2 Leathers

For additional protection, leathers are often worn over work clothing. Leather aprons, split leg aprons, sleeves and jackets are flexible enough to allow ease of movement while providing added protection from sparks and heat. Leathers should always be worn when cutting out-of-position or when cutting in tight quarters. *Figure 3-2* shows the different types of leather protective clothing.

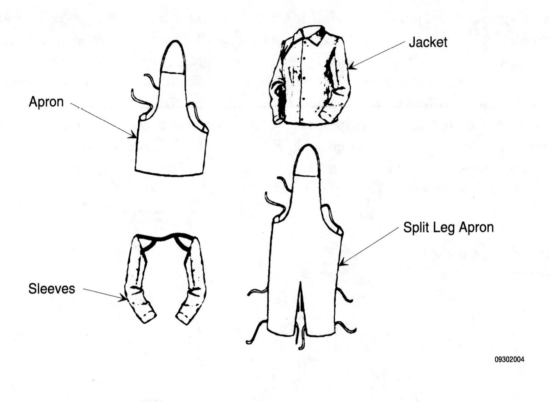

Apron

Jacket

Split Leg Apron

Sleeves

09302004

Figure 3-2. Leather Protective Clothing

3.1.3 Welding Gloves

Leather gauntlet-type gloves (commonly called welding gloves) are designed specifically for cutting and welding. They must be worn when performing any type of arc cutting to protect against hot metal spatter and the ultraviolet and infrared arc rays. Leather welding gloves are not designed to handle hot metal. Pliers, tongs or some other means should be used to handle hot metal.

CAUTION   Picking up hot metal with leather gauntlet-type (welding) gloves will burn the leather, causing it to shrivel and harden. Never handle hot metal with leather welding gloves.

Figure 3-3 shows typical leather gauntlet-type welding gloves.

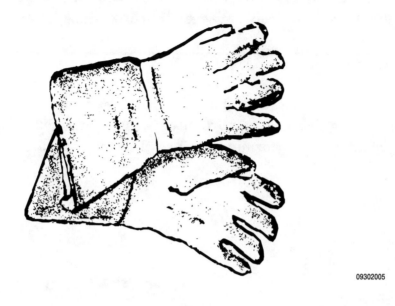

09302005

Figure 3-3. Leather Welding Gloves

3.2.0   EYE PROTECTION

Eyes are delicate and can be easily damaged. The plasma cutting arc gives off ultraviolet and infrared rays which will burn unprotected eyes. The burns caused by the arc are called arc flash. Arc flash causes blistering of the outer eye. This blistering feels like sand or grit in the eyes. It can be caused from either looking directly at the arc or from reflected glare from the arc.

The grinding and surface cleaning operations relating to metal preparation create many small particles which are propelled in all directions. In order to avoid eye injury from flying debris, eye protection must be worn.

3.2.1   Safety Glasses and Goggles

Safety glasses and goggles protect the eyes from flying debris. Safety glasses have impact-resistant lenses. Side shields should be fitted to safety glasses to prevent flying debris from entering from the side. Some safety glasses are equipped with shaded lenses to protect against glare.

WARNING! Shaded safety glasses will not provide sufficient protection from the arc. Looking directly at the arc with shaded safety glasses will result in severe burns to the eyes and face.

Safety goggles are contoured to fit the wearer's face. Goggles are a very efficient means of protecting the eyes from flying debris. They are often worn over safety glasses to provide extra protection when grinding or performing surface cleaning. *Figure 3-4* shows safety glasses and goggles.

Side Shields

Safety Glasses

Clear Safety Goggles

09302006

Figure 3-4. Safety Glasses and Goggles

3.2.2 Welding Shields

Welding shields (also called welding helmets) provide eye and face protection from arc flash, sparks and metal splash. Some shields are equipped with handles, but most are designed to be worn on the head. They either connect to helmet-like headgear or attach to a hardhat. Shields designed to be worn on the head have pivot points where they attach to the headgear. They can be raised when not needed. Welding shields are made of dark, non-flammable material. The arc is observed through a window that is either 2 by 4-1/4 inches or 4-1/2 by 5-1/4 inches. The window contains a glass filter plate and a clear glass or plastic safety lens. The safety lens is on the outside to protect the more costly filter plate from damage by spatter and debris. For additional filter plate protection a clear safety lens is sometimes also placed

WELDING THREE TRAINEE TASK MODULE 09302

on the inside of the filter plate. On most welding shields the window is fixed in the shield. However some welding shields have a hinge on the 2 by 4-1/4 inch window. The hinged window containing the filter plate can be raised, leaving a separate clear safety lens. This protects the face from flying debris during surface cleaning. *Figure 3-5* shows typical welding shields.

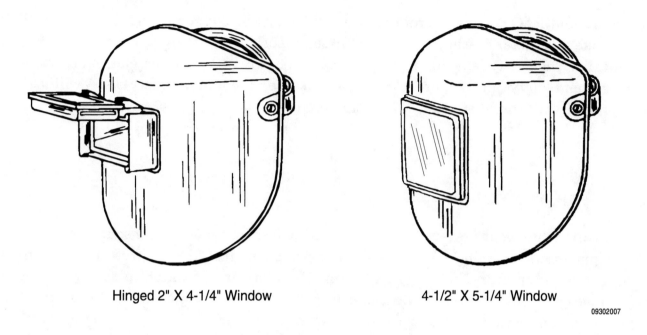

Hinged 2" X 4-1/4" Window   4-1/2" X 5-1/4" Window

09302007

Figure 3-5. Welding Shields

Filter plates come in varying shades. The shade required depends on the maximum amount of amperage to be used. The higher the amperage, the more intense the arc and the darker the filter plate must be. Filter plates are graded by numbers. The larger the number, the darker the filter plate. The American National Standards Institute (ANSI) publication *Z87.1, Practices for Occupational and Educational Eye and Face Protection,* provides guidelines for selecting filter plates. The following recommendations are based on these guidelines.

	Welding Current (Amps)	Lowest Shade Number	Comfort Shade Number
Plasma-Arc Cutting	less than 300	8	9
(PAC)	300 - 400	9	12
	400 - 800	10	14

To select the best shade lens, first start with the darkest lens recommended. If it is difficult to see, try a lighter shade lens until there is good visibility. However, do not go below the lowest recommended number.

WARNING!	Using a filter plate with a lower shade number than is recommended for the welding current (amps) being used can result in severe arc flash (burns) to the eyes.

During normal PAC cutting operations the window in the cutting shield may become dirty from smoke and metal spatter. To be able to see properly the window will have to be cleaned periodically. The safety lens and filter plate can be easily removed and cleaned in the same manner that safety glasses are cleaned. With use, the outer safety lens will become impregnated with metal spatter. When this occurs, replace it.

3.3.0 EAR PROTECTION

Ear protection may be necessary to prevent hearing loss from noise. One source of damaging noise is pneumatic chipping and scaling hammers. Ears must also be protected from flying sparks and metal spatter. This is especially important for out-of-position cutting where falling sparks and metal spatter may enter the ear canal causing painful burns. If ear protection is not used, the sparks may also cause perforated ear drums. Always wear earmuffs or earplugs for protection when performing PAC. *Figure 3-6* shows earmuffs and ear plugs.

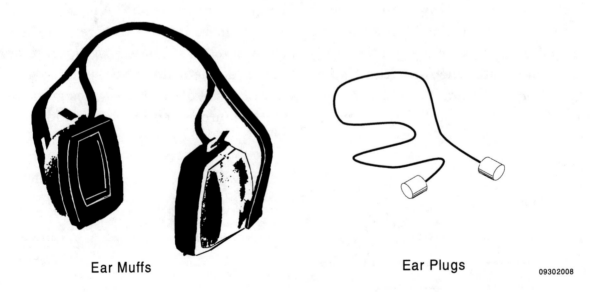

Ear Muffs

Ear Plugs

09302008

Figure 3-6. Earmuffs and Earplugs.

3.4.0 VENTILATION

The gases, dust and fumes caused by plasma-arc cutting can be hazardous if the appropriate safety precautions are not observed. The following section will define these hazards and describe the appropriate safety precautions.

3.4.1 Fume Hazard

Metals heated during PAC may give off toxic fumes and smoke. These fumes are not considered dangerous as long as there is adequate ventilation. Adequate ventilation can be a problem in tight or cramped working quarters. The following general rules can be used to determine if there is adequate ventilation:

- The cutting area should contain at least 10,000 cubic feet of air for each welder
- There should be positive air circulation
- Air circulation should not be blocked by partitions, structural barriers, or equipment

Even when there is adequate ventilation, the welder should try to avoid inhaling cutting fumes and smoke. The heated fumes and smoke generally rise straight up from the cutting torch. Observe the column of smoke and position yourself to avoid it.

If adequate ventilation cannot be ensured or if the welder cannot avoid the cutting fumes and smoke, a respirator should be used.

3.4.2 Respirator

For most PAC applications, a filter respirator offers adequate protection from dust and fumes. The filter respirator is a mask which covers the nose and mouth. The air intake contains a cloth or paper filter to remove impurities. When additional protection is required, an air-line respirator should be used. It is similar to the filter respirator, except that it provides breathing air through a hose from an external source. The breathing air can be supplied form a cylinder of compressed breathing air, or from special compressors that furnish breathing air or from special breathing air filters attached to compressed air lines.

WARNING!   Air line respirators must only be used with pure breathing air sources. Standard compressed air from oil-lubed piston compressors contains toxic oil. This compressed air can only be used for breathing if it is cleaned by a special breathing air filter placed in the line just before the respirator.

Figure 3-7 shows respirators.

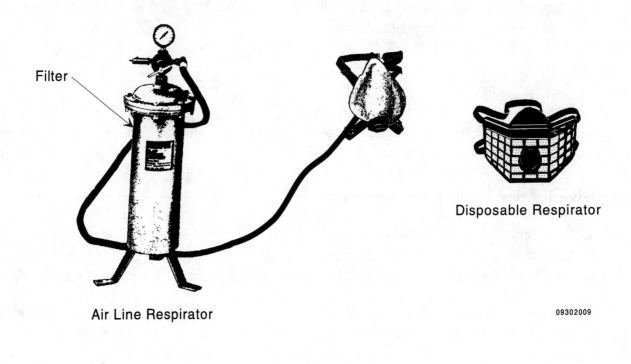

Figure 3-7. Respirators

3.5.0 ELECTRICAL SHOCK HAZARD

Plasma-arc cutting machines receive their power from alternating current sources of 115-460 volts. Contact with these voltages can produce extreme shock and possibly death. For this reason never open a power supply with power on the primary circuit. To prevent electrical shock hazard, the cutting machine must always be grounded.

A ground is an object that makes an electrical connection with the earth. Electrical grounding is necessary to prevent the electrical shock which can result from contact with a defective power supply or other electrical device. The ground provides protection by providing a path from the equipment to ground for stray electrical current created by a short or other defect. The short or other defect will cause the fuse to blow or breaker to trip and stop the current flow. Without a ground the stray electricity would be present in the frame of the equipment. When someone touched the frame the stray current would go to ground through the person.

WARNING! Death by electrocution can occur if equipment is not properly grounded.

To ensure proper electrical grounding, a power source which is using single-phase power must have a three-prong outlet plug. Power sources which use three-phase power must have a four-prong outlet plug. Power sources must only be plugged into electrical outlets which have been properly installed by licensed electricians to ensure the ground has been properly connected. *Figure 3-8* shows grounded outlet plugs.

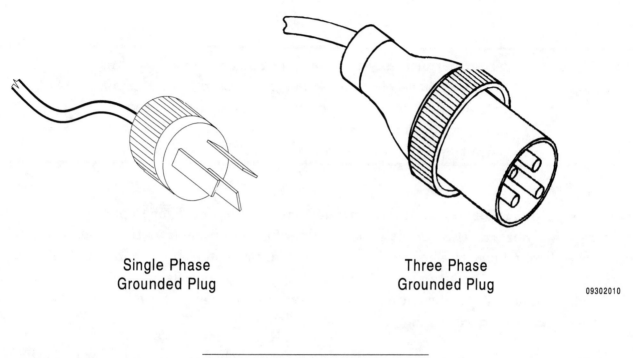

Single Phase
Grounded Plug

Three Phase
Grounded Plug

09302010

Figure 3-8. Grounded Outlet Plugs

3.6.0   AREA SAFETY

Before beginning a cutting job, the area must be checked for safety hazards.

Check the area for water. If possible, remove standing water from floors and work surfaces where cutting cables or power sources are in use in order to avoid injury from slipping or electric shock. When it is necessary to work in a damp environment, make sure that you are well-insulated by wearing rubber boots and gloves. Protect equipment by placing it on pallets above the water and shielding it from overhead leaks. To avoid electric shock, never cut with wet gloves.

Check the area for fire hazards. Cutting generates sparks that can fly ten feet or more and can drop several floors. Any flammable materials in the area must be moved or covered. Always have an approved fire extinguisher on hand before starting any cutting job.

Other workers in the area must be protected from arc flash. If possible, enclose the cutting area with screens to avoid exposing others to harmful rays or flying debris. Inform everyone in the area that you are going to be cutting so that they can exercise appropriate caution.

3.7.0   HOT WORK PERMITS AND FIRE WATCH

Most sites require the use of hot work permits and fire watches. They are necessary because the cutting operator's vision is limited and is focused on the cutting site. Because of this a fire could get out of control before the cutting operator noticed it. Severe penalties are imposed for violation of the hot work permit and fire watch standards.

WARNING!   Never perform any type of heating, cutting or welding until you have obtained a hot work permit and established a fire watch. If you are unsure of the procedure, check with your supervisor. Violation of the hot work permit and fire watch can result in fires and injury.

A hot work permit is an official authorization from the site manager to perform work which may pose a fire hazard. The permit will include information such as the time, location and type of work being done. The hot work permit system promotes the development of standard fire safety guidelines. Permits also help managers to keep records of who is working where and at what time. This information is essential in the event of an emergency or when personnel need to be evacuated.

Always have a fire watch when cutting. During a fire watch, one person (other than the PAC operator) constantly scans the work area for fires. Fire watch personnel should have ready access to fire extinguishers and alarms and know how to use them.

3.8.0   CUTTING INTO CONTAINERS

Before cutting into closed containers such as tanks, barrels and other vessels, check to see if they contain combustible and/or hazardous materials or residues of these materials. Such materials include:

- Petroleum products
- Chemicals that give off toxic fumes when heated
- Acids that could produce hydrogen gas as the result of a chemical reaction
- Any explosive or flammable residue

To identify the contents of containers check the label and then refer to the MSDS (Material Safety Data Sheet). The MSDS provides information about the chemical or material to help you determine if the material is hazardous. If the label is missing, or if you suspect the container has been used to hold materials other than what is on the label, do not proceed until the material is identified.

WARNING!   Do not heat, cut or weld on a container until its contents
                   have been identified. Hazardous material could cause the
                   container to violently explode.

If a container has held any hazardous materials, it must be cleaned before cutting takes place. Clean containers by steam cleaning, flushing with water or washing with detergent until all traces of the material have been removed.

After cleaning the container, fill it with water or an inert gas such as argon or carbon dioxide (CO_2) for additional safety. Air, which contains oxygen, is displaced from inside the container by the water or inert gas. Without oxygen combustion cannot take place. When using water, position the container to minimize the air space. When using an inert gas, provide a vent hole. *Figure 3-9* shows using water in a container to minimize the air space.

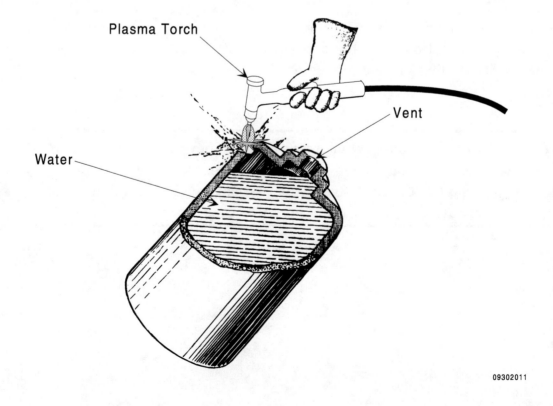

Plasma Torch

Vent

Water

09302011

Figure 3-9. Using Water in a Container to Minimize the Air Space.

3.9.0 HIGH PRESSURE GAS CYLINDER SAFETY

Cutting gases are supplied in high pressure cylinders with pressures of 2000 to 4000 psi. The cylinders come in a variety of sizes based on the cubic feet of gas they hold. The shoulder of each gas cylinder has the name of the gas stamped, labelled or stencilled on it.

Each cylinder is equipped with a bronze cylinder valve on its top. The valve has a threaded tap on one side to receive the spud of a gas regulator or control valve. Turning the cylinder valve handwheel controls the flow of gas from the cylinder. A safety plug is located on the side of the cylinder valve to allow the gas to escape if the pressure in the cylinder rises too high.

Care must be used when handling high pressure cylinders because of the extremely high pressures they contain. When not in use, cylinder valves must always be covered with their protective steel caps.

WARNING! Do not remove the protective cap unless the cylinder is secured. If a high pressure cylinder falls over and the valve breaks off, the cylinder will shoot like a rocket and could cause severe injury or death to anyone in its way.

Always transport cylinders with the protective cap installed over the valve. Transport cylinders to the work station on a hand truck or bottle cart. Use a cylinder cage to lift cylinders, never use a sling or electromagnet.

WARNING! Always handle cylinders with care. They are under high pressure and should never be dropped, knocked over or exposed to excessive heat. When moving cylinders, always be certain that the valve caps are in place.

Figure 3-10 shows transporting cylinders.

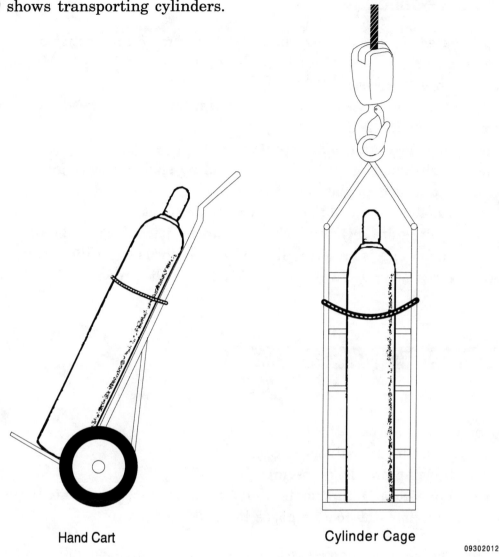

Hand Cart   Cylinder Cage

09302012

Figure 3-10. Transporting Cylinders.

SELF-CHECK REVIEW 1

1. How is plasma made from gas?
2. What are the two arc processes used for plasma-arc cutting?
3. Which arc process is commonly used for the plasma-arc cutting of metals?
4. What is an idle arc?
5. What are the three principal hazards associated with plasma-arc cutting?
6. Why are shaded safety glasses not adequate for protection from the plasma-arc?
7. Why is ear protection important when performing out-of-position plasma-arc cutting?
8. What should be protected or removed from the cutting area before cutting begins?
9. Why is a fire watch usually necessary on a job site?
10. What should be checked before transporting any high pressure gas cylinder?

ANSWERS TO SELF-CHECK 1

1. The gas is heated to an extremely high temperature in an electric arc. (1.0.0)
2. Transferred arc and nontransferred arc. (2.0.0)
3. The transferred arc process. (2.0.0)
4. The arc within the torch used to start the ionization process that does not connect with the metal work piece. (2.1.0)
5. Intense heat, arc flash and metal spatter. (3.0.0)
6. The damaging ultraviolet rays are not stopped by shaded safety glasses. (3.2.1)
7. To prevent molten metal splash from entering the ears. (3.3.0)
8. All flammable materials should be removed or protected before cutting begins because metal spatter can ignite any combustible material it contacts. (3.6.0)
9. The cutting operator's vision is too limited and is focused on the cutting site. (3.7.0)
10. The protective cap should always be installed over the valve. (3.9.0)

4.0.0   PLASMA-ARC CUTTING EQUIPMENT

Basic plasma-arc cutting equipment includes:

- Power supply-control unit
- Plasma-arc cutting torch with torch cable
- Ground (work) clamp and lead assembly
- Gases (or compressed air), flow meters and associated gas components for the plasma (orifice) gas (and shielding/cooling gases if used)

There are many different manufactures of PAC equipment. The following sections will provide general information common to most types. For specific information on controls, operation and parts, refer to the manufactures manuals for specific equipment at your site.

4.1.0   PAC POWER SUPPLY-CONTROL UNITS

Plasma-arc cutting is performed with DC current using straight polarity. The power supplies are special units designed specifically for plasma-arc cutting and are available in many sizes and configurations. The smallest and simplest PAC units are designed for light duty, manual cutting of sheet metal and light gauge plate. They typically plug into a single phase 115 volt or 230 volt AC outlet. Typical maximum cutting amperages for these light duty units range from 14 to 40 amperes. Their duty cycles are usually in the 50 to 60 percent range. The simplest units use filtered compressed air for plasma and cooling gas. They have a high frequency generator for easy arc starting. Some of these small air units even contain their own air compressor to supply gas for cooling and plasma generation.

Console controls are simple and usually include most of the following controls and indicators:

- Power On/Off switch
- Power-on (or unit-ready) indicator
- Air or gas on/off manual switch for checking and adjusting torch air or gas flow
- Output current control
- Power supply overload indicator

Figure 4-1 shows a typical light duty plasma-arc cutting unit.

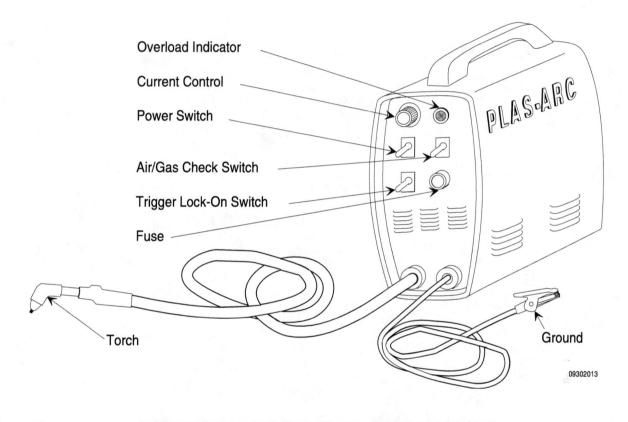

Figure 4-1, Typical Light Duty Plasma-Arc Cutting Unit.

Larger units are used for heavy duty manual cutting and general-duty mechanized cutting. They usually operate on three-phase 230/460 VAC input power, and can produce DC output currents up to 750 amperes. Their duty cycle is usually 100 percent. They can cut stainless steel and aluminum up to 2-1/2 inches thick and carbon steel to 3 inches thick. These units are always water cooled, often containing their own closed-loop torch cooling system.

The larger units often operate with several types of gases. Some use argon for plasma when idling, and then add nitrogen or hydrogen to the plasma when cutting. Some use nitrogen during idling and cut with air or oxygen.

Large units may also be used for gouging when fitted with a gouging torch or gouging tip components. These larger units sometimes use an argon-hydrogen mixture for plasma generation.

Controls on the larger units typically include a power on/off switch, output current control and ammeter, several gas flow controls and gauges, gas selector controls, local/remote switch and open circuit volt meter. *Figure 4-2* shows a typical heavy duty manual plasma-arc cutting unit.

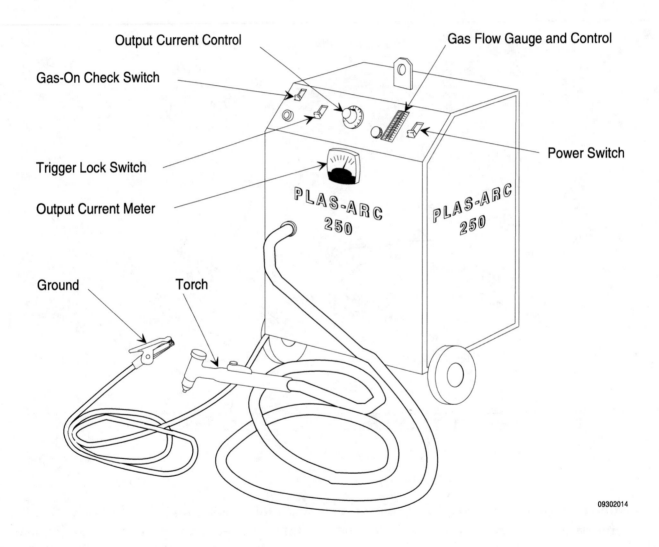

Figure 4-2, Typical Heavy Duty Manual Plasma-Arc Cutting Unit.

4.2.0   PAC TORCHES

The most common plasma-arc cutting torch utilizes the transferred arc system. In the ready mode, a start or idle arc is ignited between the electrode and the nozzle. A small amount of gas is injected into the arc chamber where it is heated to a plasma and escapes through the torch nozzle as a fine jet. However, the torch current circuit is designed so that the full voltage (potential) is not available in the torch between the electrode and nozzle. The base metal to be cut is connected to the control unit power supply by the ground clamp. This gives the base metal a higher potential than the nozzle. When the idling torch is brought near to the work piece, the arc extends to the base metal because of its higher potential. The control unit automatically increases the amperage and gas flow(s) to produce the longer cutting arc. *Figure 4-3* shows the plasma transferred arc cutting system.

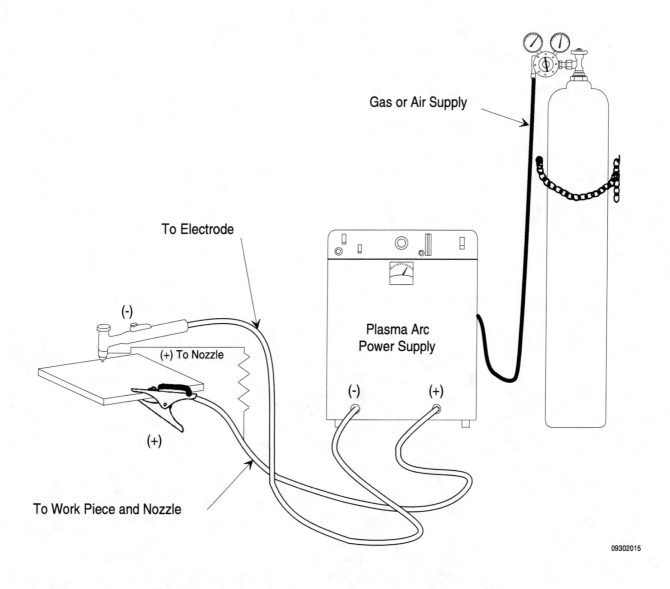

Figure 4-3, Plasma Transferred Arc Cutting System.

Plasma-arc torches may be hand held for manual cutting or machine mounted for mechanized cutting. The heaviest duty torches are usually mechanized and water cooled. Hand held torches are usually of light to medium duty and may be water cooled in the heavier duty versions. Some torches are equipped with separate passages and orifices for cooling and shielding gases. Torches have a button or lever on the torch handle to start the idle arc. A foot peddle can be also be used to start the arc. When a foot peddle is used, the switch on the control panel is set for remote operation.

The plasma-arc cutting torch contains a replaceable tungsten electrode and a gas nozzle. Gas nozzles have various sized holes for cutting different thickness metals. The torch contains coolant passages for gas or water cooling. *Figure 4-4* shows several styles of hand held PAC torches.

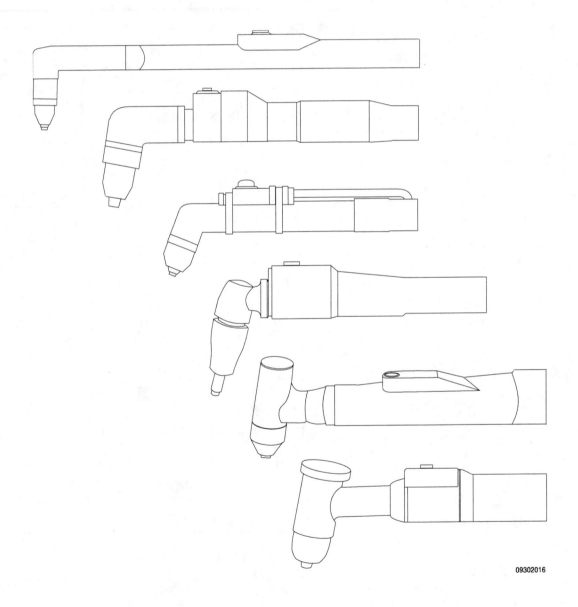

09302016

Figure 4-4, Several Styles of Hand Held PAC Torches.

Mechanized PAC torches are designed to be mounted in automated carriers. Usually they can be used in the same carriers designed for oxyfuel cutting heads. However, the oxyfuel carriers must be capable of moving at the much greater speeds of plasma-arc cutting.

Some types of PAC torches (water injection torches) inject pressurized water to create a whirlpool of water around the jet below the point where the arc narrows in the torch's arc cavity. The water swirls around the plasma jet at high velocity and compresses the jet even tighter. The narrower jet cuts with a more uniform melt rate and leaves the kerfs of stainless steel and titanium free of slag. Typically, this type of torch is mechanized.

Often several styles of shrouds, heat shields, cutting tips and electrodes are available for a given model PAC torch. These components vary in style and size to adapt to different metal types and thicknesses. Sometimes tips and parts for gouging are offered to fit some of the heavier duty torches. Electrode types and styles vary with torch manufacturers. Some use tungsten electrodes identical to those used with GTAW (TIG) torches, while others use special molded shapes. The most common parts that are changed in a torch are the electrode and the cutting tip. The electrode is changed when it becomes contaminated and the cutting tip is changed when it becomes worn (orifice washed out) or when a different type or thickness metal is to be cut. *Figure 4-5* shows various style torch replacement parts.

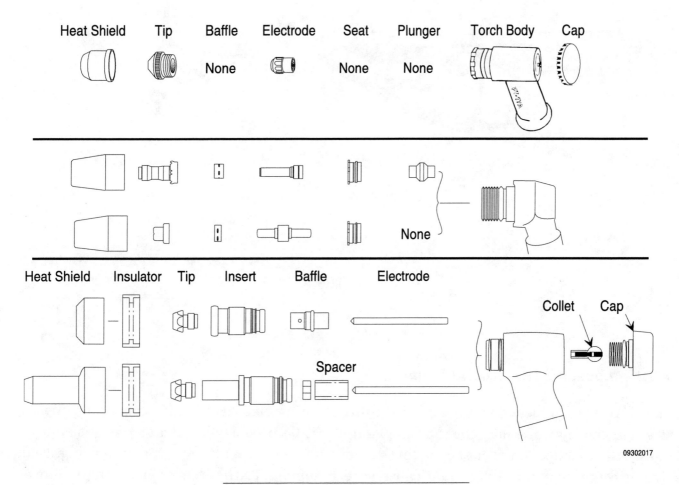

Figure 4-5, Torch Replacement Parts.

4.3.0   GROUND CLAMPS AND LEADS

The ground clamp provides the connection between the end of the ground cable and the work-piece. Ground clamps are mechanically connected to the welding cable and come in a variety of shapes and sizes. The size of a ground clamp is the rated amperage that it can carry without overheating. If a ground clamp needs to be replaced, be sure to select one that is rated at least the same as the rated capacity of the power source it will be used on. *Figure 4-6* shows examples of various styles of ground clamps.

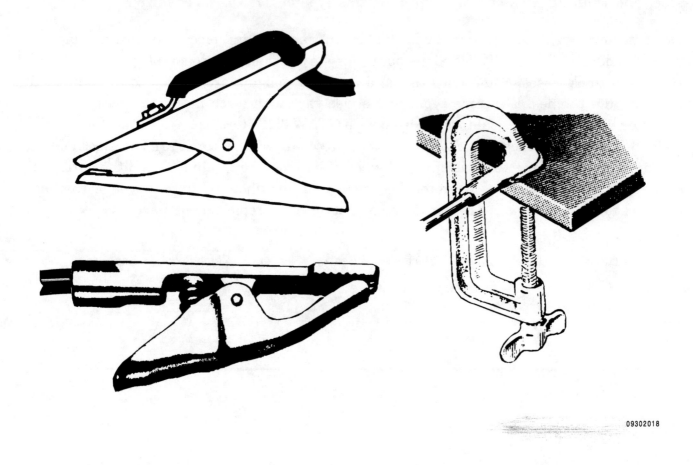

09302018

Figure 4-6. Ground Clamps.

4.3.1   Locating The Ground Clamp

When PAC is to be used to cut on equipment containing electrical or electronic components, batteries, bearings or seals, the ground clamp must be properly located to prevent damage to associated components. If the cutting current passes through any type of bearing, seal, valve or contacting surface it could cause severe damage to the item from arcing and over heating, resulting in costly replacement.

Carefully check the area to be cut and position the ground clamp so that the cutting current will not pass through any connecting surface. If in doubt, ask your supervisor for assistance before proceeding.

CAUTION   Cutting current can severely damage bearings, seals, valves or contacting surfaces. Position the ground clamp to prevent cutting current from passing through them.

Cutting current passing through electrical or electronic equipment will cause severe damage. Before cutting on any type of mobile equipment, the ground lead at the battery must be disconnected to protect the electrical system. If cutting is near the battery, the battery must be removed. Batteries produce hydrogen gas, which is extremely explosive. A spark could cause the battery to explode.

WARNING!   Do not cut near batteries. A spark could cause a battery to explode, showering the area with battery acid.

Before cutting on or near electronic or electrical equipment or cabinets, contact an electrician. The electrician will isolate the system to protect it.

CAUTION   The slightest spark of cutting current can destroy electronic or electrical equipment. Have an electrician check the equipment and, if necessary, isolate the system before cutting.

Ground clamps must never be connected to pipes carrying flammable or corrosive materials. The cutting current could cause overheating or sparks, resulting in an explosion or fire.

4.4.0   PAC GASES AND CONTROLS

All plasma-arc cutting units require one or more types of gases. The specific type of gas required and the controls necessary to set and adjust these gases vary with the type, size and manufacture of the PAC equipment.

4.4.1 PAC Gases

Several different gases and gas mixtures are used with plasma-arc cutting units. Gases used include air, nitrogen, oxygen, argon and argon mixtures, hydrogen and carbon dioxide. The simplest units use clean compressed air to cut carbon steel, stainless steel and aluminum up to about 3/16 inch thick. Heavy-duty air plasma units can cut carbon steel, stainless steel and aluminum up to about 1-3/16 inch thick. Units designed for cutting heavy plate usually require argon or a high percentage argon mixture. Some PAC units require several different gases. One gas may be used for the start or idle arc, another for cutting, and still another for shielding or cooling. General recommendations for plasma cutting gases are:

AIR: Most commonly used gas for lower current plasma cutting units. It works well for most metals from gauge thickness to one inch. It does leave an oxidized cut surface. It can also be used for plasma gouging on carbon steel.

NITROGEN: Commonly used for higher current plasma systems and for cutting materials up to three inches thick. Produces excellent quality cuts on most materials.

OXYGEN: Used where the highest quality cuts are desired on carbon steel up to one inch thick. Cut face is smooth and dross is easy to remove. It can also be used on stainless steel and aluminum but produces a rough cut face.

ARGON-HYDROGEN MIXTURES: Generally used for the cutting of stainless and aluminum. It produces a high quality clean cut face. It is also required when mechanized cutting material over three inches thick. It is an excellent gas for plasma gouging on all materials.

Specific gas requirements for a particular PAC unit or system are specified in the manufacturer's operating instructions manual and/or information tag(s) on the unit. Refer to these specification and use only the gases specified for the type of equipment being used.

4.4.2 PAC Gas Controls

Plasma-arc cutting gases are used to generate plasma, cool the torch and shield the cut against corrosion. All gases must have their pressures reduced before they can be used. The type of regulator used to control the gas pressure depends on the type of PAC equipment being used and the base metal being cut.

When compressed air is being used, a heavy-duty air filter-regulator is required to reduce the high-pressure shop air to the required PAC working pressure. Also, depending on the equipment being used, the shop air must be provided at a minimum flow rate (generally in cubic feet per hour (cfh)).

CAUTION   Check to be sure the shop air supply will provide the minimum flow rate required for the PAC equipment being used. If the air flow rate is too low it will cause poor cuts and cause the torch to overheat. Refer to the manufacturer's specifications for the required flow rate for the equipment being used.

Gases other than compressed air are normally supplied in high pressure cylinders at 1500 to 2000 psi. The high cylinder pressures are reduced by pressure regulators to a lower working pressure. A gas hose connected to the working pressure outlet of the regulator delivers the gas to the PAC console. Most heavy duty systems will require two high pressure cylinders (and two pressure regulators), one containing the cutting gas and the other a shielding or cooling gas. Some systems use a combination of compressed shop air and high pressure cylinders. Figure 4-7 shows a gas pressure regulator.

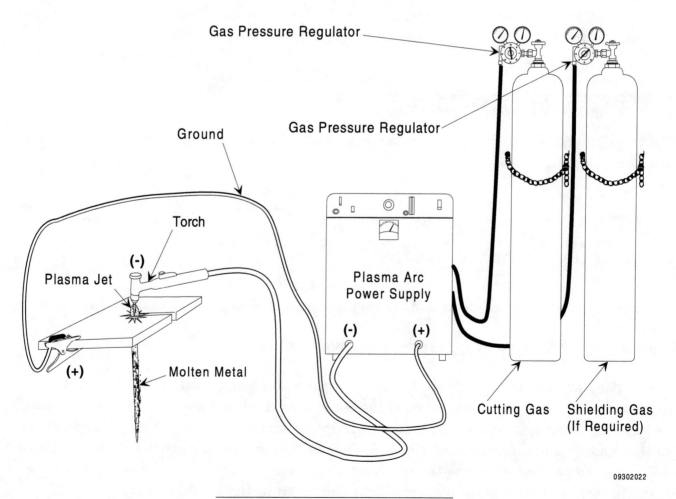

09302022

Figure 4-7. Gas Pressure Regulator.

Stop-start flow control to the torch is usually controlled by a solenoid valve which electrically opens and closes a gas valve. Solenoid valves are usually mounted on the power console and controlled by a push button on the torch. There will be one solenoid for each gas source.

5.0.0 PREPARING WORK AREA FOR PAC

The plasma-arc cutting control unit and work must be located close enough to each other for the torch cable to comfortably reach the work piece. If possible, the work piece should be located at a comfortable height and position for the torch operator. The cutting area should be well ventilated and cleared of all combustible material.

Plasma-arc cutting or gouging can spray molten metal for considerable distances, sometimes 25 feet or more. It is important that everything combustible be removed from the range of the sprayed metal. If necessary, flame-resistant shields or curtains should be erected to protect any nearby workers or equipment from ultraviolet arc rays and metal splash. It is also important to station a fire watch in the area.

Ear protection may be essential when performing PAC in out-of-position cutting to prevent metal spray from entering the operator's ears. If the site ventilation is not adequate to keep the smoke and fumes away from the operator, a respirator must be used.

6.0.0 SETTING UP PAC EQUIPMENT

PAC equipment is easy to set up because it is supplied as a complete system. However, there are some important considerations and preparations to complete before cutting should be attempted:

- The unit must have the rated power to cut the intended metal type and thickness
- The required primary power (phase, voltage and amperage) must be available for the control unit
- The required cutting tip must be installed in the torch
- The required gas or gases must be on line and at the required pressures and flow rates

6.1.0 SETTING CORRECT CUTTING AMPERAGE

The cutting amperage depends on the type of equipment being used, the type and thickness of the material being cut and the type of gas being used. Light gauge sheet steel could require as little as 7 amps, while two inch aluminum plate might require 250 amps. Always refer to the manufactures recommendations to identify the correct amperage for the equipment and job to be performed. Using the incorrect amperage will result in a poor quality cut and if the amperage is to high could cause sever damage to the torch.

6.2.0   INSTALLING GAS CYLINDERS AND SETTING GAS PARAMETERS

Several different gases may be used on the same system for cutting the same or different type and thickness metals. In many cases, the same gas or mixture is not used for both thick and thin metals of the same type. Refer to the manufacturer's instructions or tag on the unit for specific gas recommendations.

6.2.1   Installing A Gas Cylinder

When installing a gas cylinder on the PAC unit, follow these steps:

Step 1   After locating the correct gas type, move it to the unit and secure it in an upright position.

Step 2   Remove the protective valve cap.

WARNING!   Be sure the cloth used does not have any oil or grease on it. Oil or grease compressed with oxygen will explode.

Step 3   If dirt is visible in the connection seat, wipe it out with a clean dry cloth.

WARNING!   Always stand to one side of the valves when opening them to avoid injury from dirt which may be lodged in the valve.

Step 4   While standing to one side of the valve, momentarily crack open the valve to blow any dirt or debris from the valve.

Step 5   Connect the regulator to the cylinder valve and tighten with the proper wrench.

CAUTION   Do not use a pliers or a pipe wrenches to tighten the regulator's spud nut. Also be careful not to overtighten the spud nut. The spud nut is soft bronze nut and can be easily damaged.

Step 6   Open the cylinder valve all the way, and then turn the gas solenoid switch to "on" at the control panel. Gas should flow from the torch. Bleed the line for several seconds and then set the pressure or flow rate according to unit's operating instructions.

PAC unit gas pressures and/or flow rates vary with equipment design, gas type and cut depth (metal thickness). Typical gas flow rates vary from 15 cubic feet per hour (cfh) for nitrogen when cutting 0.1 inch aluminum, carbon steel or stainless steel to 62 cfh of argon and 31 cfh of nitrogen when cutting three inches of carbon steel or 2-1/2 inches of aluminum or stainless steel. Refer to the manufacturer's instructions for the correct gas pressure and flow rates.

7.0.0   OPERATING PAC EQUIPMENT

Plasma-arc cutting equipment is simple to operate. The cutting unit's operating instructions usually provide information for properly setting air or gas pressure and/or flow rates, output current, torch stand-off distance (torch tip to work distance) and torch travel speed for various metal types and thicknesses.

Note   The following steps are provided for general reference only. Refer to the manufactures operation manual for specific instructions on the PAC equipment being used.

To operate the equipment follow these steps:

Step 1   Identify the location of the primary disconnect for the electrical outlet to be used.

Step 2   Plug the PAC power supply into the outlet.

Step 3   Check that the recommended gas(es) are available. Change cylinders if they are empty. Connect the compressor's air line if compressed air is to be used.

Step 4   Turn on the PAC power supply.

Step 5   Adjust the gas regulator pressure(s) and/or flow meter(s) rates to the values recommended in the manufacturer's instructions for the type and thickness of metal being cut.

Step 6   Set the output current (amperage) control to the value recommended in the manufacturer's instructions for the type and thickness of metal being cut.

Step 7   Attach the work lead clamp to the work piece to be cut.

CAUTION   If cutting on machinery or other equipment, be sure to position the work lead clamp so that cutting current will not pass through seals, bearings, or other contacting surfaces that could be damaged from heat or arcing. Also isolate any electrical components.

Step 8   Put on the required safety equipment (hood, gloves, ear protection).

Step 9   Hold the torch at the recommended distance above the point where the cut is to begin, lower your welding hood and press the torch arc start button or foot peddle. As soon as cutting begins, move the torch forward at the recommended speed.

7.1.0   SQUARE CUTTING METAL

Square edges are cut with PAC by holding the torch at 90 degrees to the metal surface while it is advanced smoothly with no side to side movement. A straight edge or metal angle can be used to guide and steady the torch end. When cutting thicker metals, the cut does taper slightly as it narrows with depth. This can result in cuts where one or both faces are not quite square with the plate face. To compensate for this, the torch should be slightly slanted to produce a square edge on the desired piece. All the taper will then be on the scrap side. *Figure 7-1* shows compensating for the taper in thicker metal cuts.

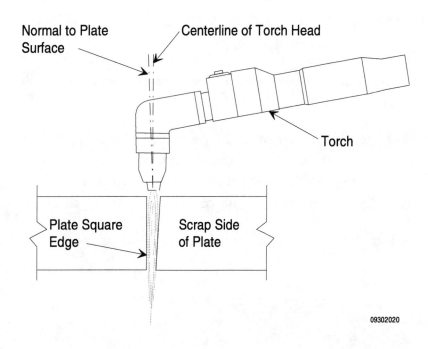

Figure 7-1, Compensating For Taper In Thicker Metal Cuts.

7.2.0 BEVEL CUTTING METAL

Bevel cuts are made with the same technique as for square cuts, except that the torch is held at the required bevel angle. A length of angle iron laid on its open face, can be used to support the torch's side at the required angle. *Figure 7-2* shows using metal angle to hold a taper angle.

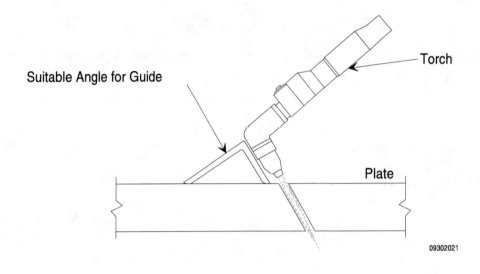

Figure 7-2, Using Metal Angle To Hold A Taper Angle.

7.3.0 PIERCING AND SLOT CUTTING IN METAL

PAC can be used to pierce and cut slots in metal in any position. Leathers are recommended when piercing or cutting metal where out-of-position orientations could cause the metal splash to fall on the torch operator.

To pierce metal, hold the torch directly over the point to be pierced and press the arc button. As soon as the jet passes completely through the work piece, move the torch head in a smooth circle (or other pattern) to produce the desired diameter or shape hole. When piercing thicker materials tip the torch slightly to prevent metal from splashing directly up into the torch.

When cutting slots, a template or straight edge can be used to precisely guide the torch tip.

WELDING THREE TRAINEE TASK MODULE 09302

Proper equipment storage and housekeeping are essential for work efficiency and safety. When finished using the PAC equipment, follow these steps to store the equipment and for good housekeeping.

Step 1   Turn off the power supply.

Step 2   Unplug and coil the power supply cable.

Step 3   Coil the torch cable.

Step 4   Close the cylinder valves and/or disconnect and coil the compressed air line.

Note   If cylinders are to be removed, be sure to remove the pressure regulators and replace the cylinder caps before releasing and moving them.

Step 5   Return the PAC equipment to its proper storage location.

Step 3   Clean off the welding bench (if used) and sweep up the slag and debris that were blown around the area.

SELF-CHECK REVIEW 2

1.   What items does a manual plasma-arc cutting control unit control?

2.   What are the most common parts that are changed in a plasma torch?

3.   What gases are used with plasma-arc cutting?

4.   What is the most commonly used gas for lower current plasma-arc cutting?

5.   What factors are used to determine the proper cutting current (output amperage)?

6.   What factors are used to determine plasma-arc cutting gas flow rates?

7.   Where can you find the detailed information to properly set the cutting variables for a plasma-arc machine?

ANSWERS TO SELF-CHECK REVIEW 2

1. Output electrical current, gas or air supply to the torch and cooling water (if used). (4.1.0)

2. The electrode and the cutting tip. (4.2.0)

3. Air, nitrogen, oxygen, argon, hydrogen, carbon dioxide and mixtures of some of these gases. (4.4.0)

4. Air. (4.4.1)

5. Equipment being used and the metal type and thickness being cut. (6.1.0)

6. Equipment design, gas type and cut depth. (6.2.1)

7. The unit's operation manual provides operating details to properly set air or gas pressure and/or flow rates, output current, torch stand-off distance (torch tip to work distance) and torch travel speed for various metal types and thicknesses. (7.0.0)

The following task is designed to test your competency with a plasma-arc cutting torch. Do not perform this cutting task until directed to do so by your instructor.

Using stainless steel or aluminum plate, lay out and cut the shape shown in *Figure 9-1*.

Note   Mild steel may be substituted if stainless steel or aluminum is not available.

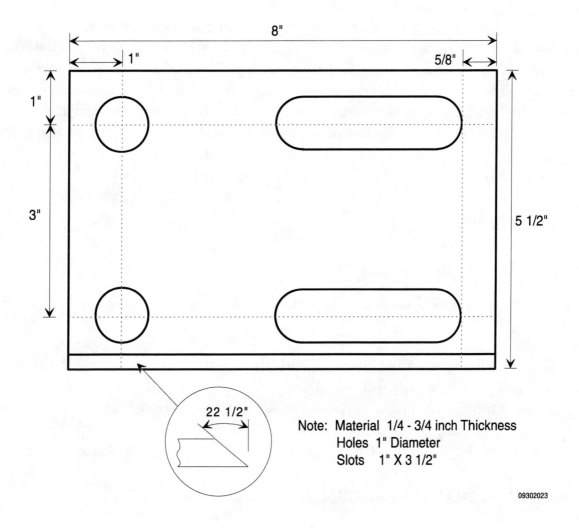

Note: Material 1/4 - 3/4 inch Thickness
Holes 1" Diameter
Slots   1" X 3 1/2"

09302023

Figure 9-1, Cutting Task.

Criteria For Acceptance:

- Outside dimensions plus or minus 1/16 inch
- Inside (holes and slots) dimensions plus or minus 1/8 inch
- Square plus or minus two degrees
- Bevel plus or minus two degrees
- Minimal amount of dross sticking to plate which can be easily removed
- Square kerf face with minimal notching not exceeding 1/16 inch deep

SUMMARY

Plasma-arc cutting is very useful for cutting and piercing many types of metals. Unlike oxyfuel cutting, it can be used to cut aluminum, magnesium, copper, nickel and stainless steel. It is usually much faster than oxyfuel cutting and produces little or no distortion or alteration zone changes. The cuts are fast and clean with little or no slag or dross. There is no chance of carbon inclusions as can happen with carbon arc cutting. Cut surfaces are usually ready for welding without further preparation. However, because the torch jet produces metal spray, good site fire-prevention practices and the wearing of the required personal protective equipment are essential. Ear protection is important when cutting out of position to prevent sparks or molten metal from entering the ears.

References

For advanced study of topics covered in this module, the following work is suggested:

Recommended Practices for Plasma-Arc Cutting, (C5.3-91), American Welding Society, Miami, FL 33135. Phone 1-800-334-9353.

Modern Welding (1988), Althouse, Turnquist, Bowditch and Bowditch, The Goodheart-Willcox Company, Inc., South Holland, IL 60473-2089. Phone 1-800-323-0440.

PERFORMANCE/LABORATORY EXERCISES

1. Set-up plasma-arc cutting equipment.
2. Perform plasma-arc cutting.

The NCCER makes every effort to keep these manuals up-to-date and free of technical errors. We appreciate your help in this process. If you have an idea for improving this manual, or if you find an error, a typographical mistake, or an inaccuracy in the *Wheels of Learning*, please write us, using this form or a photocopy. Be sure to include the exact module number, page number, a description of the problem, and the correction, if possible. We'll do our best to correct it in later editions. Thank you for your assistance.

Write: *Wheels of Learning*
National Center for Construction Education and Research
P.O. Box 141104
Gainesville, FL 32614-1104

Fax: 352-334-0932

WHEELS OF LEARNING USER UPDATE

Please let us know if you have found an inaccuracy, error, or other problem in a *Wheels of Learning* manual. Use this form or write us a letter. Please be sure to tell us the exact module name and module number, the page number, and the problem. Thanks for your help.

Craft _____ Module Name _____

Module Number _____ Page Number(s) _____

Description of Problem _____

_____

_____

(Optional) Correction of Problem _____

_____

_____

(Optional) Your Name and Address _____

_____

Weld Quality

Module 09107

NATIONAL
CENTER FOR
CONSTRUCTION
EDUCATION AND
RESEARCH

WELD QUALITY

Objectives

Upon completion of this module, the trainee will be able to:

1. Identify and explain codes governing welding.
2. Identify and explain weld imperfections and their causes.
3. Identify and explain non-destructive examination practices.
4. Identify and explain welder qualification tests.
5. Explain the importance of quality workmanship.
6. Identify and explain typical site quality organizational structures.

Prerequisites

Successful completion of the following module(s) is required before beginning study of this module:

●   SMAW Beads and Fillet Welds

Required Student Materials

None

Course Map Information

This course map shows all of the Wheels of Learning modules in the first level of the Welding curricula. The suggested training order begins at the bottom and proceeds up. Skill levels increase as a trainee advances on the course map. The training order may be adjusted by the site Training Program Sponsor.

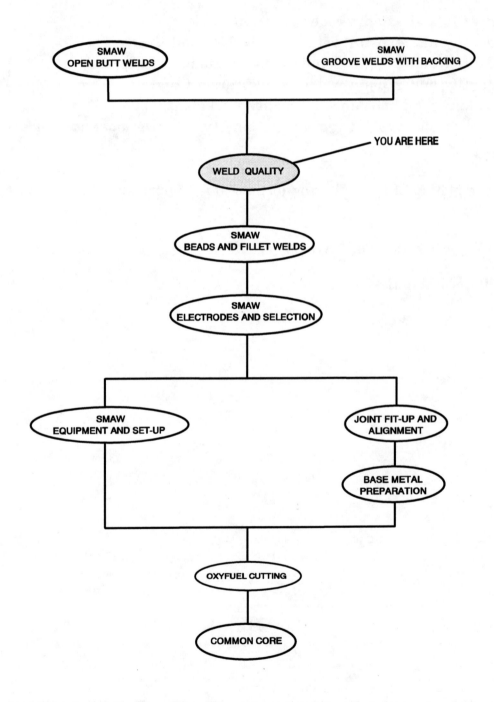

TABLE OF CONTENTS

Trade Terms Introduced in this Module

Defect: A discontinuity or imperfection that renders a part or product unable to meet minimum acceptable standards or specifications.

Discontinuity (Imperfection): A change or break in the shape or structure of a part which may or may not affect the usefulness of the item.

Homogeneity: Having a uniform structure or composition throughout.

Hardenable Materials: Metals that have the ability to be made harder by heating and then cooling.

Inclusion: Foreign matter introduced into and remaining in a weld.

Imbrittled: Metal which has been made brittle and will tend to crack with little bending.

Interpass Temperature: In a multi-pass weld, the temperature of the deposited weld metal before the next pass is deposited.

Laminations: Cracks in the base metal formed when layers separate.

Procedure Qualification: The demonstration that welds made following a specific process can meet prescribed standards.

Radiograph: Photograph made by passing x-rays or gamma rays through an object and recording the variations in density on a photographic film.

Underbead Crack: Crack in the base metal near the weld but under the surface.

Test Coupons: The metal pieces to be welded together as a test.

1.0.0 INTRODUCTION

Quality is an important aspect of the welder's job. The quality of the weld that will pass (be acceptable) for one welding application may not be acceptable for another. What is considered an acceptable weld has been established in codes and standards. There are several codes that govern welding activities, qualification requirements and tests that can be performed on weldments to determine and identify various weld imperfections.

The following sections will identify and explain the applicable codes, various weld imperfections and their causes, types of weldment testing and welder qualification test requirements. The importance of quality workmanship will also be discussed, along with typical site quality organizational structures for assuring that quality welds are achieved.

2.0.0 CODES GOVERNING WELDING

A code consists of a set of requirements covering permissible materials, service limitations, fabrication, inspection, testing procedures and qualifications of welders. Welding codes ensure that safe and reliable welded products will be produced and that persons associated with the welding operation will be safe. Clients specify in the contract which codes will be used on the project. All welding must then be performed following the guidelines and specifications outlined in that code. Since there are a number of codes and each is updated periodically, you should know which welding codes and code year apply to your job.

Codes and specifications that apply to welding safety and quality have been developed and are published by a number of nationally recognized agencies. To eliminate the necessity of writing a code for each new job, sections of these existing codes are referenced by the project contract. Agencies and societies that have established codes include:

- American Society of Mechanical Engineers (ASME)
- American Welding Society (AWS)
- American Petroleum Institute (API)
- American National Standards Institute (ANSI). This manual will not cover all the detailed requirements of each individual code or code section. Instead, general considerations that apply to weld quality will be covered.

2.1.0 AMERICAN SOCIETY OF MECHANICAL ENGINEERS (ASME)

The American Society of Mechanical Engineers has two codes: the *ASME Boiler and Pressure Vessel Code* and *B31, Code for Pressure Piping*. Both these codes are endorsed by the American National Standards Institute.

The ASME Boiler and Pressure Vessel Code (BPVC) contains eleven sections. Those sections most frequently referenced by welders are:

- Section II, "Material Specifications," which contains the specifications for acceptable ferrous (Part A) and non-ferrous (Part B) base metals and for acceptable welding and brazing filler metals and fluxes (Part C). Many of these specifications are identical to and have the same number designation as AWS specifications for welding consumables. This section is used to match base metals and filler metals.
- Section V, "Nondestructive Examination," which covers methods and standards for nondestructive examination of boilers and pressure vessels.
- Section IX, "Welding and Brazing Qualifications," which covers the qualification of welders, welding operators, brazers and brazing operators. It also covers the welding and brazing procedures that must be used for welding or brazing of broilers or pressure vessels. This section of the code is often cited by other codes and standards as the welding qualification standard.

ASME B31, Code for Pressure Piping consists of six sections. Each section gives the minimum requirements for the design, materials, fabrication, erection, testing and inspection of a particular type of piping system. In particular, B31.1, "Power Piping," covers power and auxiliary service systems for electric generation stations. B31.3 covers chemical plant and petroleum refining piping.

All sections of *ASME B31 Code for Pressure Piping* require qualification of the welding procedures and testing of welders and welding operators. Some sections require these qualifications to be performed in accordance with Section IX of the *ASME Boiler and Pressure Vessel Code*, while in others it is optional.

2.2.0 AMERICAN WELDING SOCIETY (AWS)

The American Welding Society publishes numerous documents covering welding. These documents include codes, standards, specifications, recommended practices and guides. *AWS D1.1, Structural Welding Code—Steel* is the code most frequently referenced. It covers welding and qualification requirements for welded structures of carbon and low alloy steels. It is not intended to apply to pressure vessels, pressure piping or base metals less than 1/8-inch thick.

2.3.0 AMERICAN PETROLEUM INSTITUTE (API)

The American Petroleum Institute publishes documents in all areas related to petroleum production. *API Std. 1104, Standard for Welding Pipelines and Related Facilities* applies to arc and oxyfuel gas welding of piping, pumping, transmission and distribution systems of petroleum. It presents methods for making acceptable welds by qualified welders using approved welding procedures, materials and equipment. It also presents suitable methods to ensure proper analysis of weld quality.

2.4.0 AMERICAN NATIONAL STANDARDS INSTITUTE (ANSI)

The American National Standards Institute is a private organization that does not actually prepare standards. Instead, it adopts standards that it feels are of value to the public interest. ANSI standards deal with dimensions, ratings, terminology and symbols, test methods performance and safety specifications for materials, equipment, components and products in many fields, including construction. Many of the codes used today have been adopted as an ANSI standard.

3.0.0   BASIC ELEMENTS OF WELDING CODES

All welding codes provide detailed information about qualification in three general areas. These are:

- Welding procedure qualification
- Welder performance qualification
- Welding operator qualification

Each of the above types of qualification is different and subject to different requirements.

CAUTION   The information in this manual is provided as a general
guideline only. Check with your supervisor if you are unsure of
the codes and specification requirements for your project.

3.1.0   WELDING PROCEDURE QUALIFICATION

A welding procedure is a written document that contains materials, methods, processes, electrode types, techniques and all other necessary and relevant information about the weldment. Welding procedures must be qualified before they can be used. Procedure qualification has nothing to do with the skills of the individual welder.

Welding procedure qualifications are limiting instructions written to explain how welding will be done. These limiting instructions are listed in a document known as a Welding Procedure Specification (WPS). The purpose of the WPS is to define and document in detail the variables involved in welding a certain base metal. The WPS lists the following in detail:

- The various base metals to be joined by welding
- The filler metal to be used
- The range of preheat and postweld heat treatment
- Thickness and other variables described for each welding process

WPS variables are identified as either essential or non-essential variables. Essential variables are items in the welding procedure specification that cannot be changed without requalifying the welding procedure. Non-essential variables are items in the WPS which may be changed, within a range identified by the code, but do not affect the qualification status.

For example, the following are considered essential variables in a welding procedure:

- Filler metal classification
- Material thickness
- Joint design
- Type of base metal
- Welding process

Examples of non-essential variables that may be changed without having to requalify the welding procedure include:

- Amperage
- Travel speed
- Shielding gas flow (if applicable)
- Electrode and filler wire size

CAUTION   Do not change any variable (essential or non-essential) without discussing it with your supervisor.

The WPS is qualified for use by welding test coupons (samples) and by testing the coupons in accordance with the applicable code. A test weld is made and test coupons are cut from it. The test coupons are used to make tensile tests, root bends and face bends as required by the code. The test results are then recorded on a document known as a Procedure Qualification Record (PQR). If the weldment produced by the particular procedure meets the standards of the code, the procedure becomes qualified.

Each Welding Procedure Specification must have a PQR to document the quality of the weld produced.

The methods used to qualify procedures are more detailed and thorough than those used to qualify either welders or welding operators. This is because procedures must qualify physical and metallurgical properties.

3.2.0   WELDER PERFORMANCE QUALIFICATION

Once a procedure is qualified, the welder using it must be qualified to use that procedure by passing a welding performance qualification test. Since no single performance test can qualify welders for all the different types of welding that must be done, a welder may be required to pass additional performance qualification tests. Section 6.0 of this manual, "Welder Performance Qualification Tests," discusses performance tests used to qualify the welder.

3.3.0 WELDER OPERATOR QUALIFICATION

When fully automatic welding equipment is used, the operators of the equipment must demonstrate their ability to set up and monitor the equipment to produce acceptable welds. The codes also contain qualification tests for these operators.

4.0.0 WELD DISCONTINUITIES AND THEIR CAUSES

Codes and standards define the quality requirements necessary to achieve the integrity and reliability of the weldment. These quality requirements help ensure that welded joints are capable of serving the intended function for the desired life of the weldment. Weld discontinuities can prevent a weld from meeting the minimum quality requirements.

AWS defines a discontinuity as "an interruption of the typical structure of a weldment, such as a lack of homogeneity in the mechanical, metallurgical, or physical characteristics of the material or weldment." A discontinuity is not necessarily a defect. A single excessive discontinuity or a combination of discontinuities can make the weldment defective (unable to meet minimum quality requirements). On the other hand, a weld can have one or more discontinuities and be acceptable.

Ideally, a weld should not have any discontinuities, but most will have one or more. When evaluating a weld, it is important to note the type, the size, and the location of the discontinuity. Any one of these factors, or all three, can change a discontinuity to a defect, causing the weld to be rejected during the inspection process. For example, discontinuities located where stresses exist tend to enlarge and are more detrimental than those in other locations. Surface or near-surface discontinuities may be more detrimental than similarly shaped, internal discontinuities.

The welder should be able to identify discontinuities and understand the effect they have on weld integrity. Some can be seen by visual inspection. Those that are internal to the weldment can only be seen through other testing methods. The most common weld discontinuities are:

- Porosity
- Inclusions
- Cracks
- Incomplete joint penetration
- Incomplete fusion
- Undercut
- Arc strikes
- Spatter
- Unacceptable weld profiles

4.1.0 POROSITY

Porosity is the presence of voids or empty spots in the weld metal. It is the result of gas bubbles trapped in the weld as it is being made. As the molten metal hardens, the bubbles form voids. Unless the bubbles work up to the surface of the weld before it hardens, porosity cannot be found by visual inspection.

Porosity can be grouped into four major types.

1. **Uniformly scattered** porosity may be located throughout single-pass welds or throughout several passes of multiple pass welds.
2. **Clustered porosity** is a localized grouping of pores that may result from improper starting or stopping of the welding.
3. **Linear porosity** may be aligned along a weld interface, the root of a weld or a boundary between weld beads.
4. **Piping porosity**, or wormholes, is a term for elongated gas pores. In fillet welds piping porosity normally extends from the root of the weld toward the face. Often it does not extend to the surface. When this occurs, the porosity cannot be visually detected.

Figure 4-1 shows porosity examples.

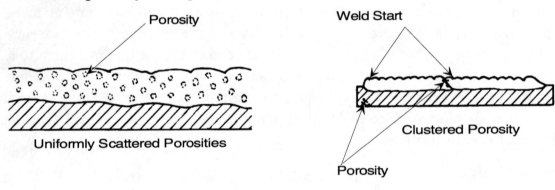

Figure 4-1. Types of Porosity

Most porosity is caused by improper welding techniques and contamination. Improper welding techniques may cause inadequate shielding gas to be formed. As a result, parts of the weld are unprotected. Nitrogen from the air or moisture in the flux or on the base metal that dissolves in the weld pool can become trapped and produce porosity.

The intense heat of the weld can decompose paint, dirt, oil or other contaminates, producing hydrogen. This gas, like nitrogen, can become trapped in the solidifying weld pool and produce porosity.

Excessive porosity has a serious effect on the mechanical properties of the joint. Although some codes permit a certain amount of porosity in welds, it is best to have as little as possible. This can be accomplished by proper cleaning of the base metal, avoiding excessive moisture in the electrode covering and using proper welding techniques.

4.2.0   INCLUSIONS

Inclusions are foreign matter trapped in the weld metal, between weld beads or between the weld metal and the base metal. Inclusions sometimes are jagged and irregularly shaped. Sometimes they form in a continuous line. This causes stresses to concentrate in one area and reduces the structural integrity (loss in strength) of the weld.

Figure 4-2 shows examples of nonmetallic inclusions

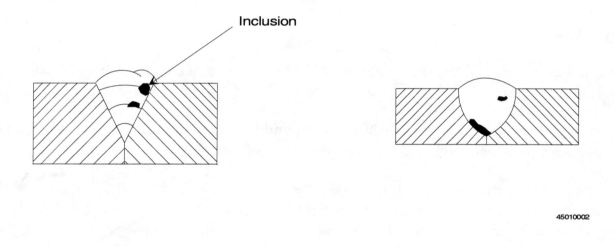

45010002

Figure 4-2. Examples of Nonmetallic Inclusions

In general, inclusions result from faulty welding techniques, improper access to the joint for welding or both. A typical example of an inclusion is slag which normally forms over a deposited weld. If the electrode is not manipulated correctly, the force of the arc will cause some of the slag particles to be blown into the molten pool. If the pool solidifies before the inclusions can float to the top they become lodged in the metal, producing a defective weld. Sharp notches in joint boundaries or between weld passes can also result in slag entrapment.

Inclusions are more likely to occur in out-of-position welding, since the tendency is to keep the molten pool small and allow it to solidify rapidly to prevent it from sagging.

With proper welding techniques and the right electrode used with the proper setting, inclusion can be avoided or kept to a minimum. Other remedies include:

- Positioning the work to maintain slag control
- Changing the electrode or flux to improve control of molten slag
- Thoroughly removing slag between weld passes
- Grinding the weld surface if it is rough and likely to entrap slag
- Removing heavy mill scale or rust on weld preparations
- Avoiding the use of electrodes with damaged coverings

4.3.0 CRACKS

Cracks are narrow breaks that may occur in the weld metal, the base metal or in the crater formed at the end of a weld bead. They are caused when localized stresses exceed the ultimate strength of the metal. Cracks are generally located near other weld or base metal discontinuities.

4.3.1 Weld Metal Cracks

Three types of cracks can occur in weld metal: transverse cracks, longitudinal cracks and crater cracks.

Transverse cracks run across the face of the weld and may extend into the base metal. They are more common in joints that have a high degree of restraint.

Longitudinal cracks are usually located in the center of the weld deposit. They may be the continuation of crater cracks or cracks in the first layer of welding. Cracking of the first pass is likely to occur if the bead is thin. If this cracking is not eliminated before the other layers are deposited, the crack will progress through the entire weld deposit.

Crater cracks have a tendency to form in the crater whenever the welding operation is interrupted. These cracks usually proceed to the edge of the crater and may be the starting point for longitudinal weld cracks. Crater cracks can be minimized or prevented by filling craters to a slightly convex shape prior to breaking the welding arc.

Figure 4-3 shows examples of weld metal cracks.

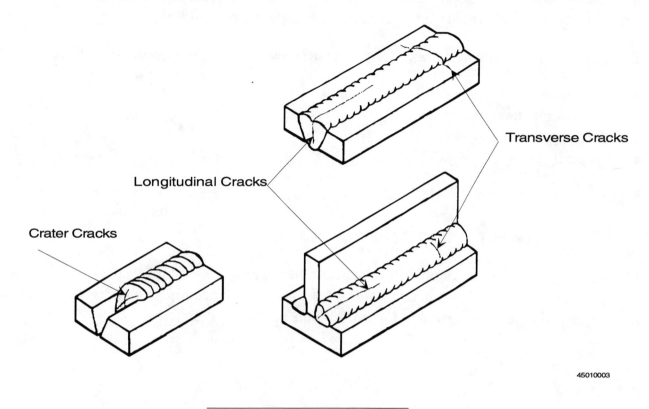

Figure 4-3. Weld Metal Cracks

Cracking of the weld metal can usually be reduced by one or more of the following actions:

- Improving the contour or composition of the weld deposit by changing the electrode manipulation or electrical conditions
- Increasing the thickness of the deposit and providing more weld metal to resist the stresses by decreasing the travel speed
- Reducing thermal stress by preheating
- Using low hydrogen electrodes
- Balancing shrinkage stress by sequencing welds
- Avoiding rapid cooling conditions

4.3.2   Base Metal Cracks

Base metal cracking usually occurs within the heat-affected zone of the metal being welded. The possibility of cracking increases when working with hardenable materials. These cracks usually occur along the edges of the weld and through the heat-affected zone into the base metal. Types of base metal cracking include underbead cracking and toe cracking.

Underbead cracks are limited mainly to steel. When they occur they are usually found at regular intervals under the weld metal do not extend to the surface. Because of this they cannot be detected by visual methods of inspection.

Toe cracks are generally the result of thermal shrinkage strains acting on a weld's heat-affected zone that has been embrittled. They sometimes occur when the base metal cannot accommodate the shrinkage strains that are imposed by welding.

Base metal cracking can usually be reduced or eliminated by one of the following means:

- Controlling the cooling rate by preheat
- Controlling heat input
- Using the correct electrode
- Properly controlling welding materials

Figure 4-4 shows examples of base metal cracks in welded joints.

4.4.0   INCOMPLETE JOINT PENETRATION

Incomplete joint penetration is a term used to describe the failure of the filler metal to penetrate and fuse with an area of the weld joint. This incomplete penetration will cause weld failure if the weld is subjected to tension or bending stresses.

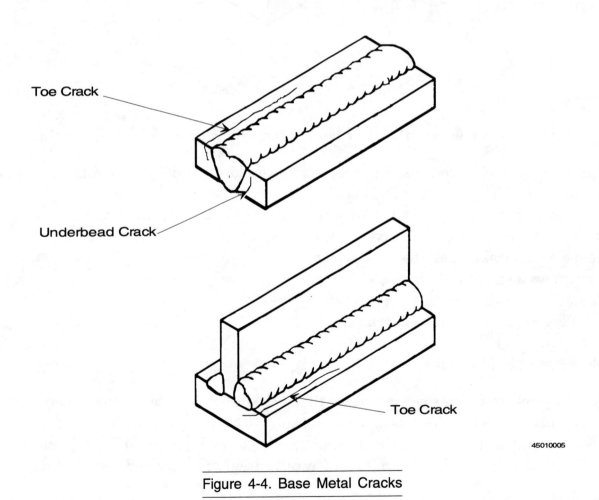

45010005

Figure 4-4. Base Metal Cracks

Incomplete joint penetration is generally associated with groove welds. It may result from insufficient welding heat, improper joint design (too much metal for the welding arc to penetrate) or poor control of the welding arc.

Figure 4-5 shows an example of incomplete joint penetration.

Insufficient heat at the root of the joint is a frequent cause of incomplete joint penetration. If the metal being joined first reaches the melting point at the surfaces above the root of the joint, molten metal may bridge across the gap between these surfaces and screen off the heat source before the metal at the root melts.

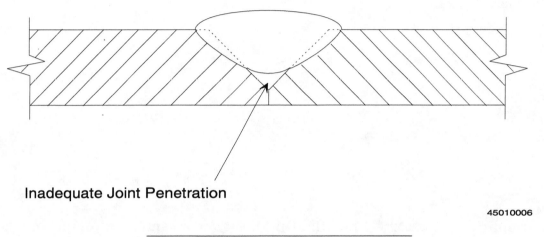

Inadequate Joint Penetration

45010006

Figure 4-5. Inadequate Joint Penetration

Improper joint design is another leading cause of incomplete penetration. If the joint is not prepared or fitted accurately, an excessively thick root face or an insufficient root gap may cause incomplete penetration. For example, incomplete penetration is likely to occur under the following conditions:

- The root face dimension is too big even though the root opening is adequate
- The root opening is too small
- The included angle of a V-groove is too small

Figure 4-6 shows typical joint designs.

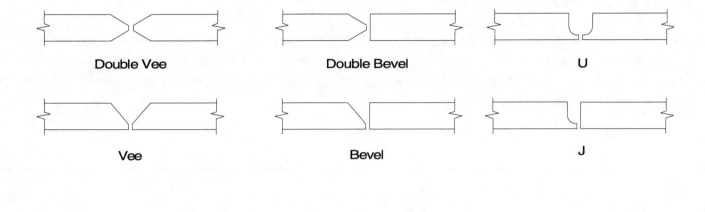

Double Vee   Double Bevel   U

Vee   Bevel   J

45010007

Figure 4-6. Typical Joint Designs

Even if the welding heat is correct and the joint design is adequate, incomplete penetration can result from poor control of the welding arc, including:

- Using an electrode that is too large
- Traveling too fast
- Using a welding current that is too low

Incomplete penetration is always undesirable in welds, especially in single-groove welds if the root of the weld is subject to either tension or bending stresses. It can lead directly to weld failure or can cause a crack to start at the unfused area.

4.5.0   INCOMPLETE FUSION

Many welders confuse incomplete joint penetration with incomplete fusion. It is possible to have good penetration without complete root fusion. Incomplete fusion is the failure of a welding process to fuse, or join together, layers of weld metal or weld metal and base metal.

Failure to obtain fusion may occur at any point in a groove or fillet weld, including the root of the weld. Often, the weld metal just rolls over onto the plate surface. This is generally referred to as overlap. In many cases the weld has good fusion at the root of the weld and plate surface, but because of poor welding technique and insufficient heat, the toe of the weld does not fuse.

Figure 4-7 shows incomplete fusion and overlap.

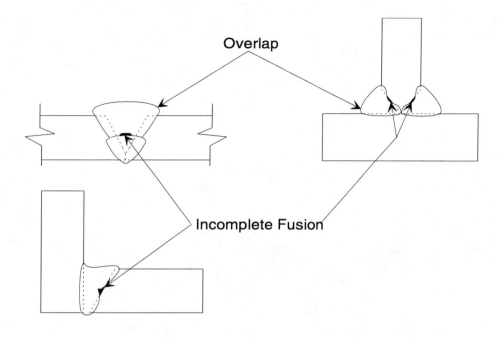

Figure 4-7. Incomplete Fusion and Overlap

Causes for incomplete fusion include the following:

- Insufficient heat as a result of low welding current, high travel speeds, or an arc gap that is too close
- Wrong size or type of electrode
- Failure to remove oxides or slag from groove faces or previously deposited beads
- Improper joint design
- Inadequate gas shielding

Incomplete fusion discontinuities affect weld joint integrity in much the same way as do porosity and slag inclusion.

4.6.0 UNDERCUT

Undercut is a groove melted into the base metal beside the weld. It is the result of the arc removing more metal from the joint face than is replaced by weld metal. On multi-layer welds, it may also occur at the point where a layer meets the wall of a groove.

Figure 4-8 shows undercutting.

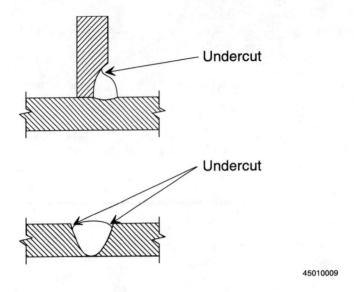

Figure 4-8. Undercutting

Undercutting is usually due to improper electrode manipulation. Other causes of undercutting include:

- Using a current adjustment that is too high
- Having an arc gap that is too long
- Failing to fill up the crater completely with weld metal

Most welds have some undercut which can be found upon careful examination. When it is controlled within the limits of the specifications and does not create a sharp or deep notch, undercut is usually not considered a weld defect. However, when it exceeds the limits, undercutting can be a serious defect since it reduces the strength of the joint.

4.7.0 ARC STRIKES

Arc strikes are small, localized points where surface melting occurs away from the joint. These spots may be caused by accidentally striking the arc in the wrong place or by faulty ground connections.

Striking an arc on base metal that will not be fused into the weld metal should be avoided. They can cause hardness zones in the base metal or can become the starting point for cracking.

Figure 4-9 shows an example of arc strikes.

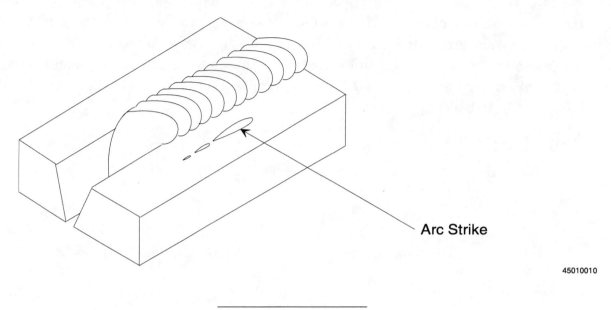

Arc Strike

45010010

Figure 4-9. Arc Strikes

4.8.0 SPATTER

Spatter is made up of very fine particles of metal on the plate surface adjoining the weld area. It is usually caused by high current, a long arc, an irregular and unstable arc or improper shielding. Spatter makes a poor appearance on the weld and base metal and can make it difficult to inspect the weld.

Figure 4-10 shows weld spatter.

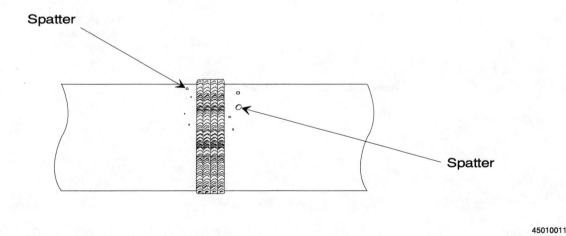

Spatter

Spatter

45010011

Figure 4-10. Weld Spatter

4.9.0   UNACCEPTABLE WELD PROFILES

The profile of a finished weld can affect the performance of the joint under load as much as other discontinuities affect it. This also applies to the profile of a single-pass or layer of a multiple-pass weld. An unacceptable profile of a single-pass of a multiple-pass weld could lead to the formation of discontinuities such as incomplete fusion or slag inclusions as the other layers are deposited. *Figure 4-11* shows acceptable and unacceptable weld profiles for both fillet and butt welds.

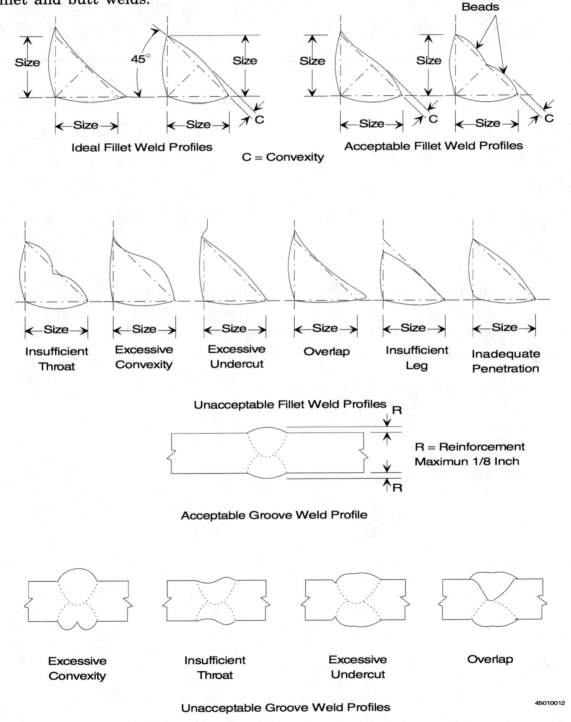

Figure 4-11. Acceptable and Unacceptable Weld Profiles

The ideal fillet has a uniform concave or convex face, although a slightly nonuniform face is acceptable. The convexity of a fillet weld or individual surface bead must not exceed 0.07 times the actual face width or individual surface bead, plus 0.06 inch.

Butt welds should be made with slight reinforcement (not exceeding 1/8 inch) and a gradual transition to the base metal at each toe. Butt welds must not have excess convexity, insufficient throat, excessive undercut or overlap. If a butt weld has any of these defects it must be repaired.

CAUTION Refer to your site's WPS (Welding Procedure Specification) for specific requirements on fillet or butt welds. The information provided here is only a general guideline. The site WPS or Quality Specifications must be followed for all welds. Check with your supervisor if you are unsure of the specifications for your application.

SELF-CHECK REVIEW 1

1. What is a welding code?
2. What are two welding codes produced by the American Society of Mechanical Engineers?
3. What does the *AWS D1.1, Structural Welding Code—Steel* cover?
4. What does the *API Standard 1104* apply to?
5. What are the essential variables in a welding procedure?
6. Can a weld have a discontinuity and still be acceptable?
7. What is porosity?
8. What causes inclusions in a weld?
9. What are the three types of base metal cracks?
10. Where does base metal cracking usually occur?
11. Insufficient heat at the root of the joint is a frequent cause of what type of discontinuity?
12. What is it called when the weld metal just rolls over onto the plate surface?
13. What is the usual cause of undercut?
14. Why should arc strikes be avoided?
15. What is the maximum reinforcement for a butt weld?

ANSWERS TO SELF-CHECK REVIEW 1

1. A code consists of a set of requirements covering permissible materials, service limitations, fabrication, inspection, testing procedures and qualifications of welders. (2.0.0)

2. The *ASME Boiler and Pressure Vessel Code* and *B31, Code for Pressure Piping*. (2.1.0)

3. It covers welding and qualification requirements for welded structures of carbon and low alloy steels. It is not intended to apply to pressure vessels, pressure piping or base metals less than 1/8-inch thick. (2.2.0)

4. It applies to arc and oxyfuel gas welding of piping, pumping, transmission and distribution systems of petroleum. (2.3.0)

5. Essential variables are items in the welding procedure specification that cannot be changed without requalifying the welding procedure. The following are considered essential variables in a welding procedure:

   - Filler metal classification
   - Material thickness
   - Joint design
   - Type of base metal
   - Welding process (3.1.0)

6. A weld can have one or more discontinuities and be acceptable. (4.0.0)

7. Porosity is the presence of voids or empty spots in the weld metal. It is the result of gas bubbles trapped in the weld as it is being made. As the molten metal hardens, the bubbles form voids. (4.1.0)

8. In general, inclusions result from faulty welding techniques, improper access to the joint for welding or both. (4.2.0)

9. Transverse cracks, which run across the face of the weld and may extend into the base metal. Longitudinal cracks, which are usually located in the center of the weld deposit. Crater cracks, which have a tendency to form in the crater whenever the welding operation is interrupted. (4.3.1)

10. Base metal cracking usually occurs within the heat-affected zone of the metal being welded. (4.3.2)

11. Incomplete joint penetration. (4.4.0)

12. Incomplete fusion, generally called overlap. (4.5.0)

13. Improper electrode manipulation. (4.6.0)

14. They can cause hardness zones in the base metal or become the starting point for cracking. (4.7.0)

15. 1/8 inch. (4.9.0)

5.0.0 NONDESTRUCTIVE EXAMINATION (NDE) PRACTICES

Nondestructive examination, sometimes referred to as nondestructive testing, is a term used for those inspection methods that allow materials to be examined without changing or destroying them. NDE methods can usually detect the discontinuities and defects described in Section 4.0.0, "Weld Discontinuities and Their Causes."

Nondestructive examination is normally performed by the site quality group as part of the site quality program. Certified welding inspectors trained in the proper test methods conduct the examinations.

There is considerable overlap in the application of nondestructive and destructive tests. Destructive tests, which destroy the weld, are frequently used on several sample weldments to supplement, confirm or establish the limits of nondestructive tests. Destructive testing is also used to provide supporting information. Once this information has been established, nondestructive examinations can be made on similar-type welds to locate all discontinuities above the critical defect size that was determined by the destructive tests.

The welder should be familiar with the basic nondestructive examination practices. These include:

- Visual inspection
- Liquid penetrant inspection
- Magnetic particle inspection
- Radiographic inspection
- Ultrasonic inspection
- Eddy current inspection

5.1.0 VISUAL INSPECTION

In visual inspection the surface of the weld and the base metal are observed for visual imperfections. Certain tools and gauges may be used during the inspection. Visual inspection is the examination method most commonly used by the welder and inspector. It is the fastest and most inexpensive method for examining a weld. However, it is limited to what can be detected by the naked eye or through a magnifying glass.

Properly done before, during after welding, visual inspection can detect over 75 percent of the discontinuities before they are found by more expensive and time-consuming nondestructive examination methods.

Prior to welding, the base metal should be examined for conditions that may cause weld defects. Dimensions, including edge preparation, should also be confirmed by measurements. If problems or potential problems are found, corrections should be made before proceeding any further.

After the parts are assembled for welding, the weld joint should be visually checked for proper root opening, and any other aspects that might affect the quality of the weld. The following should be visually examined:

- Proper cleaning
- Joint preparation and dimensions
- Clearance dimensions of backing strips ring, or consumable inserts
- Alignment and fit-up of the pieces being welded
- Welding procedures and machine settings
- Specified preheat temperature (if applicable)
- Tack weld quality

During the welding process, visual inspection is the primary method for controlling quality. Some of the aspects that should be visually examined include:

- Quality of the root pass and the succeeding weld layers
- Sequence of weld passes
- Interpass cleaning
- Root preparation prior to welding a second side
- Conformance to the applicable procedure

After the weld is completed the weld surface should be thoroughly cleaned. A thorough visual examination may disclose such surface defects as cracks, shrinkage cavities, undercuts, incomplete penetration, non-fusion, overlaps and crater deficiencies prior to inspecting for surface defects.

An important aspect of visual examination is checking the dimensional accuracy of the weld after it is completed.

Dimensional accuracy is determined by conventional measuring gauges. The purpose of using the gauges is to determine if the weld is within allowable limits as defined by the applicable codes and specifications.

Some of the more common welding gauges used are:

- Undercut gauge
- Combination butt and fillet weld gauge
- Fillet weld blade gauge set
- Pocket fillet weld gauge set

5.1.1 Undercut Gauge

An undercut gauge is used to measure the amount of undercut on the base metal. Typically, codes allow for undercut to be no more than 0.010-inch deep when the weld is transverse to the primary stress in the part that is undercut. There are several types of gauges that can be used to check the amount of undercut. These gauges have a pointed end that is pushed into the undercut. The back side of the gauge indicates the measurement in either inches or millimeters.

Figure 5-1 shows two types of undercut gauges being used.

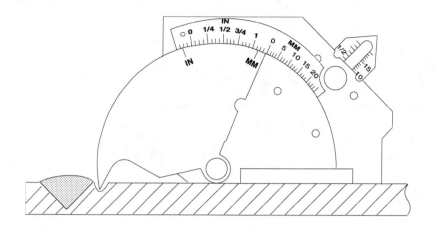

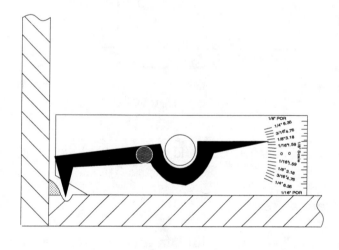

45010013

Figure 5-1. Undercut Gauges

5.1.2 Combination Butt and Fillet Weld Gauge

The combination butt and fillet weld gauge has a sliding pointer calibrated to several different scales which are used to measure the size of a convex fillet weld, the maximum convexity of a fillet weld, the maximum concavity of a fillet weld or the reinforcement of a butt weld. To use the gauge, position it as shown in *Figure 5-2* and slide the pointer to contact the base metal or weld metal. Be sure to read the correct scale for the measurement being taken.

Figure 5-2 shows using a combination butt and fillet weld gauge.

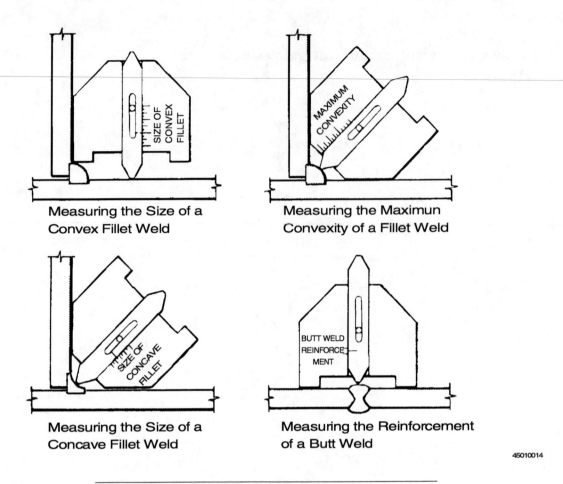

Figure 5-2. Combination Butt and Fillet Weld Gauge

5.1.3   Fillet Weld Blade Gauge Set

The fillet weld blade gauge set has seven individual gauges for measuring convex and concave fillet welds. The individual gauges are held together by a knurled nut. The seven individual blade gauges can measure eleven fillet weld sizes: 1/8, 3/16, 1/4, 5/16, 3/8, 7/16, 1/2, 5/8, 3/4, 7/8 and 1 inch and their metric equivalents, either concave or convex. The same blade size cannot be used for measuring both concave and convex fillet welds.

To use the fillet weld blade set, identify the type of fillet weld to be measured (concave or convex) and the size. Select the appropriate blade and position it. Be sure the gauge blade is flush to the base metal with the tip touching the vertical member.

Figure 5-3 shows using a fillet weld blade gauge.

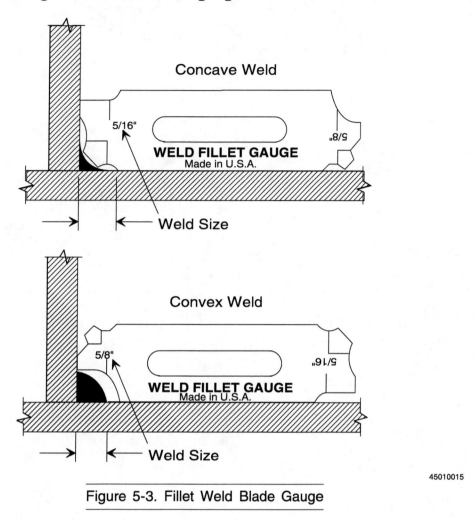

Figure 5-3. Fillet Weld Blade Gauge

5.1.4   Pocket Fillet Weld Gauge Set

The pocket fillet weld gauge is a set of two gauges, one for measuring the throat and allowable convexity of a fillet weld one for measuring the leg of a fillet weld. The gauge set will measure eight fillet weld sizes but cannot measure concave fillet welds. They are not as accurate as the fillet weld blade gauge but are often used by the welder as a pre-inspection visual check.

To use the pocket fillet weld gauge, select the appropriate gauge and position it being sure the gauge blade is flush to the base metal with the tip touching the vertical member.

Figure 5-4 shows using a pocket fillet weld gauge set.

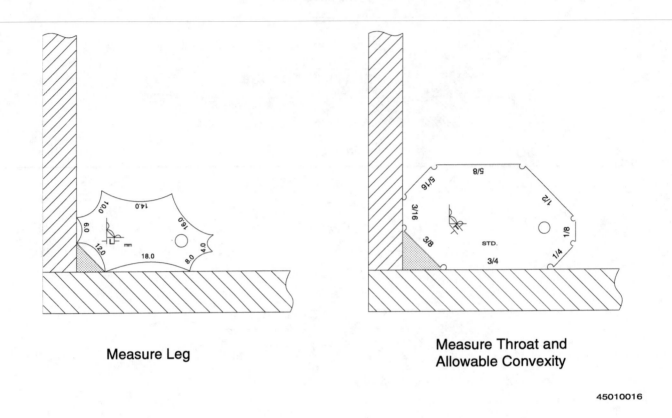

Measure Leg

Measure Throat and
Allowable Convexity

45010016

Figure 5-4. Pocket Fillet Weld Gauge Set

5.2.0 LIQUID PENETRANT INSPECTION

Liquid penetrant inspection is a nondestructive method for locating defects open to the surface. It cannot detect internal defects. The technique is based on the ability of a penetrating liquid (usually red in color) to wet the surface opening of a discontinuity and to be drawn into it. A liquid or dry powder developer (usually white in color) is then applied over the metal. If the flaw is significant, red penetrant bleeds through the white developer to indicate a discontinuity or defect.

The dye, cleaner and developer are available in spray cans for convenience. Some solvents used in the cleaners and developers contain high amounts of chlorine, a known health hazard, to make the liquids nonflammable.

WELDING ONE TRAINEE TASK MODULE 09107

The most common defects found using this process are surface cracks. Most cracks exhibit an irregular shape. The width of the bleed-out (the red dye bleeding through the white developer) is a relative measure of the depth of a crack.

Surface porosity, metallic oxides and slag will also hold penetrant and cause bleed-out. These indications are usually more circular and have less width than a crack. *Figure 5-5* shows indications of surface defects using liquid penetrant inspection.

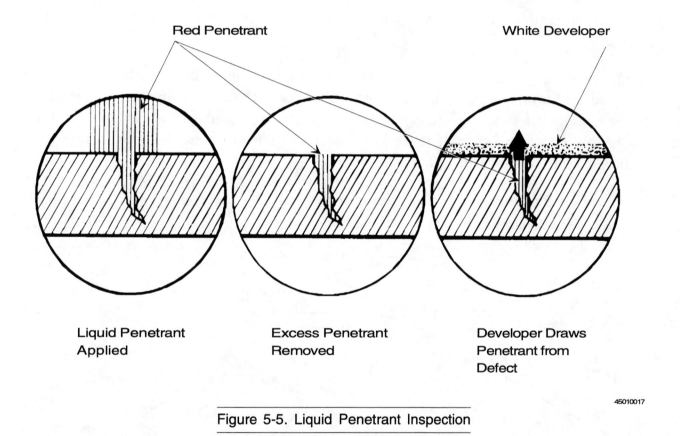

Red Penetrant   White Developer

Liquid Penetrant
Applied

Excess Penetrant
Removed

Developer Draws
Penetrant from
Defect

45010017

Figure 5-5. Liquid Penetrant Inspection

The advantages of liquid penetrant inspection are that it can find small defects not visible to the naked eye, can be used on most types of metals, is basically inexpensive and is fairly easy to use and interpret.   It is most useful to examine welds susceptible to surface cracks. Except for visual inspection, it is perhaps the most commonly used nondestructive examination method for surface inspection of nonmagnetic parts.

The disadvantages of liquid penetrant inspection are that it takes more time to use than visual inspection and that it can only find surface defects. The presence of weld bead ripples and other irregularities can also hinder interpretation of indications. Since chemicals are used, care must be taken when performing the inspection.

5.3.0 MAGNETIC PARTICLE INSPECTION

Magnetic particle inspection is a nondestructive examination method that uses electricity to magnetize the weld to be examined. After the metal is magnetized, metal particles are sprinkled on the weld surface. If there are defects in the surface or just below the surface the metal particles will be grouped into a pattern around the defect. The defect can be identified by the shape, width height of the particle pattern.

Magnetic particle inspection is used to test welds for such defects as surface cracks, lack of fusion, porosity, undercut, poor root penetration and slag inclusions. It can also be used to inspect plate edges before welding for surface imperfections. Defects can be detected at or near the surface of the weld. Defects much deeper than this are not likely to be found. Certain discontinuities exhibit characteristic powder patterns that can be identified by a skilled inspector.

For magnetic particle examination, the part to be inspected must be smooth, clean, dry and free from oil, water and excess slag. The part is then magnetized by using an electric current to set up a magnetic field within the material. The magnetized surface is covered with a thin layer of magnetic powder. If there is a defect, the powder is held to the surface at the defect because of the powerful magnetic field.

Figure 5-6 shows magnetic particle examination.

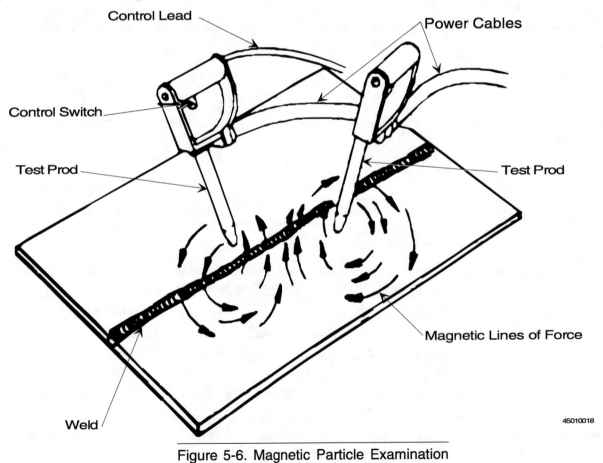

Control Lead

Power Cables

Control Switch

Test Prod

Test Prod

Magnetic Lines of Force

Weld

45010018

Figure 5-6. Magnetic Particle Examination

When this examination is used, there is normally a code or standard that governs both the method as well as the acceptance/rejection criteria of indications.

The advantages of using magnetic particle examination are that it can find small defects not visible to the naked eye and it is faster than liquid penetrant inspection.

The disadvantages of magnetic particle examination are that the materials must be capable of being magnetized, the inspector must be skilled in interpreting indications, rough surfaces can interfere with the results and it cannot find internal discontinuities located deep in the weld.

5.4.0   RADIOGRAPHIC INSPECTION

Radiography is a nondestructive examination method that uses radiation (x-ray or gamma ray) to penetrate the weld and produce an image on film. When a joint is radiographed, the radiation source is placed on one side of the weld and the film on the other. The joint is then exposed to the radiation source. The radiation penetrates the metal and produces an image on the film. The film is called a radiograph and provides a permanent record of the weld quality. Radiography should only be used and interpreted by trained, qualified personnel.

Figure 5-7 shows a radiography examination.

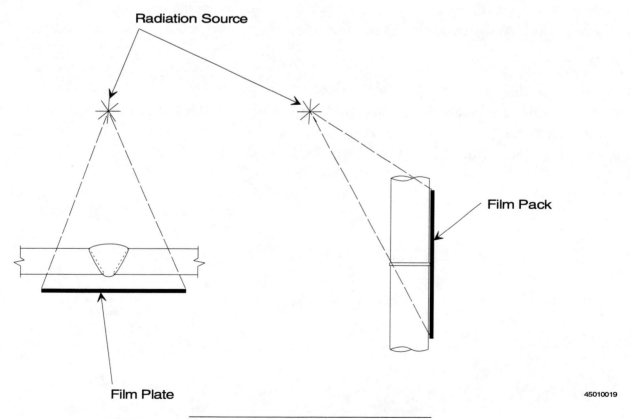

Figure 5-7. Radiography Examination

Radiographic examination can produce a visible image of weld discontinuities, either surface or subsurface, when they have differences in density from the base metal and in thickness parallel to the radiation. Surface discontinuities are better identified by visual, penetrant or magnetic particle examination.

The advantages of radiographic examination are that the film gives a permanent record of the weld quality, the entire thickness can be examined and it can be used on all types of metals.

The disadvantages of radiographic examination are that it is a slow and expensive method for inspecting welds, some joints are inaccessible to radiography and radiation of any type is very hazardous to humans. Cracks can frequently be missed if they are very small or are not aligned with the radiation beam.

5.5.0   ULTRASONIC INSPECTION

Ultrasonic inspection is a relatively low-cost nondestructive examination method that uses vibrations similar to sound waves to find surface and subsurface defects in the weld material. Ultrasonic waves are passed through the material being tested and are reflected back by any density change caused by a defect. The reflected signal is shown on a screen display (oscilloscope).

The term ultrasonic comes from the fact that these frequencies are above those heard by the human ear. The ultrasonic device operates very much like depth sounders or "fish finders."

Ultrasonic examination can be used to detect and locate cracks, laminations, shrinkage cavities, pores, slag inclusions, incomplete fusion and incomplete joint penetration, as well as other discontinuities in the weld. A trained, qualified inspector can interpret the signal on a screen to determine the approximate position, depth and size of the discontinuity.

Figure 5-8 shows a portable ultrasonic device.

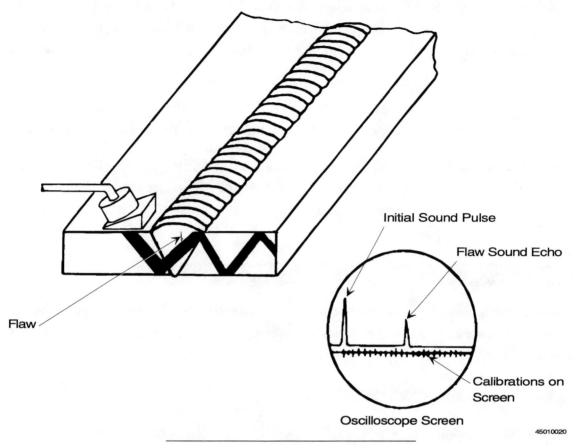

Figure 5-8. Portable Ultrasonic Device

The advantages of ultrasonic examination are that it can find defects through the thickness of the material being examined, it can be used to check materials that cannot be radiographed, it is nonhazardous to personnel or other equipment and it can detect even small defects.

The disadvantage of ultrasonic examination are that it requires a high degree of skill to properly interpret the patterns, very small or thin weldments are difficult to inspect and a permanent record is not readily obtained.

5.6.0   EDDY CURRENT INSPECTION

Eddy current examination, like magnetic particle testing, uses electromagnetic energy to detect defects in the joint. An AC coil, which produces a magnetic field, is placed on or around the part being tested. After being calibrated, the coil is moved over the part to be inspected. The coil produces a current in the metal by induction. The induced current is called an eddy current.

If a discontinuity is present in the test part, it will interrupt the flow of the eddy currents. This change can be observed on the oscilloscope.

Eddy currents only detect discontinuities near the surface of the part. The method is suitable for both ferrous and nonferrous materials it is used in testing welded tubing and pipe. It can determine the physical characteristics of a material and the wall thickness in tubing. It can check for porosity, pinholes, slag inclusions, internal and external cracks and lack of fusion.

The advantages of eddy current examination are that it can detect surface and near-surface weld defects. It is particularly useful in inspecting circular parts like pipes and tubing.

The disadvantages of eddy current examination are that eddy currents decrease with depth so defects further from the surface may go undetected. The accuracy of the examination depends in large part on the calibration of the instrument and the qualification of the inspector.

6.0.0   WELDER PERFORMANCE QUALIFICATION TESTS

The purpose of performance qualification is to measure the proficiency of individual welders. As previously discussed, codes require that welders take a test to qualify to a welding procedure.

6.1.0   WELDING POSITIONS QUALIFICATION

The welder or welding operator is qualified by welding position. Welders may be qualified to perform a welding procedure in only one or possibly all positions by taking a welding test. The qualification tests are designed to measure the welder's ability to make groove and fillet welds in different positions on plate, pipe or both. Each welding position is designated by a number and a letter (for example, 1G). These designations are standard for all codes.

The letter "G" designates a groove weld. The letter "F" designates a fillet weld. For plate welding, the positions are designated by the following numbers:

- "1" means flat position welding
- "2" means horizontal position welding
- "3" means vertical position welding
- "4" means overhead position welding

Figure 6-1 shows the plate welding positions for both fillet welds and groove welds.

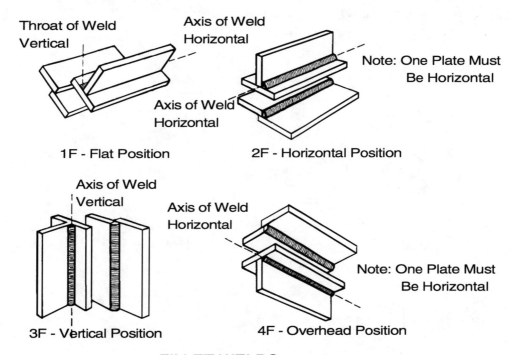

FILLET WELDS

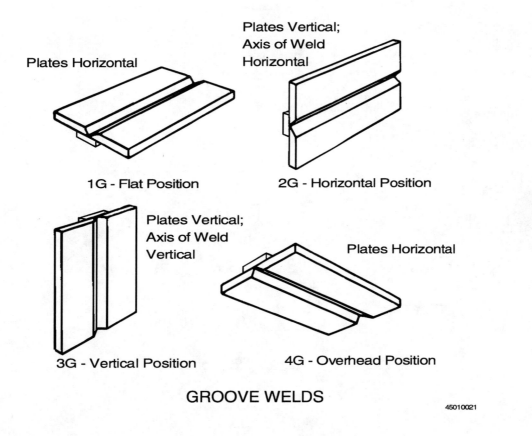

GROOVE WELDS

45010021

Figure 6-1. Welding Positions for Plate

For pipe welding there are two additional positions: 5G and 6G. Also, in the "1G" (flat groove) and "1F" (flat fillet) positions, the pipe is rotated during welding. *Figure 6-2* shows the pipe welding positions for both fillet welds and groove welds.

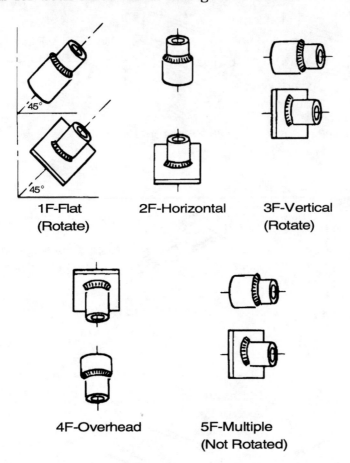

1F-Flat
(Rotate)

2F-Horizontal

3F-Vertical
(Rotate)

4F-Overhead

5F-Multiple
(Not Rotated)

Fillet Welds in Pipe

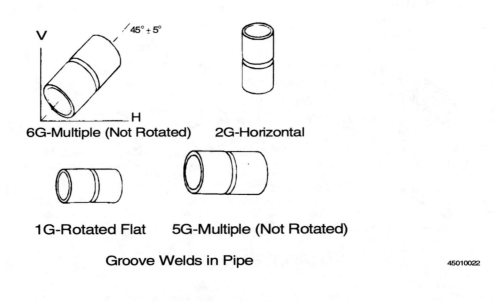

6G-Multiple (Not Rotated)

2G-Horizontal

1G-Rotated Flat

5G-Multiple (Not Rotated)

Groove Welds in Pipe

45010022

Figure 6-2. Welding Positions for Pipe

A welder who qualifies in one position does not automatically qualify to weld in all positions. However, in most cases qualification for groove welds will qualify the welder for fillet welds; qualification for pipe will qualify the welder for plate.

Note   Refer to your site requirements/code for qualifying requirements.

6.2.0   THE AWS STRUCTURAL CODE

AWS D1.1, Structural Welding Code—Steel provides information concerning the qualification of welding procedures, welders and welding operators for the types of welding done by contractors and fabricators in building and bridge construction.

Qualification for plate welding also qualifies the welder for rectangular tubing.

The mild steel electrodes used with SMAW are classified by F numbers F1, F2, F3 and F4. Qualification with an electrode in a particular F number classification will qualify the welder with all electrodes identified in that classification and in lower F number classifications. *Figure 6-3* shows the AWS F number electrode classification.

Group	AWS Electrode Classification			
F4	EXX15	EXX16	EXX18	
F3	EXX10	EXX11		
F2	EXX12	EXX13	EXX14	
F1	EXX20	EXX24	EXX27	EXX28

Figure 6-3.   AWS F-Number Electrode Classification

45010024

Material thickness is an essential variable in qualification tests under the AWS Code. Some of the tests in the code qualify the welder only up to twice the thickness of the test piece. Others qualify the welder for unlimited thicknesses.

A typical AWS welder qualification test is a V-butt weld with metal backing in the 3G and 4G positions using a F4 electrode. Passing this test qualifies the welder to weld with F4 or lower electrodes and make groove and fillet welds in all positions. *Figure 6-4* shows a typical AWS plate test.

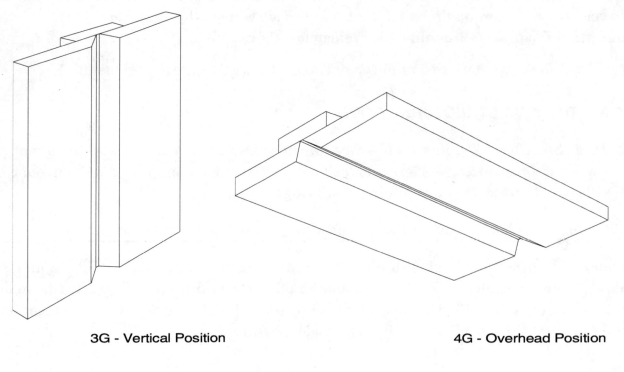

3G - Vertical Position

4G - Overhead Position

45010025

Figure 6-4 Typical AWS Plate Test

6.3.0 THE ASME CODE

Individual welders and welding operators must qualify in accordance with Section IX of the *ASME Boiler and Pressure Vessel Code* on either plate or pipe. Qualification on pipe also qualifies the welder to weld plate, but not vice-versa. Qualification with groove welds also qualifies the welder for fillet welds, but not vice-versa. It is possible, under the Code, to qualify for fillet welds only.

The typical ASME welder qualification test is to weld pipe in the 6G position using an open root. Passing this test qualifies the welder to weld pipe in all positions and plate in all positions (fillet and groove). If F3 electrodes are used all the way out, the welder only qualifies on F3 electrodes on both plate and pipe. If F3 electrodes are used for welding the root F4 electrodes are used for filler, the welder qualifies to weld pipe or plate with F3 electrodes and pipe or plate with F4 electrodes as long as backing is used. *Figure 6-5* shows a typical ASME pipe test.

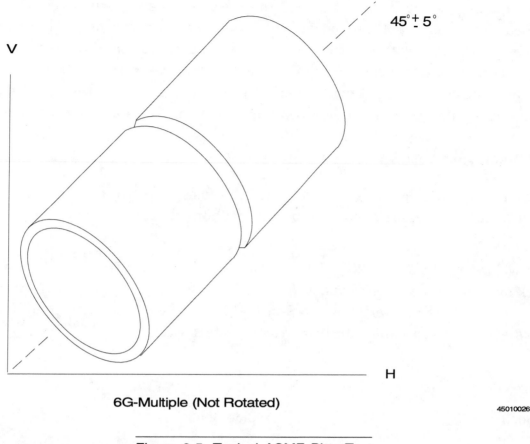

45° ± 5°

V

H

6G-Multiple (Not Rotated)

45010026

Figure 6-5. Typical ASME Pipe Test

6.4.0   WELDER QUALIFICATION TESTS

The welder becomes qualified by successfully completing a weld made in accordance with the welding procedure specification (WPS). It is a general practice for code welding to qualify welders on the groove weld tests. Passing these tests permits the welder to also weld fillet welds.

Note   Various codes and specifications require similar, though sometimes different, methods or details for qualifying welders. The applicable code or specification should be consulted for specific details and requirements. Ask your supervisor if you are unsure of which codes or specifications apply to your project.

6.4.1   Making the Test Weld

While the qualification tests are designed to determine the capability of welders, welders have failed for reasons not related to their welding ability. This is due principally to carelessness in the application of the weld and in the preparation of the test specimen. One important item is to note, prior to welding, where the test strips will be cut from the weld coupon. By doing this you can avoid potential problems such as restarts in the area of the test strips. The following sections will explain how to prepare a test specimen.

6.4.2   Removing Test Specimens

After making the qualification test weld, the test specimens are cut from the test pipe or plate by any suitable means. There are specific locations that the test specimen is taken from the pipe or plate.

For pipe welded in the 1G or 2G positions, two specimens are required. For material 3/8-inch thick and under, a face bend and a root bend are required. For 3/8-inch only, side bends can be substituted. For material over 3/8-inch thick two side bends are required. *Figure 6-6* shows where the specimens are cut from the test weld.

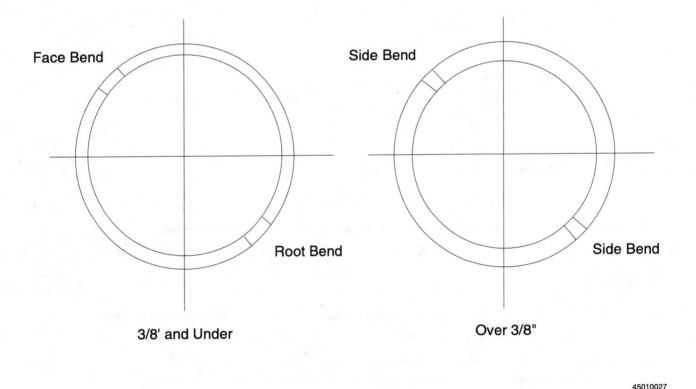

Figure 6-6. Location of Pipe Specimens for 1G and 2G Positions

For pipe welded in the 5G or 6G positions, four specimens are required. For material 1/16-up to 3/8-inch thick two face bends and two root bends are required. For material 3/8-inch thick and over, four side bends are required. *Figure 6-7* shows where the specimens are cut from the test weld.

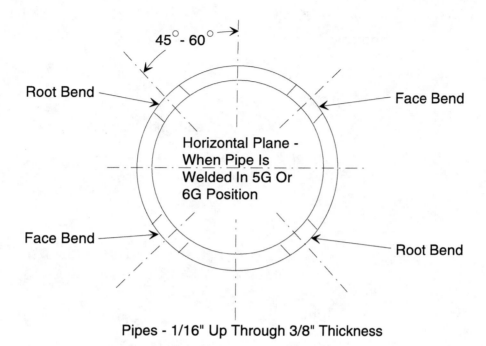

Pipes - 1/16" Up Through 3/8" Thickness

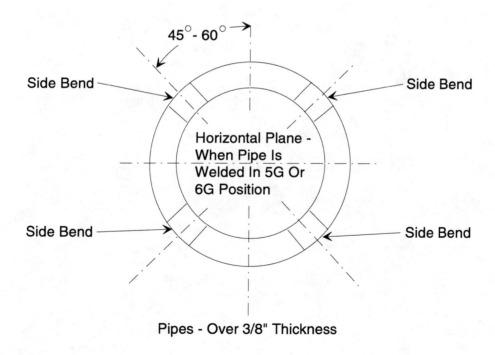

Pipes - Over 3/8" Thickness

45010028

Figure 6-7. Locations of Pipe Specimens for 5G and 6G Positions

Typical specimen locations for plate welds are shown in *Figure 6-8*. For material 3/8-inch thick a face bend and a root bend are required. For material over 3/8-inch thick two side bends are required.

Note   Tests are usually given on 3/8-inch plate for limited thickness and 1-inch plate for unlimited thickness.

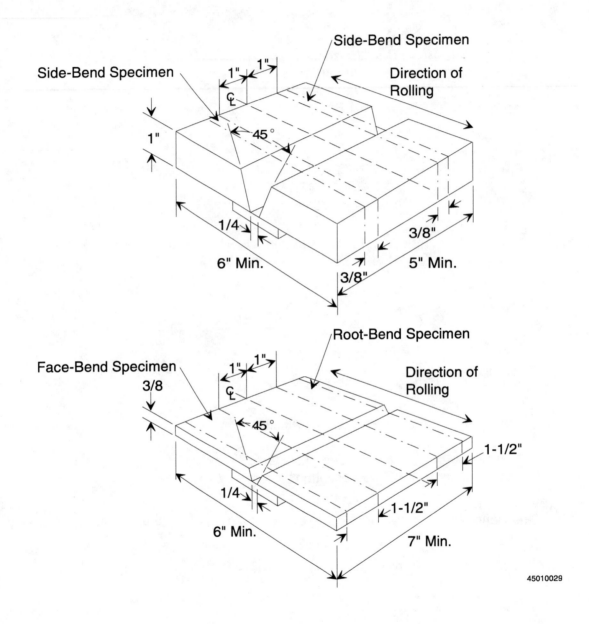

Figure 6-8 Specimen Locations for Plate Welds

6.4.3 Preparing the Specimens for Testing

After the specimen has been cut from the test piece it must be properly prepared for testing. Poor specimen preparation can cause a sound weld metal to fail. For example, a slight nick may open up under the severe bending stress of the test, causing the specimen to fail.

To properly prepare the test specimen:

- Ground or machine the surface to a smooth finish. All grinding and machining marks must be lengthwise on the sample. Otherwise they produce a notch effect, which may cause failure.
- Remove any bead reinforcement either on the face or on the root side of the weldment. This is part of the test requirement and, more important to the welder, failure to do so can cause failure of a good weld.
- Round the edges to a smooth 1/16-inch radius. This can be done with a file. Rounded edges help prevent failure caused by cracks starting at the sharp corner.
- When grinding specimens, do not water-quench them when they are hot. Quenching may create small surface cracks that become larger during the bend test.

Figure 6-9 shows a prepared test specimen.

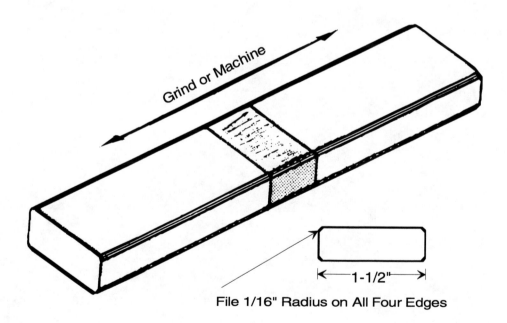

File 1/16" Radius on All Four Edges

45010030

Figure 6-9. Example of a Prepared Test Specimen

6.4.4 Testing Qualification Weld Specimens

Guided bend tests are used to evaluate groove test welds on both plate and pipe. In this method, specimens are bent into a "U" shape with a device called a "jig." The bending action places stress on the weld metal and reveals any discontinuities in the weld.

Note   The jig has a plunger and die which are dimensioned for the thickness of the specimen being bent. Refer to the welding code being used for testing for the required dimensions of the bending jig's plunger and die.

Three types of tests are performed on the jig: root bends, face bends and side bends.

- Root bends test the penetration and fusion throughout the root of the joint
- Face bends test the quality of fusion to the side walls and the face of the weld joint, porosity, slag inclusion, gas pockets other defects
- Side bends test for soundness and fusion

Face and root bend tests are used for materials up to 3/8-inch thick. Between 3/8-inch and 3/4-inch thickness, face and root bends or side bends may be used. Side bends are used for 3/4 inch and over.

Figure 6-10 shows the guided bend test method and examples of a root, face and side bend.

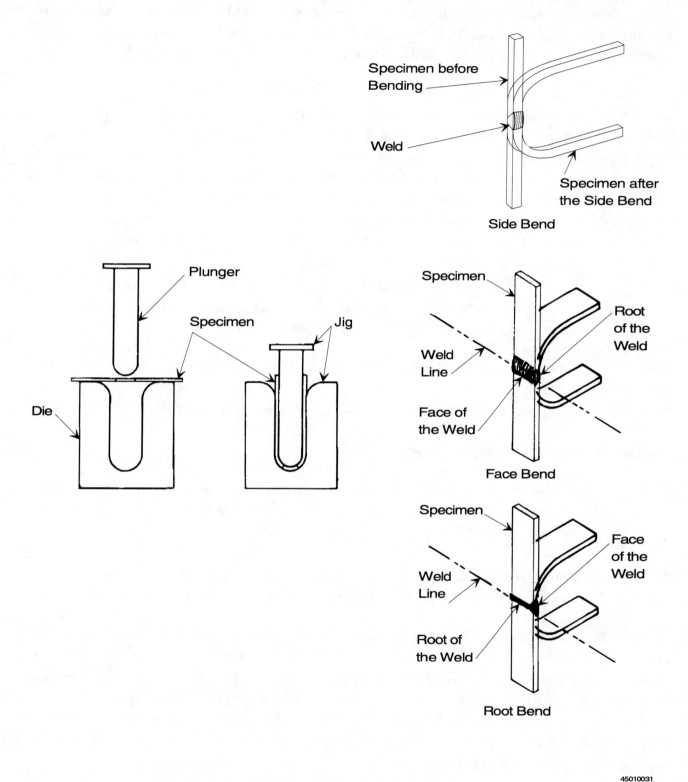

Figure 6-10. Guided Bend Test Method and Samples

45010031

After bending, the specimen is evaluated by measuring the discontinuities that are exposed. The criteria for acceptance can vary by code or site quality standards. The criteria for acceptance according to the *AWS Structural Welding Code, D1.1* states that for acceptance, the surface shall contain no discontinuities exceeding the following dimensions:

1. 1/8 inch (3.2 mm) measured in any direction on the surface.
2. 3/8 inch (9.5 mm) - sum of the greatest dimensions of all discontinuities exceeding 1/32 inch (0.8 mm), but less than or equal to 1/8 inch (3.2 mm).
3. 1/4 inch (6.4 mm) - maximum corner crack, except when that corner crack results from visible slag inclusion or other fusion type discontinuities, then 1/8 inch (3.2 mm) maximum shall apply. The specimen with corner cracks exceeding 1/4 inch (6.4 mm) with no evidence of slag inclusions or other fusion type discontinuities may be discarded a replacement test specimen from the original weldment shall be tested.

Note   In some cases a radiographic examination will be used instead of the guided bend test. This allows the entire weld to be examined and can detect small defects at any location within the weld.

When the welder passes the qualification test(s), the test results and the procedure(s) that he or she may weld to are listed on a record that is kept by the company. This record becomes part of the quality documentation and the welder becomes certified to weld to that procedure.

6.4.5   Welder Qualification Limits

If a welder fails the test, he or she may retest. An immediate retest consists of two test welds of each type failed. All the test specimens must pass this retest. A complete retest may be made if the welder has had further training or practice since the last test.

Note   Retest requirements may vary depending on site quality standards. Check your site's quality standards for specific retest requirements.

After welders have qualified they may have to requalify if they have not used the specific process for a certain time period. (This time period varies from three months under ASME to six months under AWS.) They may also be required to requalify if there is a reason to question their ability to make welds which meet the Welding Procedure Specification. Also, since welder performance qualification is limited to the essential variables of a particular procedure, any change in one or more of the essential variables requires the welder to requalify with the new procedure.

7.0.0 QUALITY WORKMANSHIP

The codes and standards discussed in this manual were written to ensure that quality welds are consistently made by the welder. Although many of the weldments fabricated will be examined and tested, not all will be examined and tested due to the cost and time this would required. However, quality workmanship is expected in every weld a welder makes.

7.1.0 TYPICAL SITE ORGANIZATION

The welder should be able to interface with appropriate site representatives to assure that quality work is achieved. To do this the site organizational structure needs to be understood. If quality problems arise the welder needs to follow the appropriate chain of command to eliminate the problems. *Figure 7-1* shows a typical site organizational structure.

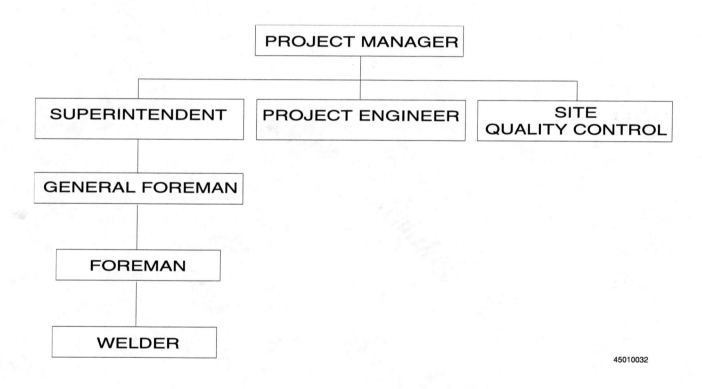

45010032

Figure 7-1. Typical Site Quality Organizational Structure

Note   Your site may differ in its organizational structure. Check with your immediate supervisor to find out the structure for your site.

7.2.0 CHAIN OF COMMAND

A welder should always follow the site chain of command. However, there may be instances when the welder should bypass the chain of command. Examples of such instances are:

- You have been directed to perform an unsafe act. If you cannot resolve the matter with your immediate supervisor, it is your responsibility to go to the general foreman, superintendent, project manager or safety engineer.
- You have been directed to perform a weld which requires certification and you are not certified to perform it. If you cannot resolve the matter with your immediate supervisor it is your responsibility to go to the general foreman, superintendent, project manager or quality engineer.

Note   Always try to resolve problems with your immediate supervisor before bypassing the chain of command.

SELF-CHECK REVIEW 2

1. What is the examination method most commonly used by the welder and inspector?
2. Typically, how much undercut will the codes allow when the weld is transverse to the primary stress in the part that is undercut?
3. Can the same blade size from the fillet weld gauge be used to measure convex and concave fillet welds?
4. Explain liquid penetrant inspection.
5. Explain magnetic particle inspection.
6. What NDE process provides a permanent record of the weld quality and can be used on all types of metals to examine there entire thickness?
7. What NDE process uses sound waves?
8. What NDE process uses electromagnetic energy to detect defects just below the surface?
9. In most cases will qualifying on groove welds qualify the welder on fillet welds?
10. What is the typical welder qualification test with the *AWS Structural Welding Code D1.1*?
11. According to Section IX of the *ASME Boiler and Pressure Vessel Code*, does qualification on pipe also qualify the welder to weld plate?
12. Why should you mark, prior to welding, where the test strips will be cut from the test coupon?
13. How many and what type of test specimens are required for pipe under 3/4-inch thick welded in the 5G or 6G positions?
14. When preparing a test specimen, what must be considered for the machining or grinding marks and why?
15. What is a guided bend test?
16. Should a welder ever bypass the chain of command? If yes, why?

ANSWERS TO SELF-CHECK REVIEW 2

1. Visual inspection. (5.1.0)
2. Typically, codes allow for undercut to be no more than 0.010-inch deep when the weld is transverse to the primary stress in the part that is undercut. (5.1.1)
3. No, there are separate blade sizes for concave and convex fillet welds. (5.1.3)
4. The technique uses a penetrating liquid (usually red in color) to wet the surface opening of a discontinuity and to be drawn into it. A liquid or dry powder developer (usually white in color) is then applied over the metal. If the flaw is significant, red penetrant bleeds through the white developer to indicate a discontinuity or defect. (5.2.0)
5. Magnetic particle inspection uses electricity to magnetize the weld to be examined. After the metal is magnetized, metal particles are sprinkled on the weld surface. If there are defects in the surface, or just below the surface, the metal particles will be grouped into a pattern around the defect. (5.3.0)
6. Radiographic inspection. (5.4.0)
7. Ultrasonic inspection. (5.5.0)
8. Eddy current inspection. (5.6.0)
9. Yes. (6.1.0)
10. A typical AWS welder qualification test is a V-butt weld with metal backing in the 3G and 4G positions using a F4 electrode. (6.2.0)
11. Yes but not vice-versa. (6.3.0)
12. To avoid potential problems such as restarts in the area of the test strips. (6.4.1)
13. Four specimens are required, two face bends and two root bends. (6.4.2)
14. All grinding and machining marks must be lengthwise on the sample. Otherwise they produce a notch effect which may cause failure. (6.4.3)
15. In a guided bend test, specimens are bent into a "U" shape with a device called a "jig." The bending action places stress on the weld metal and reveals any discontinuities in the weld. The discontinuities are measured and compared to evaluation criteria specified in the welding code being used. (6.4.4)
16. Yes, if you have been directed to perform an unsafe act or if you are being directed to perform a weld which requires certification and you are not certified to perform it. But the chain of command should be bypassed only if you cannot resolve the matter with your immediate supervisor. (7.2.0)

SUMMARY

Quality is everyone's responsibility. If the work being done cannot be defined as "quality work" it reflects on all those involved in the process. One essential trait of a craftsman is a sense of quality workmanship. Since the craftsman is generally the closest to the work he or she will have a major impact on product quality. Keeping quality in mind as you perform each step of your job will identify and correct small problems before they become major problems. This will make everyone's job easier and instill a sense of pride in what has been accomplished.

Safety, Quality, Production--Each of these items has a cost of its own, while on the project each of these items should also have proper guidelines. At the completion of the project, when all records for safety, cost, planning, scheduling, and effectiveness have been evaluated, the papers are usually filed away and never accessed again. Quality remains as possibly the most important criteria of all. It remains for all eyes to see indefinitely. As the job is viewed by the customer, blown budgets and late schedules are often forgotten as long as a quality job was received. So perhaps quality is the major reason for repeat business. Thus, how well each person performed will be graded long after the project has been laid to rest.

References

For advanced study of topics covered in this module, the following works are suggested:

Welding Handbook, Volume 5, Seventh Edition, The American Welding Society, Miami, FL, 1984. Phone 1-800-334-9353.

Welding Principles and Applications, Jeffus and Johnson, Delmar Publishers, Inc., Albany, NY, 1988. Phone 1-800-347-7707.

Modern Welding, Althouse, Turnquist, Bowditch, and Bowditch, Goodheart Willcox Publishers, South Holland, IL, 1988. Phone 1-800-323-0440.

PERFORMANCE / LABORATORY EXERCISES

None.

The NCCER makes every effort to keep these manuals up-to-date and free of technical errors. We appreciate your help in this process. If you have an idea for improving this manual, or if you find an error, a typographical mistake, or an inaccuracy in the *Wheels of Learning*, please write us, using this form or a photocopy. Be sure to include the exact module number, page number, a description of the problem, and the correction, if possible. We'll do our best to correct it in later editions. Thank you for your assistance.

Write: *Wheels of Learning*
National Center for Construction Education and Research
P.O. Box 141104
Gainesville, FL 32614-1104
Fax: 352-334-0932

WHEELS OF LEARNING USER UPDATE

Please let us know if you have found an inaccuracy, error, or other problem in a *Wheels of Learning* manual. Use this form or write us a letter. Please be sure to tell us the exact module name and module number, the page number, and the problem. Thanks for your help.

Craft _____ Module Name _____

Module Number _____ Page Number(s) _____

Description of Problem _____

_____

_____

(Optional) Correction of Problem _____

_____

_____

(Optional) Your Name and Address _____

_____